Otoliths and Their Applications in Fishery Science

Otoliths and Their Applications in Fishery Science

Editor

Josipa Ferri

MDPI • Basel • Beijing • Wuhan • Barcelona • Belgrade • Manchester • Tokyo • Cluj • Tianjin

Editor
Josipa Ferri
University of Split
Croatia

Editorial Office
MDPI
St. Alban-Anlage 66
4052 Basel, Switzerland

This is a reprint of articles from the Special Issue published online in the open access journal *Fishes* (ISSN 2410-3888) (available at: https://www.mdpi.com/journal/fishes/special_issues/otoliths_application_fishery).

For citation purposes, cite each article independently as indicated on the article page online and as indicated below:

LastName, A.A.; LastName, B.B.; LastName, C.C. Article Title. *Journal Name* **Year**, *Volume Number*, Page Range.

ISBN 978-3-0365-6612-2 (Hbk)
ISBN 978-3-0365-6613-9 (PDF)

Cover image courtesy of Josipa Ferri

Contents

About the Editor

Josipa Ferri

Josipa Ferri is an Associate Professor at the University of Split, Department of Marine Studies (Split, Croatia), where she has been working since 2006. She studied marine fisheries science at the University Department of Marine Studies (graduated in 2005) and obtained a PhD in natural sciences (2011) from the University of Split. Her research is mainly focused on ichthyology, with a special interest in fish age, growth and diet. Josipa teaches several courses on various physiological processes in fish on both undergraduate and graduate levels, and she has mentored over 50 students through their BSc and MSc thesis.

Editorial

Otoliths and Their Applications in Fishery Science

Josipa Ferri

Department of Marine Studies, University of Split, 21000 Split, Croatia; josipa.ferri@unist.hr; Tel.: +385-21558195

Otoliths are one of the most useful and important biological structures for the various studies of fish, leading to many practical applications that are not limited only to ichthyology. It is, therefore, not surprising that different approaches to otolith research are experiencing continuous growth on a global scale. To collect the latest findings on these approaches and applications, the Special Issue entitled, "Otoliths and Their Applications in Fishery Science", was created. Collected papers cover several different research areas, from the otolith macro- (e.g., shape and size) to the microstructure (e.g., increment analyses and age estimates) and microchemistry, analyzing different types of otoliths (sagittae, asterisci and lapilli) and investigating various freshwater and marine species in their larval, juvenile and adult stages that were laboratory-reared or field-collected.

In a study regarding the morphological alterations in otoliths during ontogenesis, Gao et al. [1] found that both sagittae and lapilli of the laboratory-reared *Sinogastromyzon wui* were visible upon hatching, whereas asterisci were present four days post-hatching. The results show that lapillus developmental increments were deposited daily, and the authors proposed this otolith type as the most optimal for age determination and increment deposition rate confirmation. All different types of otoliths were also investigated by Zhu et al. [2] in one of the four major, commercially important carp species in China. However, this study focused on the marking of hard structures (from lateral-line scales and rays of all fins to all otolith types) in the juvenile *Hypophthalmichthys molitrix* by using a fluorescent marker. The lapilli and lateral-line scales were marked most effectively for all sampled hard structures. In general, such markings can be used in fishery science for a wide range of applications, including the investigation of fish life-history and movement patterns, as well as determining restocking effectiveness. These valuable applications guided the authors in their new research. This time, Zhu et al. [3] exposed the juvenile crucian carp (*Carassius carassius*) to a single concentration of $SrCl_2{\cdot}6H_2O$ and evaluated the efficiency of strontium marking all fish otoliths. Sr marking signatures formed a peak in all otolith types, but again, lapilli were the most suitable for these marking observations.

The study by Jawad and Mahé [4] took a different approach and tested the potential asymmetry in asterisci. They found significant fluctuating asymmetry between the length of the common carp (*Cyprinus carpio*) and every asterisci descriptor. Considering that the level of asymmetry can modify the boundaries of stocks according to the use of left or right otoliths, the authors highlighted the importance of verifying this phenomenon before using otoliths in fisheries research. Several other studies have also pointed out the importance of otolith macrostructures. Marval-Rodríguez et al. [5] used otolith shape analysis to test possible differences between two important red snapper species of the genus *Lutjanus* in the western Atlantic, for which there is a suggestion that the two are a single species. The otolith shape revealed differences, but not between the two investigated species; rather, differences were found among their populations in the studied area. In addition, Morawicki et al. [6] revealed the spatial stock structure of the silverside *Odontesthes argentinensis*, which is an ecologically and economically important species in the SW Atlantic. The authors used a combination of elliptic Fourier descriptors, Wavelet coefficients and otolith shape indices. More proof that otolith shape analysis is effective in discriminating fish groups, especially those experiencing different environmental conditions, was presented by Moura et al. [7].

Citation: Ferri, J. Otoliths and Their Applications in Fishery Science. *Fishes* **2023**, *8*, 35. https://doi.org/10.3390/fishes8010035

Received: 15 December 2022
Accepted: 17 December 2022
Published: 3 January 2023

Using the traditional otolith shape descriptors, it was discovered that the European eel population in the Minho River (NW Iberian Peninsula) shows complete discrimination between the two main types of habitats (tributaries and estuaries).

Generally, otoliths have mainly been used in age and growth studies, and the significance of this application in fishery science continues today, as reflected by the number of related papers in this Special Issue. Given that the information on age and growth is essential for stock assessments and the development of management plans, Huang et al. [8] presented several analyses on the two important croakers (*Pennahia macrocephalus* and *Atrobucca nibe*) that are under high fishing pressure in SW Taiwan. The authors validated the periodicity of ring formation, examined three age determination methods and calculated the representative growth models using a multimodel approach. The validation of aging is crucial because it directly influences the correct determination of age, as proved by Rodriguez-Marin et al. [9]. The results obtained in their study provide evidence that the annulus formation in the otoliths of Atlantic bluefin tuna are completed later in the calendar year than previously thought. As a conclusion, it is necessary to delay the date of the current July 1 adjustment criterion to November 30. On the other hand, García-Fernández et al. [10] focused on the usefulness and potential of otolith daily growth analyses. The authors analyzed otoliths of adult *Merluccius merluccius* females from the Galician waters and showed that the daily growth of females decreases during their spawning period. Good environmental conditions are key for successful growth, and Denis et al. [11] highlighted the importance of small estuaries that are crucial habitats for supporting higher growth during a fish's growth phase. Particularly, they investigated eels in small estuaries along the French coast of the eastern English Channel.

Several papers on fish age and growth have presented the first data on certain areas of distribution of investigated species. Christoffersen et al. [12] provided new information about the quillback rockfish in the northern part of its range (SE Alaska) and detected similar growth patterns and rates that were previously identified in the southern area (Salish Sea). Moreover, Ferri and Brzica [13] investigated the age and growth of the European barracuda, *Sphyreana sphyreana*, in the eastern Adriatic for the first time. The authors also measured otolith length, width and mass to test the utility of these morphometrics as predictors of age in investigated fish. The results showed that counting otolith annuli produced a better estimation of age than proposed linear models based on relationships between observed fish age and otolith morphometrics. Using otolith morphometrics in aging is just one way to facilitate the challenging and time-consuming reading of otoliths. Over the last few decades, various autonomous techniques have been proposed to automate this tedious activity. In this context, Ordoñez et al. [14] used otolith images from the Greenland halibut to train a convolutional neural network (CNN) for the automatic predicting of fish age. The study demonstrated that applying a CNN model trained on images from one lab does not lead to a suitable performance when predicting fish ages from otolith images from another lab. The authors detected that this was due to a problem known as dataset shift, which can be handled using domain adaptations. In addition, Politikos et al. [15] presented DeepOtolith, an open-source artificial intelligence (AI) platform that provides a web system with a simple interface that automatically estimates fish age by combining otolith images with a CNN. This platform currently contains classifiers for only three fish species; however, the authors highlighted that more species will be included as soon as more related work on aging is tested.

Finally, several papers analyzed otolith microchemistry and emphasized applications of this research area in: (i) describing demographics of an anadromous fish's life-history; (ii) identifying whether fish exhibit homing and return to natal streams to spawn; and (iii) assessing population connectivity. Roloson et al. [16] examined two hypotheses that anadromous brook trout (*Salvelinus fontinalis*) are more likely to arise from sea-run mothers, and that freshwater entry timing makes them vulnerable to pesticide-induced fish kills; meanwhile, Shrimpton et al. [17] estimated the fidelity to natal streams of the Columbia-origin Kokanee (*Oncorhynchus nerka*) in the Williston Reservoir. The work by Jiang et al. [18]

assessed the population connectivity of *Coilia nasus*. Results strongly suggest that there are two original natal populations in the Qiantang and Changjiang Rivers, whereas the population in the Yellow Sea has little connectivity with that of the Qiantang River, but has a supplementary relationship with that of the Changjiang River.

Authors of all these papers dealt with different questions and tested various hypotheses, but they were all motivated by specific applications of otoliths in fishery science. Hopefully, these papers will inspire many more researchers and lead to compelling new advances in otolith research.

Acknowledgments: I would like to take this opportunity to thank all the authors for their contributions to this Special Issue; the reviewers for their effort and valuable time, comments and suggestions during the review process; and the editorial team of the journal *Fishes* for their hard work, effective cooperation and support.

Conflicts of Interest: The author declares no conflict of interest.

References

1. Gao, M.; Wu, Z.; Huang, L.; Tan, X.; Li, M.; Huang, H. Growth and Microstructural Features in Otoliths of Larval and Juvenile *Sinogastromyzon wui* (F. Balitoridae, River Loaches) of the Upper Pearl River, China. *Fishes* **2022**, *7*, 57. [CrossRef]
2. Zhu, Y.; Jiang, T.; Chen, X.; Liu, H.; Phelps, Q.; Yang, J. A Pilot Study Assessing a Concentration of 100 mg/L Alizarin Complexone (ALC) to Mark Calcified Structures in *Hypophthalmichthys molitrix*. *Fishes* **2022**, *7*, 66. [CrossRef]
3. Zhu, Y.; Jiang, T.; Chen, X.; Liu, H.; Phelps, Q.; Yang, J. Inter-Otolith Differences in Strontium Markings: A Case Study on the Juvenile Crucian Carp *Carassius carassius* (Linnaeus, 1758). *Fishes* **2022**, *7*, 112. [CrossRef]
4. Jawad, L.; Mahé, K. Fluctuating Asymmetry in Asteriscii Otoliths of Common Carp (*Cyprinus carpio*) Collected from Three Localities in Iraqi Rivers Linked to Environmental Factors. *Fishes* **2022**, *7*, 91. [CrossRef]
5. Marval-Rodríguez, A.; Renán, X.; Galindo-Cortes, G.; Acuña-Ramírez, S.; Jiménez-Badillo, M.; Rodulfo, H.; Montero-Muñoz, J.; Brulé, T.; De Donato, M. Assessing the Speciation of *Lutjanus campechanus* and *Lutjanus purpureus* through Otolith Shape and Genetic Analyses. *Fishes* **2022**, *7*, 85. [CrossRef]
6. Morawicki, S.; Solimano, P.; Volpedo, A. Unravelling Stock Spatial Structure of Silverside *Odontesthes argentinensis* (Valenciennes, 1835) from the North Argentinian Coast by Otoliths Shape Analysis. *Fishes* **2022**, *7*, 155. [CrossRef]
7. Moura, A.; Dias, E.; López, R.; Antunes, C. Regional Population Structure of the European Eel at the Southern Limit of Its Distribution Revealed by Otolith Shape Signature. *Fishes* **2022**, *7*, 135. [CrossRef]
8. Huang, S.; Chang, S.; Lai, C.; Yuan, T.; Weng, J.; He, J. Length-Weight Relationships, Growth Models of Two Croakers (*Pennahia macrocephalus* and *Atrobucca nibe*) off Taiwan and Growth Performance Indices of Related Species. *Fishes* **2022**, *7*, 281. [CrossRef]
9. Rodriguez-Marin, E.; Busawon, D.; Luque, P.; Castillo, I.; Stewart, N.; Krusic-Golub, K.; Parejo, A.; Hanke, A. Timing of Increment Formation in Atlantic Bluefin Tuna (*Thunnus thynnus*) Otoliths. *Fishes* **2022**, *7*, 227. [CrossRef]
10. García-Fernández, C.; Domínguez-Petit, R.; Saborido-Rey, F. The Use of Daily Growth to Analyze Individual Spawning Dynamics in an Asynchronous Population: The Case of the European Hake from the Southern Stock. *Fishes* **2022**, *7*, 208. [CrossRef]
11. Denis, J.; Mahé, K.; Amara, R. Abundance and Growth of the European Eels (*Anguilla anguilla* Linnaeus, 1758) in Small Estuarine Habitats from the Eastern English Channel. *Fishes* **2022**, *7*, 213. [CrossRef]
12. Christoffersen, C.; Shiozawa, D.; Suchomel, A.; Belk, M. Age and Growth of Quillback Rockfish (*Sebastes maliger*) at High Latitude. *Fishes* **2022**, *7*, 38. [CrossRef]
13. Ferri, J.; Brzica, A. Age, Growth, and Utility of Otolith Morphometrics as Predictors of Age in the European Barracuda, *Sphyraena sphyraena* (Sphyraenidae): Preliminary Data. *Fishes* **2022**, *7*, 68. [CrossRef]
14. Ordoñez, A.; Eikvil, L.; Salberg, A.; Harbitz, A.; Elvarsson, B. Automatic Fish Age Determination across Different Otolith Image Labs Using Domain Adaptation. *Fishes* **2022**, *7*, 71. [CrossRef]
15. Politikos, D.; Sykiniotis, N.; Petasis, G.; Dedousis, P.; Ordoñez, A.; Vabø, R.; Anastasopoulou, A.; Moen, E.; Mytilineou, C.; Salberg, A.; et al. DeepOtolith v1.0: An Open-Source AI Platform for Automating Fish Age Reading from Otolith or Scale Images. *Fishes* **2022**, *7*, 121. [CrossRef]
16. Roloson, S.; Knysh, K.; Landsman, S.; James, T.; Hicks, B.; van den Heuvel, M. The Lifetime Migratory History of Anadromous Brook Trout (*Salvelinus fontinalis*): Insights and Risks from Pesticide-Induced Fish Kills. *Fishes* **2022**, *7*, 109. [CrossRef]
17. Shrimpton, J.; Breault, P.; Turcotte, L. Fidelity to Natal Tributary Streams by Kokanee Following Introduction to a Large Oligotrophic Reservoir. *Fishes* **2022**, *7*, 123. [CrossRef]
18. Jiang, T.; Liu, H.; Hu, Y.; Chen, X.; Yang, J. Revealing Population Connectivity of the Estuarine Tapertail Anchovy *Coilia nasus* in the Changjiang River Estuary and Its Adjacent Waters Using Otolith Microchemistry. *Fishes* **2022**, *7*, 147. [CrossRef]

MDPI

Article

Growth and Microstructural Features in Otoliths of Larval and Juvenile *Sinogastromyzon wui* (F. Balitoridae, River Loaches) of the Upper Pearl River, China

Minghui Gao [1], Zhiqiang Wu [2,*], Liangliang Huang [2,3,*], Xichang Tan [4], Mingsi Li [2] and Haibo Huang [2]

1 College of Life Science and Technology, Guangxi University, Nanning 350000, China; gmhxxzj@126.com
2 College of Environmental Science and Engineering, Guilin University of Technology, Guilin 541000, China; lms980330@163.com (M.L.); eleveshhb@126.com (H.H.)
3 Coordinated Innovation Center of Water Pollution Control and Water Security in Karst Area, Guilin University of Technology, Guilin 541004, China
4 Bureau of Hydrology and Water Resources, Pearl River Conservancy Commission of Ministry of Water Resources, Guangzhou 510000, China; rjimtxc@hotmail.com
* Correspondence: wuzhiqiang@glut.edu.cn (Z.W.); llhuang@glut.edu.cn (L.H.)

Citation: Gao, M.; Wu, Z.; Huang, L.; Tan, X.; Li, M.; Huang, H. Growth and Microstructural Features in Otoliths of Larval and Juvenile *Sinogastromyzon wui* (F. Balitoridae, River Loaches) of the Upper Pearl River, China. *Fishes* **2022**, *7*, 57. https://doi.org/10.3390/fishes7020057

Academic Editor: Filipe Martinho

Received: 5 January 2022
Accepted: 26 February 2022
Published: 1 March 2022

Abstract: Otolith growth and microstructural features of fish are essential to the understanding of the early fish lifecycle. This paper assesses the features of otoliths from laboratory-reared larval and juvenile *Sinogastromyzon wui* (*S. wui*, 0 to 25 days post-hatching) that were obtained as eggs from the Shilong Reach of Xijiang River between April and August 2021. We observed the development of the three pairs of otoliths (lapilli, sagittae, and asterisci) and compared the shape changes and growth of the lapilli and sagittae, as well as the timing and deposition rate of increments of the lapilli. The lapilli and the sagittae were visible on hatching, whereas the asterisci were present at four days post-hatching (dph). The shape of the sagitta changed more obviously than that of the lapillus, and a strong correlation was observed between sagitta shape changes and fish ontogenesis. The otolith shape greatly modulated during the post-flexion larval stage (Post-FLS), it corresponded with the formation period of individual fins. Analysis of the microstructural features indicated that lapilli were the optimal otolith for age determination and increment deposition rate confirmation. Using regression analysis of the known age and the number of lapillus daily increments, we demonstrated that the lapillus developmental increments were deposited daily, and the first increment formed at two days post-hatching. Our conclusions support employing the lapillus increment deposition rate and the time of the first daily increments in the determination of the age of wild larval and juvenile *S. wui*.

Keywords: ontogeny; teleost anatomy; development; fish morphology; Fourier

1. Introduction

Otoliths are acellular biomineralized concretions of calcium carbonate and other minor elements (Na, Sr, K, S, N, Cl, and P), generated on a protein matrix in vertebrates' inner ears [1]. These structures are mainly used for sound reception and balance orientation [2,3]. The evidence of daily otolith growth was discovered and described by Panella in the 1970s [4]. In subsequent studies, the otolith daily increment deposition was shown to commonly occur in the early life stages of Osteichthyes [5,6].

The fish developmental rule involves resource dynamic assessment and fishery management [7,8]. During the early fish lifecycle, otoliths develop with the growing fish. The otolith growth model is based on the otolith radius (OR) and the total length (TL), which are used to infer the developmental rate of the larval and juvenile fish [9]. The daily increment width is influenced by water temperature and prey and can be used to study the spawning populations in varying seasons [10,11]. In addition, the otolith shape alters as it grows. Therefore, otolith shape analysis is often used to distinguish fish species or populations [12,13].

Otoliths are one of the most studied elements of the teleost fish anatomy, because they represent a permanent record of life history [14]. Otolith development is modulated by both endogenous (ontogeny, physiology, and feeding habits) and exogenous elements (water depth, temperature, salinity, and substrate), and the otolith microstructure possesses a certain degree of species-specificity [15,16]. Hence, the otolith microstructure is often used to examine early life events such as hatching, first-feeding, and habitat alteration [17,18]. In addition, the central nuclear characteristics can be used to distinguish between varying fish populations [19], while the width of the otolith increments can be used to analyze the growth stage of fish in early life history [20].

Sinogastromyzon wui comes from the Balitoridae family and Homalopterinae subfamily. It is among the most prevalent species in the upper reaches of the Pearl River [21]. *S. wui* spawns drifting eggs, which develop in drifting water and it is a member of Ostariophysi, whose inner ear (and sagittae, in particular) is highly modified during their ontogenesis. In recent years, with the dams' construction, the river's hydrology has altered significantly. The reproduction and population recruitment of *S. wui* are key indices for the evaluation of fisheries' resources. However, there are limited studies on the early growth of *S. wui*. Herein, we explored the morphological alterations in otoliths during ontogenesis and examined a potential association between age and shape formation. Furthermore, we confirmed the deposition time of the first increment and identified the deposition frequency. Our conclusions will provide important benefits to fishery resource protection and assessment.

2. Materials and Methods

2.1. Fish Rearing and Sampling

Our fish egg samples were collected from the Shilong Reach of the Xijiang River (109°31′30″ E, 23°52′21″ N) between April and August 2021. The collection was performed daily during the spawning season using Jiang nets (total length 5 m; rectangular iron opening/mouth 1.0 m × 1.5 m and a mesh net size of 0.5 mm attached to a 0.8 m × 0.4 m × 0.4 m filter collection bucket), which were placed vertically below the surface of the water and against the current. Each sampling period lasted 1 h (6–7 a.m.).

After collection, the eggs were sorted in the laboratory. *S. wui* eggs were readily recognized by their non-viscous pale yellow texture, and they were of a radius between 0.4 and 0.5 cm. Each sample collection rendered 50–80 *S. wui* eggs, which were then maintained in 1.5 L Zug bottles. Once the eggs hatched, the larvae were transferred to 30 L water pools. The water in the Zug bottles and pools was river water from the sampling site, and an air distributor was added to provide sufficient oxygen. The water temperature was maintained between 19 and 28 °C. The photoperiod was equal to the natural length of the day (23°52′ N). Following 2 dph, the larvae were fed with cooked chicken egg yolks three times a day. After half an hour, any unconsumed yolks were suctioned out of tanks, with the help of a water cleaner (WEIPRO, TC3500, Guangzhou, China), to keep the water clean.

2.2. Otolith Preparation

While rearing, all samples were separated into four ontogenetic stages: the preflexion larval stage (Pre-FLS), the flexion larval stage (FLS), the post-flexion larval stage (Post-FLS), and the juvenile stage (JS) [22]. Otolith samples were collected daily during Pre-FLS and FLS; then, that otolith samples were selected once every 2 days during Post-FLS and JS. In total, 320 individuals ranging from 5.66 to 13.93 mm were sampled (10 in every age group used for shape research and 10 for validating increment deposition frequency).

Before dissection, the larvae and juveniles were anesthetized with lidocaine hydrochloride (30 mg/L), and the TL was measured with vernier calipers. All of the otoliths (lapilli, sagittae, and asterisci) from the experimental fish were selected using a microscope (Motic, SMZ-168). The otoliths were first cleaned with absolute ethyl alcohol and, subsequently, encapsulated on a glass slide with colorless enamel resin. Finally, the otoliths were photographed under a microscope (Olympus, BX53) with a camera (MicroPublisher 5.0 Real-Time Viewing).

2.3. Assessment of Otolith Development

Given that asterisci were not present upon hatching, this study only examined lapilli and sagittae growth. Otoliths are generally divided into four growth areas; namely, anterior, posterior, dorsal, and ventral (Figure 1) [23]. The growth rate of each area was expressed by the otolith radius of that area. Moreover, the otolith radius was measured using a computer image analysis system (computer imaging identification system for the analysis of otolith microstructure of fish) [24].

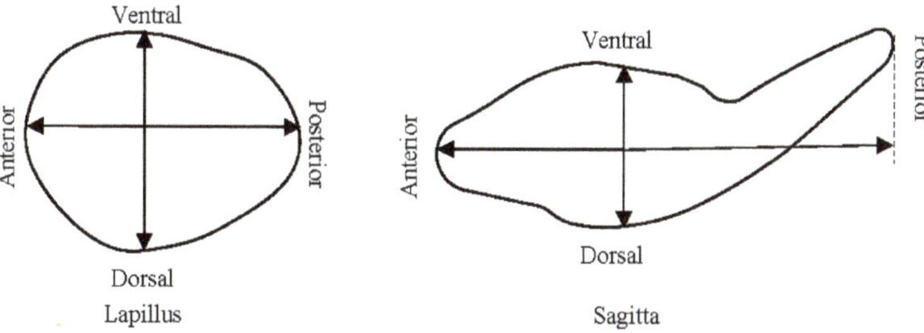

Figure 1. The measured radius of each growth area of lapillus and sagitta.

To better elucidate otolith shape alterations with age, parameters closely linked to 140 groups of lapilli and sagittae (once every 2 days during the larval and juvenile stages) were obtained using elliptic Fourier analysis [25,26]. The shapes of lapilli and sagittae were described as a two-dimensional projection carrying elements termed harmonics. Next, each otolith was represented by four elliptic Fourier coefficients, which mathematically normalized its shape, without any alterations to the otolith size, position, or rotation. Following this, the Fourier power spectrum was calculated for each otolith, which determined the appropriate quantity of harmonics for each otolith using a 99.99% accurate contour definition [27]. Our results suggested 20 harmonics, with each harmonic carrying 4 coefficients, namely, a, b, c, and d (80 Fourier coefficients in total). However, since the initial three elliptic Fourier coefficients of the first harmonics were adjusted to constant values (a = 1, b = 0, and c = 0), they were disregarded, giving us a final count of 77 Fourier coefficients. Standardization and coefficient calculations were carried out in SHAPE 1.3 software [28].

2.4. Daily Increment Verification and Spacing Assessment

We excluded sagittae from the daily increment assessment due to its highly fragile nature. Instead, we employed 25 dph left lapilli, each of which was encapsulated on a glass slide with colorless enamel resin and buffed using 3000-grit sandpapers to expose the central nucleus. Images were captured with a digital camera, as described before. Lapillus counting was conducted twice by three separate individuals. Error rates were calculated for each counter. Readings were only included in the analysis if the error rate remained <5% [26].

2.5. Statistical Analysis

To determine whether the otolith shape differed between ontogenetic stages, a canonical discriminant analysis (CDA) on the principal component scores in the scores file (*.pcs) output by PrinComp was performed in SPSS 2020. The Wilks' Lambda of each discriminant function was used to assess the CDA performance, ranging from 0 (total discrimination) to 1 (no discrimination).

The linear relationship between the known age and daily increment quantity was used to determine the deposition frequency. A *t*-test was used to test the difference between the slope of the linear function and 1. The increment frequency was verified when the t between the slope and 1 was insignificant. The association between the otolith radius (OR) and age (days) was fitted to the following logistic growth model: $OR_t = (OR_\infty \times OR_0)/((OR_\infty - OR_0)\, e^{-kt} + OR_0)$, gompertz growth model: $OR_\infty \times (OR_\infty/OR_0)\hat{}e^{-kt}$, or linear growth model: $OR_t = kt + OR_0$.

In both cases, t represented age in days, OR_t represented the OR stage t, k represented the instantaneous growth rate, and OR_0 represented the initial OR. The parameters were derived using Origin 2018 (OriginLab, Northampton, MA, USA). In addition, Origin 2018 was used to fit the relationship between otolith length and total length.

3. Results

3.1. Ontogenetic Development

The collected eggs were mostly in the tailbud and caudal fin-appearing stage. After 5–8 h of incubation at 19–28 °C, the eggs began to develop otoliths. At this time, the auditory sac formed an oval shape and two otoliths (lapilli and sagittae) were visible inside. Following a 7–10 h incubation period, the eggs began to hatch. The Pre-FLS was from 0 to 3 dph. Nutrition was provided via yolk sac, and the notochord was not upturned. The FLS was from 3 to 6 dph. At this point, the notochord was upturned, and nutrition was provided via exogenous sources. Moreover, the caudal fin was almost completely formed, and the dorsal fin started to develop at 6 dph. The Post-FLS was from 6 to 13 dph, at which point, the pectoral, dorsal, and pelvic fins were visible. Lastly, the JS was from 14 to 25 dph, when the whole body became covered in scales. The transition age between the Pre-FLS and the FLS was at 4 dph, between the FLS and the Post-FLS at 6 dph, and between the Post-FLS and the JS at 13.5 dph. Otoliths of 2–3 groups were selected from the samples at each stage; the sample population, predicted otolith age (increment number), predicted daily developmental rate, and TLs were calculated. The data are presented in Table 1.

Table 1. Sampling details for each stage of larval and juvenile *Sinogastromyzon wui*, including the sample size (N), otolith age estimates (increment count), daily growth rate estimate (DGR), and TL.

DPH	N	Increment Count (Mean ± SD, Days)	DGR (Mean ± SD, μm·day^{-1})	TL (Mean ± SD, mm)
2	10	1 ± 1	3.02 ± 0.26	6.02 ± 0.25
3	10	2 ± 1	2.64 ± 0.31	6.68 ± 0.26
4	10	3 ± 1	3.23 ± 0.49	6.75 ± 0.31
7	10	6 ± 2	4.49 ± 0.54	6.92 ± 0.40
13	10	12 ± 2	3.46 ± 0.58	8.46 ± 0.73
17	10	16 ± 2	4.99 ± 0.34	9.95 ± 0.67
23	10	22 ± 3	4.38 ± 0.43	11.31 ± 1.12

Only the lapilli and sagittae were visible after hatching, with the asterisci emerging at 4 dph. All otolith pairs underwent anatomical alterations during ontogenesis (Figure 2). Lapilli were round in shape at hatching (0 dph, TL = 6.05 ± 0.25 mm) and maintained their round shape from Post-FLS (13 dph, TL = 8.46 ± 0.73 mm) until the end of JS (25 dph, TL = 12.03 ± 0.92 mm), when the otolith developed into an ovoid.

The shape of the sagitta altered significantly during development. From the newly hatched larval stage to the Pre-FLS (0 dph, TL = 6.59 ± 0.28 mm), the shape of the sagitta changed from nearly round to oval. Subsequently, from the FLS (7 dph, TL = 6.92 ± 0.40 mm) until the Post-FLS (13 dph, TL = 8.46 ± 0.73 mm), it was shaped like a sickle. During the JS, the sagitta continued to develop but was more stable in terms of structure. The asteriscus was renal when it appeared (4 dph, TL = 6.68 ± 0.31 mm).

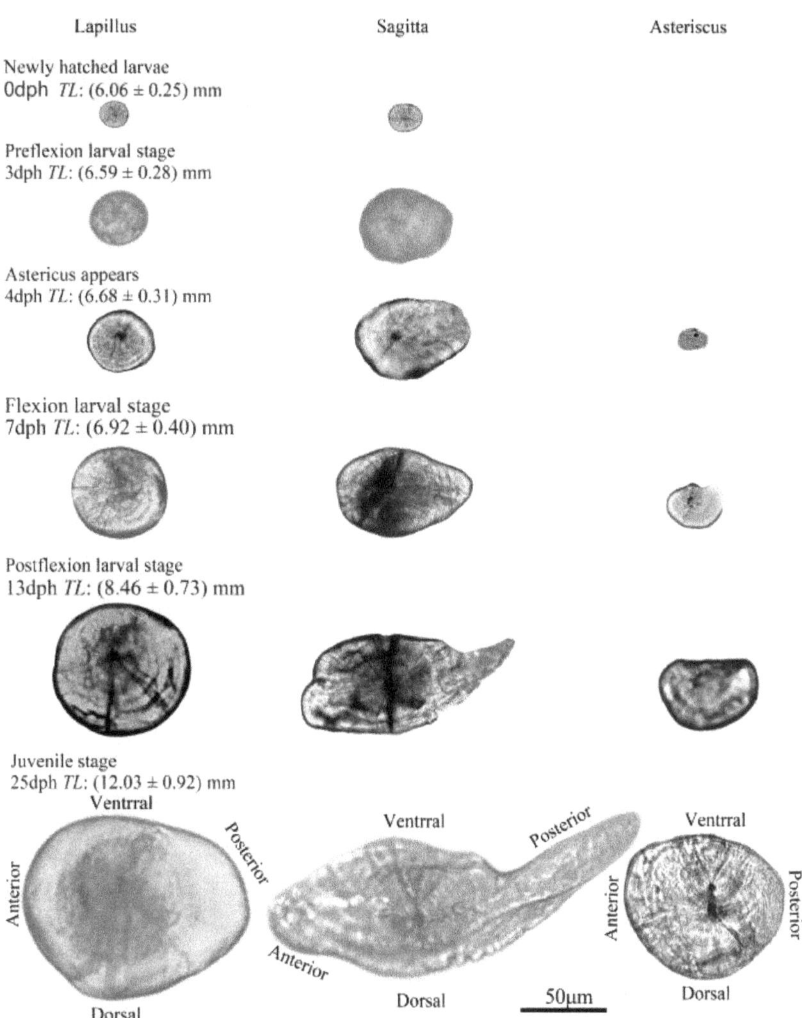

Figure 2. Otolith development in *Sinogastromyzon wui*.

3.2. Lapilli and Sagittae Shape Analysis

Our shape analyses demonstrated that both lapillus and sagitta structures underwent alterations with age. The age distinctions were more pronounced in sagittae than lapilli. Based on the lapilli CDA, the age-related changes were not distinguishable (Wilks' $\lambda > 0.05$; LD1 77.1%, LD2 13.5%) with the two discriminant functions of CDA (Figure 3a). In contrast, the sagittae CDA showed that the age distinctions were obvious (Wilks' $\lambda < 0.05$; LD1 98.3%) using the first discriminant function of CDA, which stratified ages into four classes (1 and 3 days; 5 days; 9, 13, and 17 days; 25 days) (Figure 3b).

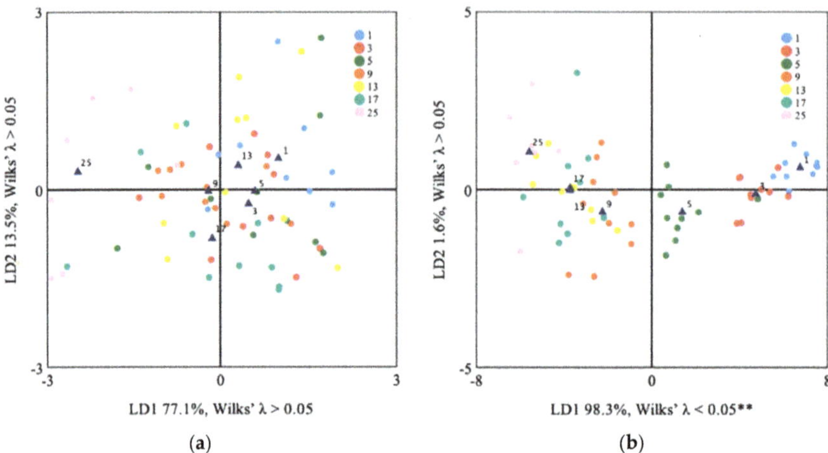

Figure 3. Canonical discriminant analyses performed on the Fourier coefficients of lapilli (**a**) and sagittae (**b**) from *Sinogastromyzon wui* of each age class. **, $p < 0.01$.

3.3. Otolith Growth

In this study, the logarithmic growth model displayed a very good fit for lapilli growth (Figure 4a), and the results showed no significant difference in the growth of each otolith area ($p > 0.05$). In contrast, the linear and gompertz growth model fit the sagitta growth remarkably (Figure 4b), and the growth in each area showed a significant difference ($p < 0.05$).

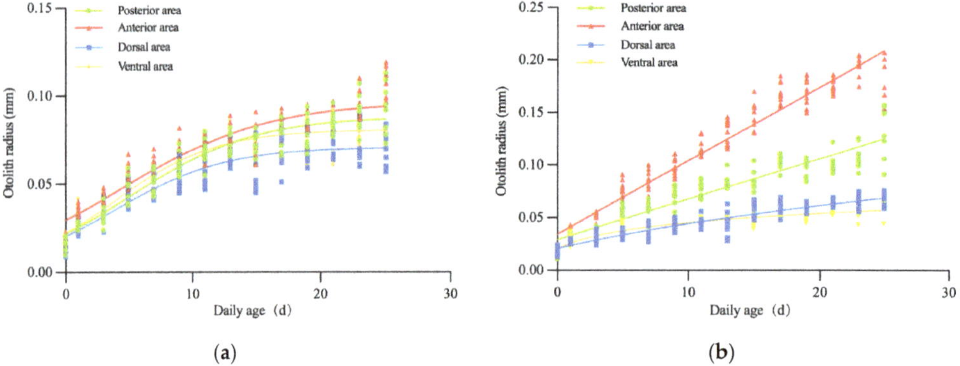

Figure 4. The relationship between daily age and otolith radius of lapilli (**a**) and sagittae (**b**) of *Sinogastromyzon wui*: (**a**) posterior area, $OR_t = -0.045 + 0.04 \times \ln(t + 4.563)\,(R^2 = 0.9086)$; anterior area, $OR_t = -0.013 + 0.033 \times \ln(t + 2.72)\,(R^2 = 0.9090)$; dorsal area, $OR_t = -0.001 + 0.022 \times \ln(t + 1.806)\,(R^2 = 0.8370)$; and ventral area, $OR_t = -0.0025 + 0.025 \times \ln(t + 1.631)\,(R^2 = 0.8299)$; (**b**) posterior area, $OR_t = 0.004t + 0.029\,(R^2 = 0.8838)$; anterior area, $OR_t = 0.0074t + 0.034\,(R^2 = 0.9449)$; dorsal area, $OR_t = 0.022e^{(-0.021x)}\,(R^2 = 0.8479)$; and ventral area $OR_t = 0.023e^{(-0.124x)}\,(R^2 = 0.8223)$.

The otolith length of lapilli (OL_L) was larger than the otolith length of sagittae (OL_S) at larval hatching, but OL_S was larger than OL_L when the body length of larvae reached 6.02 ± 0.25. Correlations between TL and OL_L or OL_S were fitted to the following exponential growth model: $OL_L = 4.494e^{4.694TL}$ ($R^2 = 0.786$, n = 150) (Figure 5) and $OL_S = 5.089e^{2.548TL}$ ($R^2 = 0.845$, n = 150) (Figure 5).

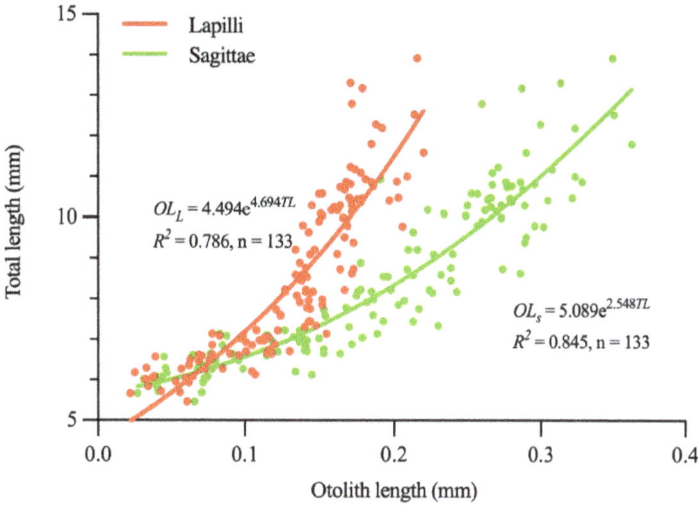

Figure 5. The relationship between otolith length of lapilli or sagittae and total length of *Sinogastromyzon wui*.

3.4. The Lapilli Otolith Microstructure

The otolith core area had one central primary primordia (Figure 6). Following incubation at 21–26 °C water temperature, the first daily increment of lapilli formed at 2 dph. The primordial core radius and the first developmental increment (F1) was between 11.0 and 14.4 µm in length, with an average of 13.5 ± 1.1µm (n = 10). Moreover, the width of the daily increment fluctuated. The daily increment width before Post-FLS (7–12 dph) widened (mean, <5.9 µm). However, past the breaking point, the increment width narrowed and then widened again as juveniles reached 20 dph, when it gradually reduced to the otolith edge (Figure 7).

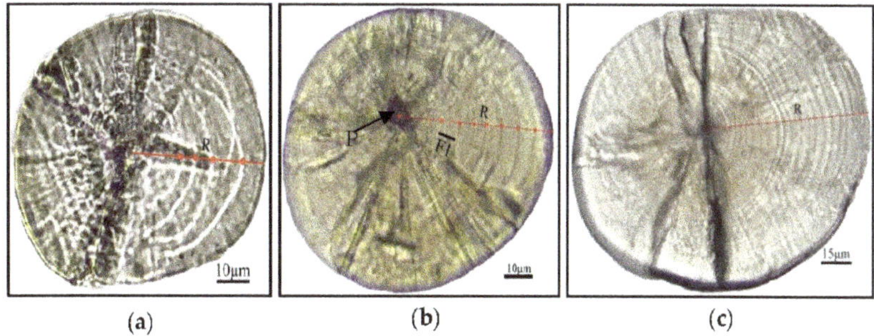

Figure 6. Lapilli otoliths of Sinogastromyzon wui. (**a**) The lapillus otolith of larva at 5 dph; showing the radius (R) along which increments were read and measured. (**b**) The lapillus otolith of larva at 9 dph, showing the primordium (P) and the first increment (F1). (**c**) The lapillus otolith of larva at 23 dph, showing the radius (R) along which increments were read.

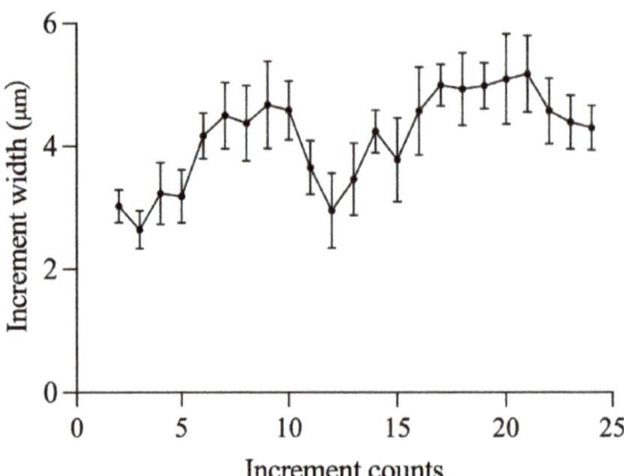

Figure 7. Daily increment widths (mean ± SD) of the lapilli.

3.5. Verification of the Daily Lapilli Increment Formation

Here, we verified the daily increment formation by testing 150 lapilli. To accomplish this, we fitted the correlation between growth increment numbers (N) and larval and juvenile ages (A) to the function as follows:

$$A = 0.97\ N - 1.08 \quad (1)$$

4. Discussion

This study revealed that the lapilli morphology did not significantly alter from the nearly round to ovoid shape, while the sagittae morphology underwent significant changes from nearly a round shape to a sickle. The otolith morphological development is similar to many reported Cypriniformes [26,29,30]. Moreover, otolith morphology is species-specific and is under the regulation of both genetic and environmental factors [31]. For instance, the lapilli of *Macropodus opercularis* develop from an oval shape in the larval stage to prismatic in the JS [32]. Moreover, the sagittae of *Liza haematocheila* develop from an oval shape in the larval stage to a leaf-like shape in the JS [33]. These data indicate that the otolith morphology in the JS exhibits a certain degree of species-specificity, which can be very helpful in species identification during the early stages of fish development [34].

The CDA results revealed no obvious correspondence between lapillus shape development and varying stages of larval and juvenile development. Lapilli were always nearly round during this period, which was consistent with the analysis results of the same growth rate in each lapillus region. However, in terms of the sagitta, there was a significant correlation between the shape development and varying stages of larvae and juveniles. During the Pre-FLS, the growth rate of each sagittal region was the same, and the shape remained round. During the FLS, the anterior and posterior regions began to grow faster than the ventral and dorsal regions; hence, the shape became an ellipse. During the Post-FLS, the anterior and posterior regions continued to grow rapidly, and the shape became a sickle. By the JS, the sagitta shape became quite stable. Therefore, the sagitta shape was more suitable than the lapillus shape for predicting the developmental stages of *S. wui* larvae and juveniles.

Herein, we made novel verifications of the otolith daily increment formation rate and initial increment development in *S. wui* larvae and juveniles. The first increment was obvious at 2 dph. This was consistent with the observation that *Myxocyprinus asiaticus* forms the first increment at 2 dph [29] but was different from other reports that stated as follows: *Oreochromis aureus* and *Cyprinus carpio* form the first increment before hatching [35,36]; *Acanthopagrus schlegeli* and *Theragra chalcogramma* form the first increment at hatching [37,38]; *Cirrhinus molitorella* forms the first increment at 3 dph [39]. The formation of the first daily increment is species-specific, possibly due to the regulation of ontogenetic and environmental factors such as yolk absorption and first-feeding [39–42]. To avoid deviation in larval and juvenile age calculation, it is essential to confirm the initial daily increment deposition time separately [3]. In this study, we further validated the daily lapillus increment formation rates. We revealed that the larval and juvenile age can be calculated by D + 1, which is consistent with reports of most fish [39,43]. The conclusions from this study can be used for the reproductive duration and early growth study of *S. wui*.

The microstructures and growth features of otolith should be extensively studied to better elucidate the critical events and stages of larvae and juveniles [23]. Among the previously examined factors are hatch check [44], first-feeding [35], and development of numerous accessory growth centers [45]. In this study, we only detected one growth center. There was no clear hatching check or sign of first-feeding. We observed fluctuating alterations in daily increment width. For example, the width gradually increased during the FLS and the early JS, which corresponded to the exogenous and active ossification transitions, respectively. Based on other studies, in terms of growth, the association between body length (BL) and otolith radius (OR) can be either linear [45,46] or nonlinear [26,35]. In this study, an exponential growth curvilinear was employed to fit the BL–OL correlation. We observed a shift in the relationship at the end of the larval stage (OL, 0.20–0.25 mm). The same result was also observed in *Cirrhinus molitorella* [35], *Siniperca chuatsi* [45], *Strangomera bentincki,* and *Engraulis ringens* [47], indicating the likelihood that this shift is brought on by endogenous factors such as genetics, onto-genetics, and physiology.

5. Conclusions

These results provide previously unavailable ontogeny and otolith growth information for *S. wui* that can increase an understanding of its biology. In the Xijiang River, the results of this study can be used to assess the dynamics of *S. wui* recruitment and the relationship between their ontogeny and key environmental factors such as river flow, temperature, and feed, which may provide insights into the relationship between changes in river environment and overall population dynamics. Moreover, as *S. wui* is the dominant species in this river, the migration of its spawning grounds may better reveal the impact of dam closure on fish spawning in the upstream section of the river.

Author Contributions: Conceptualization, methodology, software and writing—original draft preparation, M.G.; formal analysis and validation X.T.; investigation, H.H.; resources and data curation, M.L.; writing—review and editing, supervision, Z.W.; project administration and funding acquisition, L.H. All authors have read and agreed to the published version of the manuscript.

Funding: This research was funded by the National Natural Science Foundation of China, grant number "32060830" and "U20A2087".

Institutional Review Board Statement: The study was conducted according to the guidelines of the Declaration of Helsinki and approved by the Institutional Review Board of Animal Experimental Ethics committee of Guangxi University (GXU-2022-1 and 10 April 2021).

Data Availability Statement: Data from this study are available from the corresponding author upon request (Z.W.: wuzhiqiang@glut.edu.cn).

Acknowledgments: We are grateful to the "Chunhui Planning" Project from the Ministry of Education, China. We also thank all the editors and reviewers for providing constructive comments on the present work.

Conflicts of Interest: The authors declare no conflict of interest.

References

1. D'Iglio, C.; Natale, S.; Albano, M.; Savoca, S.; Famulari, S.; Gervasi, C.; Lanteri, G.; Panarello, G.; Spanò, N.; Capillo, G. Otolith Analyses Highlight Morpho-Functional Differences of Three Species of Mullet (Mugilidae) from Transitional Water. *Sustainability* **2021**, *14*, 398. [CrossRef]
2. Panfili, J.; De Pontual, H.; Troadec, H.; Wrigh, P.J. *Manual of Fish Sclerochronology*; Ifremer-IRD: Brest, France, 2002; p. 466.

3. Lundberg, Y.W.; Xu, Y.; Thiessen, K.D.; Kramer, K.L. Mechanisms of otoconia and otolith development. *Dev. Dyn.* **2015**, *244*, 239–253. [CrossRef]
4. Panella, G. Fish otoliths: Daily growth layers and periodical patterns. *Science* **1971**, *173*, 1124–1127. [CrossRef] [PubMed]
5. Cieri, M.D.; McCleave, J.D. Validation of daily otolith increments in glass-phase American eels *Anguilla rostrata* (Lesueur) during estuarine residency. *J. Exp. Mar. Biol. Ecol.* **2001**, *257*, 219–227. [CrossRef]
6. Joh, M.; Takatsu, T.; Nakaya, M.; Higashitani, T.; Takahashi, T. Otolith microstructure and daily increment validation of marbled sole (*Pseudopleuronectes yokohamae*). *Mar. Biol.* **2005**, *147*, 59–69. [CrossRef]
7. Higgins, R.M.; Diogo, H.; Isidro, E.J. Modelling growth in fish with complex life histories. *Rev. Fish Biol. Fish.* **2015**, *25*, 449–462. [CrossRef]
8. Van Poorten, B.T.; Walters, C.J. How can bioenergetics help us predict changes in fish growth patterns? *Fish. Res.* **2016**, *180*, 23–30. [CrossRef]
9. Campana, S.E. *How Reliable Are Growth Back-Calculations Based on Otoliths?* NRC Research Press: Ottawa, ON, Canada, 1990; Volume 47.
10. Hwang, S.D.; Song, M.H.; Lee, T.W.; McFarlane, G.A.; King, J.R. Growth of larval Pacific anchovy Engraulis japonicus in the Yellow Sea as indicated by otolith microstructure analysis. *J. Fish Biol.* **2006**, *69*, 1756–1769. [CrossRef]
11. Takahashi, M.; Watanabe, Y.; Kinoshita, T.; Watanabe, C. Growth of larval and early juvenile Japanese anchovy, *Engraulis japonicus*, in the Kuroshio-Oyashio transition region. *Fish. Oceanogr.* **2001**, *10*, 235–247. [CrossRef]
12. Li, W.; Zhang, C.; Tian, Y.; Liu, Y.; Liu, S.; Tian, H.; Cao, C. Otolith Shape Analysis as a Tool to Identify Two Pacific Saury (*Cololabis saira*) Groups from a Mixed Stock in the High-Seas Fishing Ground. *J. Ocean Univ. China* **2021**, *20*, 402–408. [CrossRef]
13. Souza, A.T.; Soukalová, K.; Děd, V.; Šmejkal, M.; Moraes, K.; Říha, M.; Muška, M.; Frouzová, J.; Kubečka, J. Otolith shape variations between artificially stocked and autochthonous pikeperch (*Sander lucioperca*). *Fish. Res.* **2020**, *231*, 105708. [CrossRef]
14. D'Iglio, C.; Albano, M.; Famulari, S.; Savoca, S.; Panarello, G.; Di Paola, D.; Perdichizzi, A.; Rinelli, P.; Lanteri, G.; Spano, N.; et al. Intra- and interspecific variability among congeneric Pagellus otoliths. *Sci. Rep.* **2021**, *11*, 16315. [CrossRef] [PubMed]
15. Hüssy, K. Otolith shape in juvenile cod (*Gadus morhua*): Ontogenetic and environmental effects. *J. Exp. Mar. Biol. Ecol.* **2008**, *364*, 35–41. [CrossRef]
16. Fischer, P. Otolith microstructure during the pelagic, settlement and benthic phases in burbot. *J. Fish Biol.* **1999**, *54*, 1231–1243. [CrossRef]
17. Zhang, Z.; Beamish, R.J.; Riddell, B.E. *Differences in Otolith Microstructure between Hatchery-Reared and Wild Chinook Salmon (Oncorhynchus Tshawytscha)*; NRC Research Press: Ottawa, ON, Canada, 1995; Volume 52.
18. Campana, S.E.; Thorrold, S.R. Otoliths, increments, and elements: Keys to a comprehensive understanding of fish populations? *Can. J. Fish. Aquat. Sci.* **2001**, *58*, 30–38. [CrossRef]
19. Campana, S.E.; Neilson, J.D. *Microstructure of Fish Otoliths*; NRC Research Press: Ottawa, ON, Canada, 1985; Volume 42.
20. Wu, L.; Liu, J.S.; Wang, X.L.; Zhang, G.; Zhang, Z.Y.; Murphy, B.R.; Xie, S.G. Identification of individuals born in different spawning seasons using otolith microstructure to reveal life history of *Neosalanx taihuensis*. *Fish. Sci.* **2011**, *77*, 321–327. [CrossRef]
21. Gao, M. *Fish Resource in the Early Life Stages and Its Relation to Environmental Factors in Laibin Section of Xijiang River*; Guilin University of Technology: Guilin, China, 2018.
22. Kendall, A.W.; Ahlstrom, E.H.; Moser, H.G. *Early Life History Stages of Fishes and Their Characters*; Allen Press Inc.: Lawrence, KS, USA, 1984.
23. Taiming, Y.; Jiaxiang, H.; Ting, Y.; Liulan, Z.; Zhi, H. Study on the otolith development and the formation of increments in larvae and juvenile of Chuanchia labiosa. *Acta Hydrobiol. Sin.* **2014**, *38*, 764–771. [CrossRef]
24. Zhu, Q.; Xia, L.; Chang, J. Computer identification on otolith microstructure of fish. *Acta Hydrobiol. Sin.* **2002**, *26*, 600–604.
25. Vignon, M. Ontogenetic trajectories of otolith shape during shift in habitat use: Interaction between otolith growth and environment. *J. Exp. Mar. Biol. Ecol.* **2012**, *420–421*, 26–32. [CrossRef]
26. Bounket, B.; Gibert, P.; Gennotte, V.; Argillier, C.; Carrel, G.; Maire, A.; Logez, M.; Morat, F. Otolith shape analysis and daily increment validation during ontogeny of larval and juvenile European chub Squalius cephalus. *J. Fish Biol.* **2019**, *95*, 444–452. [CrossRef]
27. Crampton, J.S. *Elliptic Fourier Shape Analysis of Fossil Bivalves: Some Practical Considerations*; John Wiley & Sons: Hoboken, NJ, USA, 1995.
28. Iwata, H.; Ukai, Y. SHAPE: A computer program package for quantitative evaluation of biological shapes based on elliptic Fourier descriptors. *J. Hered.* **2002**, *93*, 384–385. [CrossRef] [PubMed]
29. Song, Z.; Fu, Z.; Li, J.; Yue, B. Validation of daily otolith increments in larval and juvenile Chinese sucker, *Myxocyprinus asiaticus*. *Environ. Biol. Fishes* **2007**, *82*, 165–171. [CrossRef]
30. Gao, M.H.; Wu, Z.Q.; Huang, L.L.; Tan, X.C.; Liu, H.; Rad, S. Otolith shape analysis and growth characteristics in larval and juvenile Squalidus argentatus. *Environ. Biol. Fishes* **2021**, *104*, 937–945. [CrossRef]
31. Wilson, R.R., Jr. Depth-related changes in Sagitta morphology in six macrourid fishes of the Pacific and Atlantic oceans. *Copeia* **1985**, *1985*, 1011–1017. [CrossRef]
32. Zhao, T.; Chen, G.; Lin, X. Otolith ontogeny and increment of larval Macropodus opercularis. *J. Fish. Sci. China* **2010**, *17*, 1364–1370.

33. Ji, Y.; Zhao, F.; Yang, Q.; Ma, R.; Yang, G.; Zhang, T.; Zhuang, P. Sagittal otolith morphology and the relationship between its mass and the age of Liza haematocheila in the Yangtze Estuary, China. *Chin. J. Appl. Ecol.* **2018**, *29*, 953–960.
34. Campana, S.; Stevenson, D.K. Otolith Microstructure Examination and Analysis. *Copeia* **1992**, 1197. [CrossRef]
35. Karakiri, M.H.C. Preliminary notes on the formation of daily increments in otoliths of Oreochromis aureus. *J. Appl. Ichthyol.* **1989**, *5*, 53–60. [CrossRef]
36. Smith, B.B.W.; Keith, F. Validation of the ageing of 0+ carp (*Cyprinus carpio* L.). *Mar. Freshw. Res.* **2003**, *54*, 1005–1008. [CrossRef]
37. Huang, W.-B.; Chiu, T.-S. Daily Increments in Otoliths and Growth Equation of Black Porgy, Acanthopagrus schlegeli, larvae. *Acta Zool. Taiwanica* **1997**, *8*, 121–130.
38. Akira, N.; Juro, Y. *Age and Growth of Larval and Juvenile Walleye Pollock, Theragra Chalcogramma (Pallas), as Determined by Otolith Daily Growth Increments*; Elsevier: Amsterdam, The Netherlands, 1984; Volume 82.
39. Huang, Y.; Chen, F.; Tang, W.; Lai, Z.; Li, X. Validation of daily increment deposition and early growth of mud carp *Cirrhinus molitorella*. *J. Fish Biol* **2017**, *90*, 1517–1532. [CrossRef]
40. Brothers, E.B.; Mathews, C.P.; Lasker, R. Daily growth increments in otoliths from larval and adult fishes. *Fish. Bull.* **1976**, *74*, 1–8.
41. Jenkins, G.P. Age and growth of co-occurring larvae of two flounder species, Rhombosolea tapirina andAmmotretis rostratus. *Mar. Biol.* **1987**, *95*, 157–166. [CrossRef]
42. Lagardère, F.; Troadec, H. Age estimation in common sole Solea solea larvae: Validation of daily increments and evaluation of a pattern recognition technique. *Mar. Ecol. Prog. Ser.* **1997**, *155*, 223–237. [CrossRef]
43. Durham, B.W.; Wilde, G.R. Validation of Daily Growth Increment Formation in the Otoliths of Juvenile Cyprinid Fishes from the Brazos River, Texas. *N. Am. J. Fish. Manag.* **2008**, *28*, 442–446. [CrossRef]
44. Ding, C.; Chen, Y.; He, D.; Tao, J. Validation of daily increment formation in otoliths for Gymnocypris selincuoensis in the Tibetan Plateau, China. *Ecol. Evol.* **2015**, *5*, 3243–3249. [CrossRef]
45. Song, Y.; Cheng, F.; Zhao, S.; Xie, S. Ontogenetic development and otolith microstructure in the larval and juvenile stages of mandarin fish Siniperca chuatsi. *Ichthyol. Res.* **2018**, *66*, 57–66. [CrossRef]
46. Souza, A.; Soukalová, K.; Děd, V.; Šmejkal, M.; Blabolil, P.; Říha, M.; Jůza, T.; Vašek, M.; Čech, M.; Peterka, J.; et al. Ontogenetic and interpopulation differences in otolith shape of the European perch (*Perca fluviatilis*). *Fish. Res.* **2020**, *230*, 105673. [CrossRef]
47. Molina-Valdivia, V.; Landaeta, M.F.; Castillo, M.I.; Alarcón, D.; Plaza, G. Short-term variations in the early life history traits of common sardine Strangomera bentincki and anchoveta Engraulis ringens off central Chile. *Fish. Res.* **2020**, *224*. [CrossRef]

Article

A Pilot Study Assessing a Concentration of 100 mg/L Alizarin Complexone (ALC) to Mark Calcified Structures in *Hypophthalmichthys molitrix*

Yahua Zhu [1], Tao Jiang [2], Xiubao Chen [2], Hongbo Liu [2], Quinton Phelps [3] and Jian Yang [1,2,*]

[1] Wuxi Fisheries College, Nanjing Agricultural University, Wuxi 214081, China; 2020213001@stu.njau.edu.cn
[2] Key Laboratory of Fishery Ecological Environment Assessment and Research Conservation in Middle and Lower Reaches of the Yangtze River, Freshwater Fisheries Research Center, Chinese Academy of Fishery Sciences, Wuxi 214081, China; jiangt@ffrc.cn (T.J.); chenxb@ffrc.cn (X.C.); liuhb@ffrc.cn (H.L.)
[3] Department of Biology, Missouri State University, Springfield, MO 65897, USA; quintonphelps@missouristate.edu
* Correspondence: jiany@ffrc.cn; Tel.: +86-510-8555-7823

Abstract: The effectiveness of chemical compounds for marking hard tissues in juvenile silver carp (*Hypophthalmichthys molitrix*) is not well known. We analyzed the use of alizarin complexone (ALC) as a fluorescent marker to mark the various hard structures of juvenile silver carp. Experimental fish (~2 months old) were randomly assigned to either control or marking groups, which were immersed in 0 or 100 mg/L ALC solutions, respectively, for 2 days. The otoliths, fin rays, and scales of the fish were then sampled, visualized using fluorescence microscopy, and evaluated after 10 days. The ALC treatment was effective for marking certain hard structures and the marking color was affected by the light source. There were no obvious differences in the marking efficiency of rays from pectoral, dorsal, ventral, anal, and caudal fins, but the lapilli and lateral line scales were marked most effectively from the sampled otolith and scale types, respectively. Our findings indicate that ALC immersion and fin ray and scale sampling can be used for the effective marking and non-lethal evaluation of hard structures in juvenile silver carp.

Keywords: marking; alizarin complexone; hard tissue; fin ray; otolith; scale; silver carp (*Hypophthalmichthys molitrix*)

Citation: Zhu, Y.; Jiang, T.; Chen, X.; Liu, H.; Phelps, Q.; Yang, J. A Pilot Study Assessing a Concentration of 100 mg/L Alizarin Complexone (ALC) to Mark Calcified Structures in *Hypophthalmichthys molitrix*. *Fishes* **2022**, *7*, 66. https://doi.org/10.3390/fishes7020066

Academic Editor: Josipa Ferri

Received: 29 January 2022
Accepted: 14 March 2022
Published: 16 March 2022

Publisher's Note: MDPI stays neutral with regard to jurisdictional claims in published maps and institutional affiliations.

1. Introduction

The rapid decline in global fishery resources has increased the demand for restocking efforts to supplement wild fish populations [1]. For example, wild populations of the silver carp (*Hypophthalmichthys molitrix* Valenciennes 1844), which is one of the four major commercially important carp species in China [2], have been considerably reduced over the past few decades by overfishing, water pollution, habitat loss, and the construction of hydroelectric facilities [3]. Consequently, this fish has recently become an important restocking species for natural resource enhancement.

The marking of hard structures in fish can be used in fisheries ecology for a wide range of applications, including the investigation of life-history characteristics, population dynamics, and movement patterns [4,5], as well as determining restocking effectiveness [6]. Lethally sampled otoliths and non-lethally sampled hard tissues (e.g., fin rays, scales) have received increasing attention as a strategy for assessing fish marking [4,7,8].

The selection of a marking strategy should consider the factors of mark quality [4], as well as time- and cost-effectiveness [5], and the survival rate of marked fish should be highlighted in the development of novel marking techniques [9]. The effectiveness of marking strategies depends on multiple aspects, including life-history stage, marking chemical concentration, fish immersion time, and the method of application [10,11], which can be classified as spraying [12,13], injection [14], dietary supplement [15], or immersion [11].

Fluorescent marking is considered a high-quality strategy for marking hard structures (e.g., otoliths, scales, fin rays) in fish [16]; the resulting chemical marking is easily distinguished from naturally occurring variations in calcified structures [17]. Among the various fluorescent-marking compounds, alizarin complexone (ALC) has recently been successfully applied in studies on several fish species [7,15,18]. However, relatively little is known about the effectiveness of chemical compounds (e.g., ALC) for marking hard tissues in juvenile (the life-cycle stage that is most suitable for mass marking) silver carp [8].

Accordingly, the present study aimed to evaluate ALC immersion as a strategy for marking the hard structures of juvenile silver carp, with a focus on marking structures that can be sampled non-lethally (i.e., scales and fin rays). This evaluation provides a scientific basis for a better understanding of the intra- and inter-hard structure differences of fluorescent marking, as well as the feasibility and operability of this marking technology for other teleost fish.

2. Materials and Methods

2.1. Experimental Fish

Juvenile silver carp (~2 months old) (Figure 1) were purchased from the Changzhou Professional Nursery Breeding Base, Changzhou City, Jiangsu Province, China, and maintained in a glass aquarium (45 cm width × 100 cm length × 50 cm height) filled with aerated tap water in a laboratory. Every day during the temporary culture process, food residues and feces were cleaned from the aquarium bottom, and approximately one-third of the aquarium's volume was replaced with clean water.

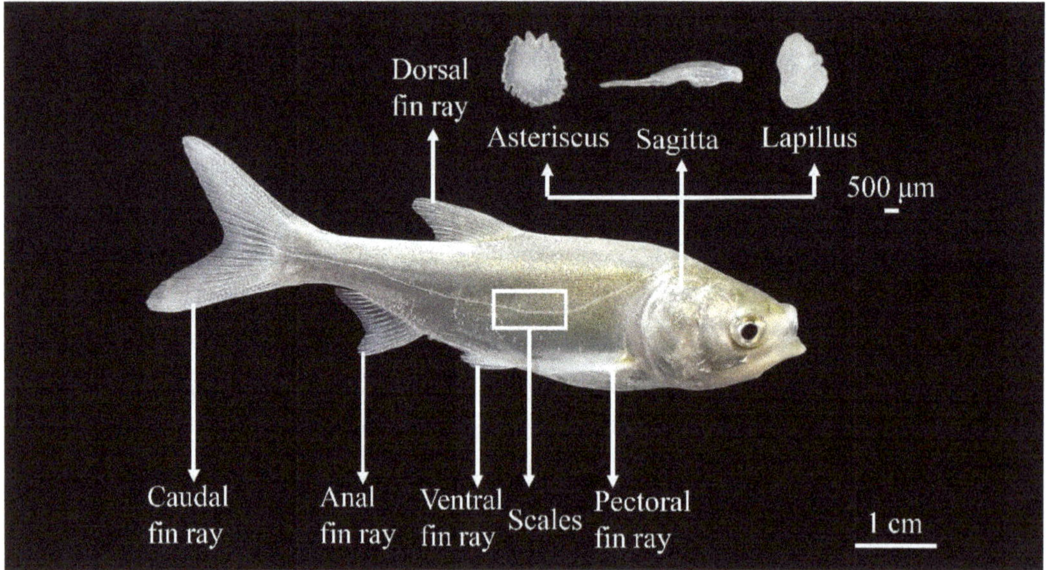

Figure 1. Location of targeted hard structures in the juvenile silver carp (*Hypophthalmichthys molitrix*).

2.2. Marking Procedure

The experiment was conducted at the Nanquan Base of the Freshwater Fisheries Research Center of the Chinese Academy of Fishery Sciences in Wuxi City, Jiangsu Province, China. After approximately three weeks of temporary culture, juvenile fish (five individuals per group) were transferred to marking solutions (0 and 100 mg/L ALC; Bioengineering Co., Ltd., Shanghai, China), which had been prepared using tap water that had been previously aerated for 48 h. After adding 1 g ALC powder into 10 L of aerated tap water, the mixture was slowly stirred with a glass rod for 30 min to accelerate its dissolution and then left for one day to dissolve completely before starting the experiment. After the fish were maintained in the marking solutions for 48 h, the treatment groups were individually transferred to a plastic basin, where they were rinsed in water for 10 min and then subjected to another clean-water-rinsing step. This step was repeated three times to ensure that there was no solution residue on the surface of the fish. The two treatment groups were then separately transferred to two clean glass aquaria (50 cm width × 50 cm length × 40 cm height) for post-marking culture. After 10 days, the fish were sampled, sealed in plastic bags, and frozen (−20 °C) until processing.

2.3. *Hard Tissue Sampling*

Hard tissues (Figure 1) were collected from the fish using a dissecting microscope (NV10, Shanghai Precision Instruments Co., Ltd., Shanghai, China). Lateral line scales and non-lateral line scales (≥5 each) were collected from each fish using forceps, washed in deionized water to remove the majority of the surface impurities and films, and then gently flattened on glass slides using coverslips. The first pectoral, dorsal, ventral, anal, and caudal fin rays were sampled from each fish using dissecting needles and scissors, and all three otolith types were extracted. Both fin rays and otoliths were washed in deionized water to remove the majority of surface impurities and films, dehydrated using anhydrous ethanol, dried, and then embedded in clear nail polish, without grinding or polishing. All samples were stored in the dark and were maintained at temperatures of < 25 °C to prevent the degradation of the chemical markings [19].

2.4. *Marking Evaluation*

Within two weeks of sampling, all hard tissue samples, without any additional grinding, polishing, or sectioning preparation, were visualized using an Olympus BX51 fluorescence microscope (Olympus Co., Tokyo, Japan) (Table 1), which was equipped with an Olympus XC10 digital camera and the "Stream Start" software included with the device (Olympus Co., Tokyo, Japan). The exposure time and gain were set to 20 ms and 5.0 dB, respectively.

Table 1. The combined wavelength of the dichromatic mirror and filter used for visual ALC marker detection.

Light Source	Wavelength (nm)		
	Excitation Filter	**Suppression Filter**	**Dichromatic Mirror**
Brightfield light	–	–	–
Blue excitation light	460–495	510–550	505
Green excitation light	530–550	575–625	570

Referring to the studies by Liu et al. [7] and Taylor et al. [20], marking effectiveness in the present study was categorized into four grades, based on observation under brightfield (BF; i.e., transmitted light), blue (WBS), and green (WGS) excitation lights: 0, no marking (unacceptable); 1, faint marking (unacceptable); 2, moderate marking (acceptable); or 3, strong marking (acceptable). The mode values of marking grades were used as the quality scores of hard tissues from each fish specimen.

2.5. *Statistical Analysis*

Body length and wet mass data are presented as mean ± standard error (SE) values. The significance of differences in the mean body length and wet mass values of the control and ALC-marked groups were evaluated using one-way analysis of variance (ANOVA) in SPSS 23.0 (IBM, Armonk, NY, USA). Statistical significance was set at $p < 0.05$.

3. Results

3.1. Effects of ALC on Fish Survival and Growth

No mortality was observed during the experiment. After the 10-day growth experiment, there were no significant differences observed in either the average body length of the ALC-marked and control groups (64.94 ± 5.51 mm and 67.13 ± 5.27 mm, respectively; $N = 5$ per group) or their body weights (3.58 ± 0.93 g and 4.14 ± 1.26 g, respectively; $N = 5$ per group) based on the results of one-way ANOVAs ($p > 0.05$).

3.2. Otolith Marking

Unlike all the otoliths of the five fish in the control group without any marker rings, all three otolith types of the five juvenile silver carp in the marked group exhibited clear marker rings under visible, blue excitation, and green excitation light, with a marking rate of 100% (Figure 2, Table 2). However, the marking rings were more complete in the lapilli and asterisci than in the sagittae, and marker ring color varied, depending on the light source. For example, under visual light, the marking rings appeared red, with those of the lapilli and sagittae being clearer than those of the asterisci. Meanwhile, under the blue excitation light, the marking rings appeared orange-red, which strongly contrasted with the green coloration of unmarked areas, while under the green excitation light, the marking rings were bright red, whereas the unmarked areas were a less bright red.

Table 2. Effect of a light source on the detection (mode marking grade) of alizarin complexone marking in hard tissues from five juvenile silver carps in the marked group.

Hard Tissue Type	Sample Size	Marking Grade		
		BF	WBS	WGS
Otolith (sagitta)	5	3	3	3
Otolith (lapillus)	5	3	3	3
Otolith (asteriscus)	5	2	2	2
Fin ray (pectoral)	5	1	3	3
Fin ray (dorsal)	5	1	3	3
Fin ray (ventral)	5	1	3	3
Fin ray (anal)	5	1	3	3
Fin ray (caudal)	5	1	3	3
Scales (above lateral line)	25	0	2	2
Scales (on lateral line)	25	0	3	3
Scales (below lateral line)	25	0	2	2

BF: brightfield/transmitted light; WBS: blue excitation light; WGS: green excitation light.

3.3. Fin Ray Marking

Under different light sources and using the fluorescence microscope, fin ray marking efficiency varied (Figure 3, Table 2). Fin ray marking was poorly detectable under visible light but was more evident under the blue and green excitation lights. Under the blue excitation light, the marked areas appeared a faint red, whereas the unmarked areas appeared green, while under the green excitation light, the entire ray appeared red, with brighter fluorescence in the marked areas. There were no obvious differences between the marking efficiencies of the five types of fin rays.

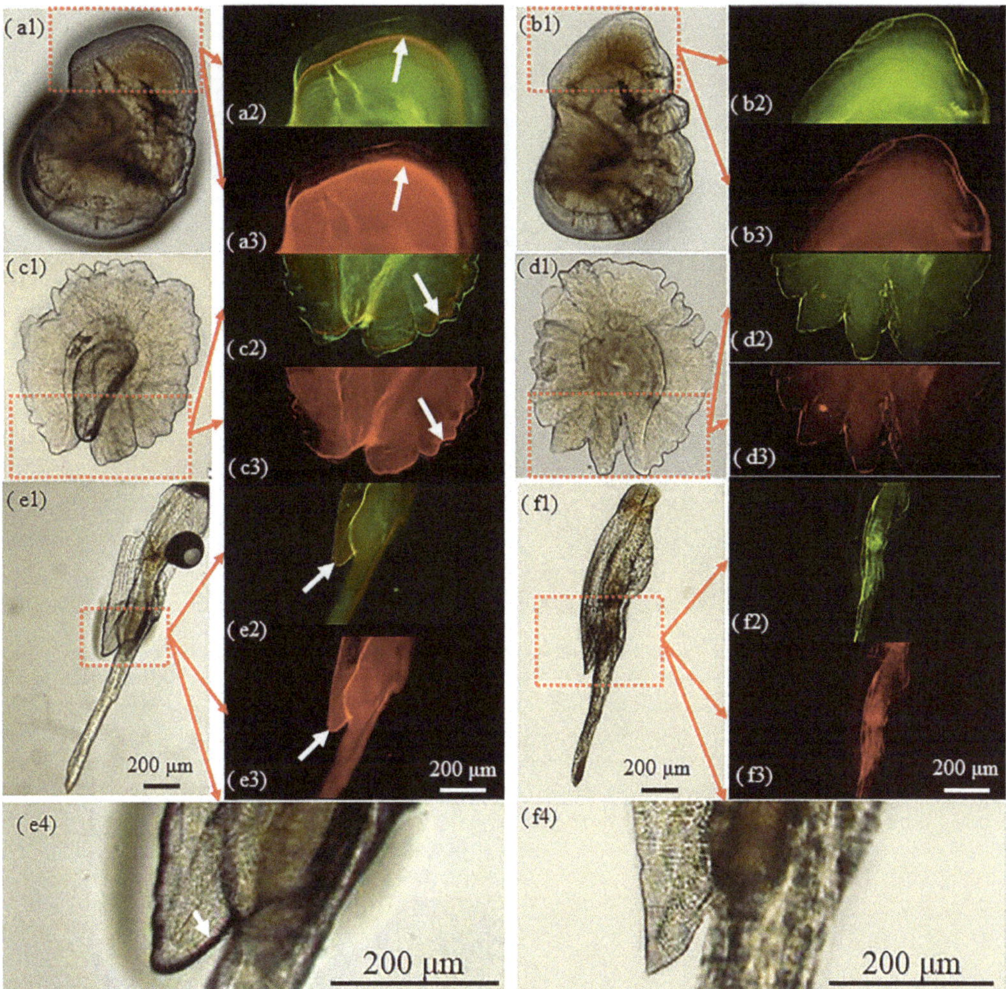

Figure 2. The typical patterns of marking when alizarin complexone (ALC) is used on otoliths (lapillus, sagitta, and asteriscus) from juvenile silver carp, *Hypophthalmichthys molitrix*. Note: (**a1**), ALC-marked lapilli under transmitted light; (**a2**), ALC-marked lapilli under blue excitation light; (**a3**), ALC-marked lapilli under green excitation light; (**b1**), control lapilli under transmitted light; (**b2**), control lapilli under blue excitation light; (**b3**), control lapilli under green excitation light; (**c1**), ALC-marked asterisci under transmitted light; (**c2**), ALC-marked asterisci under blue excitation light; (**c3**), ALC-marked asterisci under green excitation light; (**d1**), control asterisci under transmitted light; (**d2**), control asterisci under blue excitation light; (**d3**), control asterisci under green excitation light; (**e1**), (**e4**) (enlarged view), ALC-marked sagittae under transmitted light; (**e2**), ALC-marked sagittae under blue excitation light; (**e3**), ALC-marked sagittae under green excitation light; (**f1**), (**f4**) (enlarged view), control sagittae under transmitted light; (**f2**), control sagittae under blue excitation light; (**f3**), control sagittae under green excitation light. The white arrows point to the fluorescent marks.

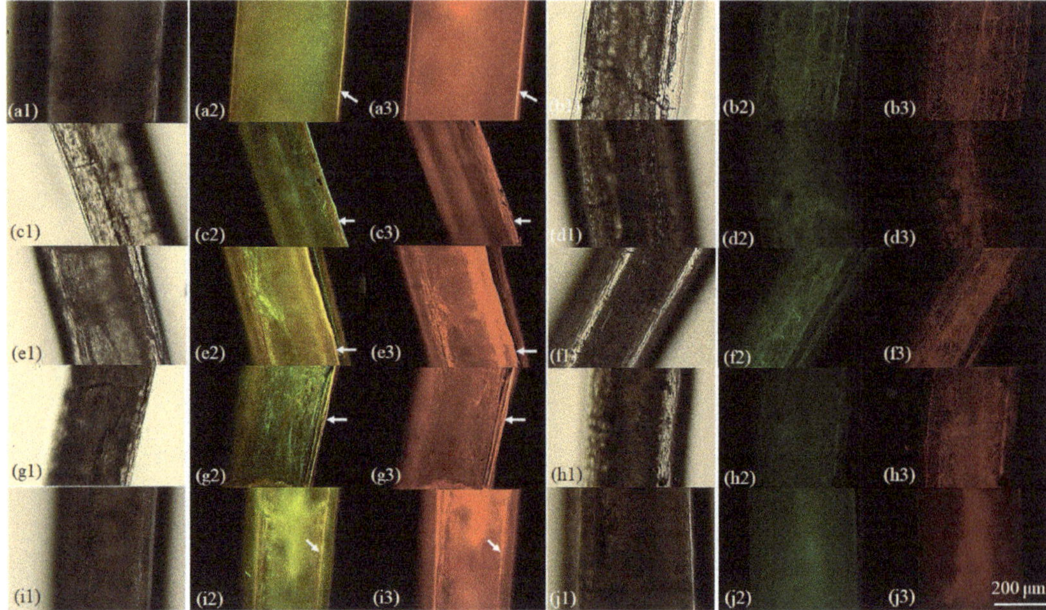

Figure 3. The typical patterns of marking when alizarin complexone (ALC) is used for marking fin rays in juvenile silver carp, *Hypophthalmichthys molitrix*. Note: (**a1**), ALC-marked pectoral fin rays under transmitted light; (**a2**), ALC-marked pectoral fin rays under blue excitation light; (**a3**), ALC-marked pectoral fin rays under green excitation light; (**b1**), control pectoral fin rays under transmitted light; (**b2**), control pectoral fin rays under blue excitation light; (**b3**), control pectoral fin rays under green excitation light; (**c1**), ALC-marked dorsal fin rays under transmitted light; (**c2**), ALC-marked dorsal fin rays under blue excitation light; (**c3**), ALC-marked dorsal fin rays under green excitation light; (**d1**), control dorsal fin rays under transmitted light; (**d2**), control dorsal fin rays under blue excitation light; (**d3**), control dorsal fin rays under green excitation light; (**e1**), ALC-marked ventral fin rays under transmitted light; (**e2**), ALC-marked ventral fin rays under blue excitation light; (**e3**), ALC-marked ventral fin rays under green excitation light; (**f1**, control ventral fin rays under transmitted light; (**f2**), control ventral fin rays under blue excitation light; (**f3**), control ventral fin rays under green excitation light; (**g1**), ALC-marked anal fin rays under transmitted light; (**g2**), ALC-marked anal fin rays under blue excitation light; (**g3**), ALC-marked anal fin rays under green excitation light; (**h1**), control anal fin rays under transmitted light; (**h2**), control anal fin rays under blue excitation light; (**h3**), control anal fin rays under green excitation light; (**i1**), ALC-marked caudal fin rays under transmitted light; (**i2**), ALC-marked caudal fin rays under blue excitation light; (**i3**), ALC-marked caudal fin rays under green excitation light; (**j1**), control caudal fin rays under transmitted light; (**j2**), control caudal fin rays under blue excitation light; (**j3**), control caudal fin rays under green excitation light. The white arrows point to the fluorescent marks.

3.4. Marking of Scales

The marking efficiency of the scales also varied under different light sources using the fluorescence microscope (Figure 4, Table 2). Fin ray marking was unapparent under visible light but was more evident under the blue and green excitation lights. Under the blue excitation light, the marked areas exhibited a weak red fluorescence, whereas the unmarked areas appeared green, and under green excitation light, the entire scale appeared red, with brighter fluorescence in the marked areas. The marking efficiency of the lateral line scales was greater than that of scales above and below the lateral line.

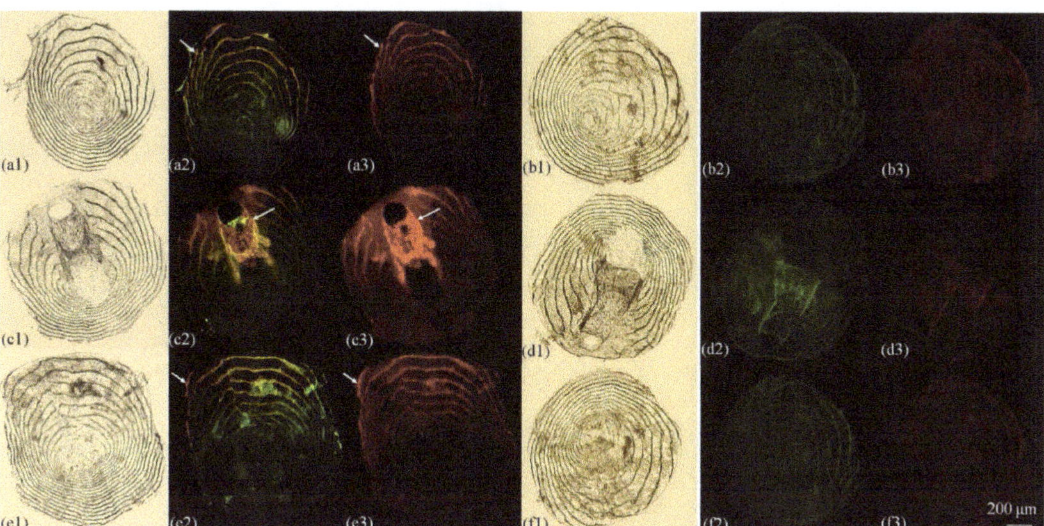

Figure 4. The typical patterns of marking when alizarin complexone (ALC) is used for marking scales in juvenile silver carp *Hypophthalmichthys molitrix*. Note: (**a1**), ALC-marked scales above the lateral line under transmitted light; (**a2**), ALC-marked scales above the lateral line under blue excitation light; (**a3**), ALC-marked scales above the lateral line under green excitation light; (**b1**), control scales above the lateral line under transmitted light; (**b2**), control scales above the lateral line under blue excitation light; (**b3**), control scales above the lateral line under green excitation light; (**c1**), ALC-marked scales on the lateral line under transmitted light; (**c2**), ALC-marked scales on the lateral line under blue excitation light; (**c3**), ALC-marked scales on the lateral line under green excitation light; (**d1**), control scales on the lateral line under transmitted light; (**d2**), control scales on the lateral line under blue excitation light; (**d3**), control scales on the lateral line under green excitation light; (**e1**), ALC-marked scales below the lateral line under transmitted light; (**e2**), ALC-marked scales below the lateral line under blue excitation light; (**e3**), ALC-marked scales below the lateral line under green excitation light; (**f1**), control scales below the lateral line under transmitted light; (**f2**), control scales below the lateral line under blue excitation light; (**f3**), control scales below the lateral line under green excitation light. The white arrows point to the fluorescent marks.

4. Discussion

4.1. Variation of Marking Effectiveness among Hard Structures

Otoliths grow in all directions and build concentric layers from the core [21,22]. Fin rays grow in an additive, incremental manner, with the oldest layers at the core and the newest layers at the outer edge [23]. Scales consist of a distinct osseous upper layer and an underlying proteinaceous basal layer. After initial deposition from the focus (similar to the core), scales grow incrementally to create bony scale circuli [23]. As mentioned above, all hard tissue samples in the present study were not ground, polished, or sectioned. Therefore, in Figures 2–4 of this study, the whole surface of the growth layer marked by ALC during the immersion process of fish is shown, and the boundary lines of the layer in the hard structures (especially the otolith and fin rays) were the brightest in the fluorescent images. Nevertheless, due to differences in the chemical compositions of hard structures in fish, marking effectiveness was expected to vary within and between the hard structures sampled in the present study, namely, otoliths (sagittae, lapilli, and asterisci), scales (below, on, and above the lateral line), and rays from the pectoral, dorsal, ventral, anal, and caudal fins. As expected, under visible light, ALC markings were more visible in the otoliths than in the fin rays or scales, and the effectiveness of ALC marking also varied among the three otolith types. Asterisci are the largest and thinnest otoliths in cyprinid fish and are more transparent; age could not be determined based on the arrangement of the rings [24]. Under both visible and excitation lights, markings were more apparent in the lapilli and sagittae than in the asterisci, which is why asterisci are usually not selected as the focus of research in marking experiments. However, for the sagitta, the entire otolith shape is arrow-type, which is susceptible to external forces and fractures during sampling. Due to the shape of the sagitta, the marked and external rings of the otoliths were close and, thus, difficult to discriminate during microscopy. In contrast, in terms of the morphology of the otolith and the difficulty of the sampling operation, lapilli are oval-shaped, with a more regular external morphology, a clearer ring arrangement, and more distinguishable marking areas. Thus, the lapilli may be the most appropriate otolith type for marking detection, which agrees with the findings of previous studies [24–27].

The effectiveness of fluorescence marking has also been reported to vary among tissue types when using other fluorescent dyes. For example, Elle et al. [28] used calcein to mark rainbow trout (*Oncorhynchus mykiss* Walbaum 1792) and reported that the fluorescence retention of otoliths was greater than that of fin rays. In some cases, such as in the measurement of fish age using oxytetracycline-marked recaptured goliath groupers (*Epinephelus itajara* Lichtenstein 1822), the results obtained from both otoliths and dorsal spines are consistent [29] and, therefore, external tissues (i.e., dorsal spines) can be used for non-lethal age determination. In contrast, dorsal ray sections were so small that readers tended to have lower confidence in age-reading. For scales, as observed under visible light in the present study, studies on other species [30] (e.g., *Aristichthys nobilis* Richardson 1845) marked with calcein and alizarin red S have also reported that marker chemicals are ineffective for marking scales.

The results of the present study demonstrated the intra- and inter-hard structure differences in fluorescent marking, i.e., there were no obvious differences in the marking efficiency of rays from the pectoral, dorsal, ventral, anal, and caudal fins, but the lapilli and lateral line scales of the sampled otolith and scale types, respectively, were marked most effectively. Notably, although otoliths are considered optimal marking structures as they consist of calcium carbonate minerals that are metabolically inert and are generally not resorbed [31], lethal sampling is a major disadvantage. Fin rays and scales can offer a non-lethal alternative to otoliths [32]. Nevertheless, the use of both calcium phosphate structures limits resorption [16,23] and loss of marks by sunlight [16]. Furthermore, scales are well known for being easily peeled off and regenerated [23]. Therefore, the aforementioned advantages and disadvantages among these hard tissues are of critical concern for future applications.

It is noteworthy that the marked areas in all three types of hard structures in ALC-immersed carp were all brighter than those of unmarked structures in the control carp when observed under blue excitation light and, especially, green excitation light (Figures 2–4). Similar cases have been found in other species with other fluorescent markers [7,30,33]. Moreover, Katakura et al. [34] demonstrated that otolith, scale, and dorsal fin ray marks were identified in captive juvenile walleye pollock (*Theragra chalcogramma* Pallas 1814) one year after they were marked with ALC. This evidence suggests that this ALC marking method could be used in the future for identification between marked fish and unmarked nature fish during stock enhancement. Unexpectedly, in this study, almost all the scale circuli appeared to be marked by ALC (Figure 4). The mechanism of this phenomenon is still unknown and should be explored in the future. Nevertheless, it could be assumed that ALC dye may be deposited into the circuli directly from the immersion solution (as in the case reported by Able et al. [35]) and indirectly from the skin layer through scale development (as in the case reported by Katakura et al. [34]).

4.2. Optimum Marker Concentration

The ideal chemical marker concentration for fluorescent marking by immersion varies among fish types and even among life-history stages. For example, ALC has been reported to significantly affect hatching. Long-lasting, distinct otolith marks were produced from the immersion of Baltic cod (*Gadus morhua* Linnaeus 1758) yolk-sac larvae in ≥ 50 mg/L ALC for 24 h. The mortality of eggs and larvae was low during the marking procedure. The hatching success of ALC-marked embryos was significantly reduced, and hatching was delayed with increasing ALC concentrations [36]. Other fluorescent dye markers may also affect fish health [37]. Indeed, results similar to those in the present study reported the immersion of bighead carp (*Aristichthys nobilis* Richardson 1845), grass carp (*Ctenopharyngodon idella* Valenciennes 1844), and black rockfish (*Sebastes schlegelii* Girard 1856) in calcein and alizarin red S solution, and the optimum concentrations for detection of the marker rings in each tissue were not the same [30,33,38].

In the present study, an ALC concentration of 100 mg/L resulted in the successful marking of hard structures in juvenile silver carp. However, under visible light, the marking was more effective in otoliths than in either fin rays or scales. This indicated that the optimal ALC concentrations for marking otoliths, fin rays, and scales were different. Similarly, Lü et al. [8] reported that the immersion of juvenile silver carp in calcein and alizarin red S at concentrations of 50–200 mg/L and 150–300 mg/L, respectively, was ineffective for marking non-lateral line scales. In the present study, the same level of detection was achieved using a lower concentration (100 mg/L) of ALC. Therefore, ALC might be more suitable than calcein or alizarin red S for the fluorescent marking of hard structures in juvenile silver carp. However, if using higher concentrations can improve marking, particularly in fin rays and scales when observed under visible light, while still ensuring fish safety, then the use of higher concentrations should be considered.

5. Conclusions

The findings of the present study demonstrate that immersion in ALC (100 mg/L) can be used for the fluorescent marking of otoliths (lapilli, sagittae, and asterisci), scales (lateral and non-lateral lines), and rays from the pectoral, dorsal, ventral, anal, and caudal fins in juvenile silver carp. Marking effectiveness varied when evaluated using different light sources. However, except in the case of fin rays and scales (the marking of which was not evident when observed under visible light), the marking of all three hard structure types could be observed under visible light, blue excitation light, and green excitation light. The lapilli were the most successful target among the three otolith types, and the marking of lateral line scales was more effective than that of the non-lateral line scales. In contrast, no obvious differences were observed in the marking of rays among the five fin types. Future research should focus more on the practical applications of non-lateral fin ray and scale structures and the development of strategies to avoid their limitations, e.g., material turnover, resorption, loss, or regeneration. The present study provides a method for marking the hard structures of juvenile silver carp and advances the current understanding of ALC marking targets and efficiency, which is invaluable for assessing the success of fish restocking programs involving teleost fishes.

Author Contributions: Conceptualization, J.Y., Y.Z., T.J. and Q.P.; methodology, Y.Z., T.J., X.C., H.L. and J.Y.; investigation, Y.Z., T.J., X.C., H.L. and J.Y.; resources, Y.Z., T.J. and X.C.; funding acquisition, J.Y.; writing—original draft, Y.Z., T.J., X.C., H.L. and Q.P.; writing—review and editing, J.Y.; supervision, J.Y. All authors have read and agreed to the published version of the manuscript.

Funding: This work was supported by the Central Public-Interest Scientific Institution Basal Research Fund, FFRC, CAFS (Grant code 2019JBFM06), the China Central Governmental Research Institutional Basic Special Research Project from Public Welfare Funds, CAFS (Grant code 2021GH08).

Institutional Review Board Statement: The study was conducted according to the guidelines of the Declaration of Helsinki, and approved by the Ethics Committee of Animal Care and Use Committee of the Freshwater Fisheries Research Center at the Chinese Academy of Fishery Sciences. The analysis was carried out following the Guidelines for the Care and Use of Laboratory Animals set by the Animal Care and Use Committee of the Freshwater Fisheries Research Center (2003WXEP61). All operations were carried out with field permit no. 20181AC1128.

Data Availability Statement: Data that support the findings of this study are available from the corresponding author upon reasonable request (Y.J.: jiany@ffrc.cn).

Acknowledgments: We thank Liangcai Pan, Ajun Sun, and Guanming He for their technical support in silver carp aquaculture.

Conflicts of Interest: The authors declare no conflict of interest.

References

1. Warren-Myers, F.; Dempster, T.; Swearer, S.E. Otolith mass marking techniques for aquaculture and restocking: Benefits and limitations. *Rev. Fish Biol. Fish.* **2018**, *28*, 485–501. [CrossRef]
2. Xia, Y.; Li, X.; Yang, J.; Zhu, S.; Wu, Z.; Li, J.; Li, Y. Elevated temperatures shorten the spawning period of silver carp (*Hypophthalmichthys molitrix*) in a large subtropical river in China. *Front. Mar. Sci.* **2021**, *8*, 708109. [CrossRef]
3. Lu, G.; Wang, C.; Zhao, J.; Liao, X.; Wang, J.; Luo, M.; Zhu, L.; Bernatzhez, L.; Li, S. Evolution and genetics of bighead and silver carps: Native population conservation versus invasive species control. *Evol. Appl.* **2020**, *13*, 1351–1362. [CrossRef] [PubMed]
4. Curtis, J.M.R. Visible implant elastomer color determination, tag visibility, and tag loss: Potential sources of error for mark-recapture studies. *N. Am. J. Fish. Manag.* **2006**, *26*, 327–337. [CrossRef]
5. Phinney, D.E.; Miller, D.M.; Dahlberg, M.L. Mass-marking young salmonids with fluorescent pigment. *Trans. Am. Fish. Soc.* **1967**, *96*, 157–162. [CrossRef]
6. Lejk, A.M.; Radtke, G. Effect of marking salmo *Trutta lacustris* L. Larvae with alizarin red S on their subsequent growth, condition, and distribution as juveniles in a natural stream. *Fish. Res.* **2021**, *234*, 105786. [CrossRef]
7. Liu, Q.; Zhang, X.M.; Zhang, P.D.; Nwafili, S.A. The use of alizarin red S and alizarin complexone for immersion marking Japanese flounder *Paralichthys olivaceus* (T.). *Fish. Res.* **2009**, *98*, 67–74. [CrossRef]
8. Lü, H.; Fu, M.; Dai, S.; Xi, D.; Zhang, Z.-X. Experimental evaluation of calcein and alizarin red S for immersion marking of silver carp *Hypophthalmichthys molitrix* (Valenciennes, 1844). *J. Appl. Ichthyol.* **2016**, *32*, 83–91. [CrossRef]
9. Bangs, B.L.; Falcy, M.R.; Scheerer, P.D.; Clements, S. Comparison of three methods for marking a small floodplain minnow. *Anim. Biotelem.* **2013**, *1*, 18. [CrossRef]
10. Partridge, G.J.; Ginbey, B.M.; Woolley, L.D.; Fairclough, D.V.; Crisafulli, B.; Chaplin, J.; Prokop, N.; Dias, J.; Bertram, A.; Jenkins, G.I. Development of techniques for the collection and culture of wild-caught fertilised snapper (*Chrysophrys auratus*) eggs for stock enhancement purposes. *Fish. Res.* **2017**, *186*, 524–530. [CrossRef]
11. Partridge, G.J.; Jenkins, G.I.; Doupé, R.G.; De Lestang, S.; Ginbey, B.M.; French, D. Factors affecting mark quality of alizarin complexone-stained otoliths in juvenile black bream *Acanthopagrus butcheri* and a prescription for dosage. *J. Fish Biol.* **2009**, *75*, 1518–1523. [CrossRef] [PubMed]
12. Gaines, P.C.; Martin, C.D. Feasibility of dual-marking age-0 Chinook salmon for mark–recapture studies. *N. Am. J. Fish. Manag.* **2004**, *24*, 1456–1459. [CrossRef]
13. Schumann, D.A.; Koupal, K.D.; Hoback, W.W.; Schoenebeck, C.W. Evaluation of sprayed fluorescent pigment as a method to mass-mark fish species. *Open Fish Sci. J.* **2013**, *6*, 41–47. [CrossRef]
14. Caraguel, J.-M.; Barreau, T.; Brown-Vuillemin, S.; Iglésias, S.P. In vivo staining with alizarin for ageing studies on Chondrichthyan fishes. *Aquat. Living Resour.* **2020**, *33*, 1. [CrossRef]
15. Watanabe, K.-I.; Ohta, K. Efficacy of a fluorescent tagging technique on otolith and scales using alizarin complexone in artificial seawater for Red Sea bream and Japanese flounder juveniles. *Aquac. Sci.* **2009**, *57*, 627–630. [CrossRef]
16. Lü, H.; Fu, M.; Zhang, Z.; Su, S.; Yao, W. Marking fish with fluorochrome dyes. *Rev. Fish. Sci. Aquac.* **2020**, *28*, 117–135. [CrossRef]
17. Mansur, L.; Catalán, D.; Plaza, G.; Landaeta, M.F.; Ojeda, F.P. Validations of the daily periodicity of increment deposition in rocky intertidal fish otoliths of the South-Eastern Pacific Ocean. *Rev. Biol. Mar. Oceanogr.* **2013**, *48*, 629–633. [CrossRef]
18. Cottingham, A.; Hall, N.G.; Loneragan, N.R.; Jenkins, G.I.; Potter, I.C. Efficacy of restocking an estuarine-resident species demonstrated by long-term monitoring of cultured fish with alizarin complexone-stained otoliths. A case study. *Fish. Res.* **2020**, *227*, 105556. [CrossRef]
19. Stötera, S.; Degen-Smyrek, A.K.; Krumme, U.; Stepputtis, D.; Bauer, R.; Limmer, B.; Hammer, C. Marking otoliths of Baltic cod (*Gadus morhua* Linnaeus, 1758) with tetracycline and strontium chloride. *J. Appl. Ichthyol.* **2019**, *35*, 427–435. [CrossRef]
20. Taylor, M.D.; Fielder, D.S.; Suthers, I.M. Batch marking of otoliths and fin spines to assess the stock enhancement of *Argyrosomus japonicus*. *J. Fish Biol.* **2005**, *66*, 1149–1162. [CrossRef]
21. Romanek, C.S.; Gauldie, R.W. A Predictive model of otolith growth in fish based on the chemistry of the endolymph. *Comp. Biochem. Physiol.* **1996**, *114*, 71–79. [CrossRef]
22. Fisher, M.; Hunter, E. Digital imaging techniques in otolith data capture, analysis and interpretation. *Mar. Ecol. Prog. Ser.* **2018**, *598*, 213–231. [CrossRef]
23. Tzadik, O.E.; Curtis, J.S.; Granneman, J.E.; Kurth, B.N.; Pusack, T.J.; Wallace, A.A.; Hollander, D.J.; Peebles, E.B.; Stallings, C.D. Chemical archives in fishes beyond otoliths: A review on the use of other body parts as chronological recorders of microchemical constituents for expanding interpretations of environmental, ecological, and life-history changes. *Limnol. Oceanogr.* **2017**, *15*, 238–263. [CrossRef]

24. Seibert, J.R.; Phelps, Q.E. Evaluation of aging structures for silver carp from midwestern U.S. rivers. *N. Am. J. Fish. Manag.* **2013**, *33*, 839–844. [CrossRef]
25. Ding, C.; He, D.; Chen, Y.; Jia, Y.; Tao, J. Otolith microstructure analysis based on wild young fish and its application in confirming the first annual increment in Tibetan *Gymnocypris selincuoensis*. *Fish. Res.* **2020**, *221*, 105386. [CrossRef]
26. Durham, B.W.; Wilde, G.R. Validation of daily growth increment formation in the otoliths of juvenile Cyprinid fishes from the Brazos River, Texas. *N. Am. J. Fish Manag.* **2008**, *28*, 442–446. [CrossRef]
27. Gao, M.H.; Wu, Z.Q.; Huang, L.L.; Tan, X.C.; Liu, H.; Rad, S. Otolith shape analysis and growth characteristics in larval and juvenile *Squalidus argentatus*. *Environ. Biol. Fishes* **2021**, *104*, 937–945. [CrossRef]
28. Elle, S.F.; Koenig, M.K.; Meyer, K.A. Evaluation of calcein as a mass mark for rainbow trout raised in outdoor hatchery raceways. *N. Am. J. Fish. Manag.* **2011**, *30*, 1408–1412. [CrossRef]
29. Brusher, J.H.; Schull, J. Non-lethal age determination for juvenile goliath grouper *Epinephelus itajara* from Southwest Florida. *Endang. Species Res.* **2009**, *7*, 205–212. [CrossRef]
30. Lü, H.; Fu, M.; Xi, D.; Yao, W.-Z.; Su, S.-Q.; Wu, Z.-L. Experimental evaluation of using calcein and alizarin red S for immersion marking of bighead carp *Aristichthys nobilis* (Richardson, 1845) to assess growth and identification of marks in otoliths, scales and fin rays. *J. Appl. Ichthyol.* **2015**, *31*, 665–674. [CrossRef]
31. Wood, R.S.; Chakoumakos, B.C.; Fortner, A.M.; Gillies-Rector, K.; Frontzek, M.D.; Ivanov, I.N.; Kah, L.C.; Kennedy, B.; Pracheil, B.M. Quantifying fish otolith mineralogy for trace-element chemistry studies. *Sci. Rep.* **2022**, *12*, 2727. [CrossRef] [PubMed]
32. Clarke, A.D.; Telmer, K.H.; Shrimpton, J.M. Elemental analysis of otoliths, fin rays and scales: A comparison of bony structures to provide population and life-history information for the Arctic grayling (*Thymallus arcticus*). *Ecol. Freshw. Fish* **2010**, *16*, 354–361. [CrossRef]
33. Lü, H.; Zhang, X.; Xi, D.; Gao, T. Use of calcein and alizarin red S for immersion marking of black rockfish *Sebastes schlegelii* Juveniles. *Chin. J. Oceanol. Limnol.* **2014**, *32*, 88–98. [CrossRef]
34. Katakura, S.; Ohta, M.; Jin, M.; Sakurai, Y. Otolith-marking experiments of juvenile walleye pollock *Theragra chalcogramma* using oxytetracycline, alizarin complexone, and alizarin red S. *Aquac. Sci.* **2003**, *51*, 327–335. [CrossRef]
35. Able, K.W.; Sakowicz, G.P.; Lamonaca, J.C. Scale formation in selected fundulid and cyprinodontid fishes. *Ichthyol. Res.* **2009**, *56*, 1–9. [CrossRef]
36. Meyer, S.; Sørensen, S.R.; Peck, M.A.; Støttrup, J.G. Sublethal effects of alizarin complexone marking on Baltic cod (*Gadus morhua*) eggs and larvae. *Aquaculture* **2012**, *324*, 158–164. [CrossRef]
37. Jurgelėnė, Ž.; Montvydienė, D.; Stakėnas, S.; Poviliūnas, J.; Račkauskas, S.; Taraškevičius, R.; Skrodenytė-Arbačiauskienė, V.; Kazlauskienė, N. Impact evaluation of marking *Salmo trutta* with alizarin red S produced by different manufacturers. *Aquat. Toxicol.* **2022**, *242*, 106051. [CrossRef]
38. Lü, H.; Chen, H.; Fu, M.; Peng, X.; Xi, D.; Zhang, Z. Experimental evaluation of calcein and alizarin red S for immersion marking grass carp *Ctenopharyngodon idellus*. *Fish. Sci.* **2015**, *81*, 653–662. [CrossRef]

 fishes

MDPI

Article

Inter-Otolith Differences in Strontium Markings: A Case Study on the Juvenile Crucian Carp *Carassius carassius* (Linnaeus, 1758)

Yahua Zhu [1], Tao Jiang [2], Xiubao Chen [2], Hongbo Liu [2], Quinton Phelps [3] and Jian Yang [1,2,*]

1 Wuxi Fisheries College, Nanjing Agricultural University, Wuxi 214081, China; 2020213001@stu.njau.edu.cn
2 Laboratory of Fishery Microchemistry, Freshwater Fisheries Research Center, Chinese Academy of Fishery Sciences, Wuxi 214081, China; jiangt@ffrc.cn (T.J.); chenxb@ffrc.cn (X.C.); liuhb@ffrc.cn (H.L.)
3 Department of Biology, Missouri State University, Springfield, MO 65897, USA; quintonphelps@missouristate.edu
* Correspondence: jiany@ffrc.cn; Tel./Fax: +86-510-8555-7823

Abstract: The release of hatchery-reared fish fry for restocking is important for the enrichment of fishery resources; however, the effective evaluation of the success rate of marking such fish is challenging. We exposed juvenile crucian carp (*Carassius carassius*) to a single concentration of $SrCl_2 \cdot 6H_2O$ for 5 d and evaluated the efficiency of Sr marking of the fish otoliths (sagittae, asterisci, and lapilli) using an electron probe micro-analyzer. Sr marking signatures formed a peak in all otolith types, with a marking success rate of 100%. The ratio of Sr to Ca in the lapilli and sagittae was higher than that in the asterisci. It took 2 d from the beginning of immersion to the deposition of Sr on the lapilli and sagittae, and the time delay for asterisci was 1 d. For the lapilli and sagittae, it took 16 d to terminate Sr marking and fully recover to the pre-marking Sr level, whereas it was 12 d for the asterisci. The application of the Sr dose had no effect on the survival or growth of the carp. This study demonstrated that the lapilli are the most suitable otolith type for Sr marking observations in crucian carp and provides a theoretical basis and technical support for carp restocking using the Sr marking approach.

Keywords: crucian carp; otolith; strontium; marking efficiency; restocking; time delay

Citation: Zhu, Y.; Jiang, T.; Chen, X.; Liu, H.; Phelps, Q.; Yang, J. Inter-Otolith Differences in Strontium Markings: A Case Study on the Juvenile Crucian Carp *Carassius carassius* (Linnaeus, 1758). *Fishes* **2022**, *7*, 112. https://doi.org/10.3390/fishes7030112

Academic Editor: Josipa Ferri

Received: 15 April 2022
Accepted: 10 May 2022
Published: 15 May 2022

1. Introduction

The decline in fishery resources is a global problem that has driven a high demand for restocking using hatchery-reared larvae and juveniles to supplement wild fish populations [1]. The compensatory recruitment of aquatic organisms is a critical approach to enriching fishery resources, protecting biodiversity, maintaining aquatic ecosystem balances, and improving water quality. Hatchery fish (e.g., salmon) have been released in numbers as high as several million to several billion individuals in some countries around the Pacific Ocean [2]. In China, approximately 150 billion commercial and endangered aquatic organisms (e.g., Cypriniformes, Acipenseriformes, and Pleuronectiformes fish) were released for restocking from 2016 to 2020. To measure the success of restocking efforts in addressing decreasing populations, hatchery-reared individuals must be distinguishable from naturally spawned fish, potentially for many years after compensatory recruitment. Scientifically marking and effectively evaluating the mark–recapture efficiency are still key technical difficulties in the validation of restocking performance worldwide [1].

Fish otoliths are widely used to reveal biological and ecological patterns related to movement and habitat use [3–6]. Salinity, temperature, and ambient trace element concentrations are physicochemical factors that affect otolith chemical composition [7–10]. As the elements in the otolith are not reabsorbed, the otolith forms a permanent record of the chemical environment to which a fish has been exposed throughout its life, known as the otolith chemical signature [7]. Otolith microchemistry can provide detailed life history information and be used to identify fish from different spawning sites [11,12]. Moreover, it is a potential tool for stock identification [13]. Marking can occur naturally [14,15] or via the deliberate manipulation of specific chemicals to create recognizable signatures in the hard structures of fish, especially the otoliths [16]. Therefore, otolith marking using chemicals, including tetracycline hydrochloride [17], oxytetracycline [18], calcein [19], alizarin complexone [20], alizarin red S [21,22], strontium chloride ($SrCl_2$) [17], enriched stable isotopes of barium (Ba) and strontium (Sr) [23,24], and lanthanide elements [25], is believed to be one of the most useful methods for evaluating the effects of mark–recapture. Strontium marking is considered the most suitable method for creating a single mark in the otolith of fish [1].

To the best of our knowledge, most marking studies have focused on a specific pair of otoliths. Notably, the different types of otoliths have potential uses for different research purposes. Specifically, asterisci, the largest of the three pairs of otoliths in cyprinid fish, have usually been used for age determination [26,27]; the lapilli are reported to be superior to the other otoliths for the daily aging of cyprinid species owing to their well-defined daily increment structure [28], and the sagittae are usually used as the experimental materials in migratory fish studies to assess stock connectivity [29,30] and determine movements and life-history patterns [31]. Only one study has used all three types of otoliths to investigate otolith biometry–body length relationships in four fish species [32], and no studies have examined the exact differences among the three types of otoliths (sagittae, asterisci, and lapilli) or evaluated the differences in Sr marking efficiencies among them. To fill this gap, the present study assessed otolith Sr marks in a typical cyprinid fish, crucian carp *Carassius carassius* (Linnaeus, 1758). The objectives of this study were to explore the influence of marking on the growth and survival of the experimental fish, analyze the differences in marking effects among the lapilli, asterisci, and sagittae in this carp species; determine the optimal otolith type for Sr marking, and investigate the feasibility and operability of this technology for teleostean fish.

2. Materials and Methods

2.1. Experimental Materials

The study was carried out at the Southern District Aquaculture Base of the Freshwater Fisheries Research Center of the Chinese Academy of Fishery Sciences in Wuxi City, Jiangsu Province, China. The fish used in the experiments were healthy juvenile *C. carassius* (Figure 1), artificially bred at the aforementioned aquaculture base. In this experiment, we randomly selected 20 fish for each of the marking and control groups. The $SrCl_2 \cdot 6H_2O$ compound (analytical reagent grade) was purchased from Shanghai Fusheng Industrial Co., Ltd. (Shanghai, China). An aqueous solution of 60 mg L^{-1} (i.e., Sr^{2+} nominal concentration of 19.85 mg L^{-1}) was prepared, with tap water aeration applied for more than 48 h, as a backup. The exact Sr^{2+} content, measured using an inductively coupled plasma mass spectrometer (Agilent 7500ce, Agilent, Santa Clara, CA, USA), was 19.20 mg L^{-1}, almost the same as the nominal Sr^{2+} concentration in the solution.

The study was divided into three experimental phases as follows: the temporary feeding phase, the Sr immersion marking phase, and the post-marking culture phase. All experiments were carried out in an immersion exposure system. Experimental fish, which were approximately 90-days-old, were cultured in one glass aquaria (45 cm width × 100 cm length × 50 cm height) containing aerated water. These fish were kept unfed 24 h prior to the Sr immersion marking phase, randomly divided into two groups, and immersed for 5 d in two special marking glass aquaria (20 cm width × 30 cm length × 30 cm height) containing 10 L of aerated water (the concentration of $SrCl_2 \cdot 6H_2O$ was 0 mg L^{-1} and 60 mg L^{-1}, respectively). In addition, no food was provided throughout the periods of the Sr immersion marking phase. After the immersion was completed, the marked and control fish were randomly placed in two glass aquaria of the same specification (50 cm width × 50 cm length × 40 cm height) with 100 L of aerated water for post-marking culture, and the fish were fed to satiation with commercial pellets of Fish Formula Feed 126, mainly containing ca. 28% crude protein, 15% ash, 12.5% moisture, 12% crude fiber, 3% crude fat, 1.2% lysine, and 0.6% phosphorus (Wuxi Tongwei Biological Technology Co., Ltd., Wuxi, China). Fish were fed twice per day at the same time each day, and feed residues were removed daily from each aquarium. During this period, the water was kept continuously oxygenated under natural light, and the water temperature ranged from 23 to 32 °C. Samples were taken every 5 d, with five fish sampled each time, and the experiment ended after all fish were sampled.

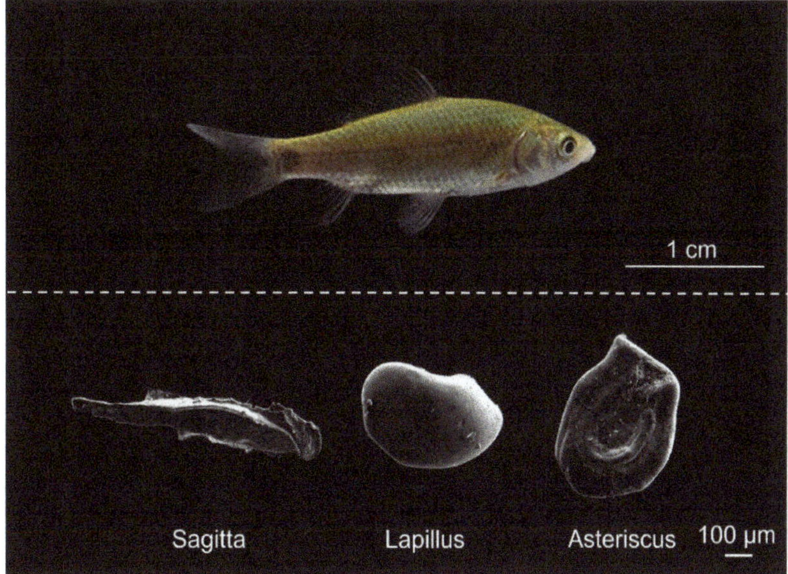

Figure 1. The three types of otoliths of juvenile crucian carp.

2.2. Extraction and Detection of Otoliths

After measuring the weight (g) and total length (mm) of the sampled fish, sagittae in the cytosacular sac, asterisci in the bottle sac, and lapilli in the elliptical sac were removed under a stereoscopic microscope (NV10, Shanghai Precision Instruments Co., Ltd., Shanghai, China; Figure 1). The otolith was first rinsed with ultrapure water that was purified using an Elga Purelab Water System (CLXXXDIM2, ELGA, High Wycombe, UK) and then dried with 99.7% ethanol. Samples were embedded in Epofix Resin and Epofix Hardener (Epofix, Struers, Rødovre, Denmark). They were grinded with sandpaper (1200 and 2000 meshes) until the core was exposed, followed by polishing the surface of the otolith using an automated polishing machine (LaboPol-35, Struers, Rødovre, Denmark) with a polishing solution (OP-S NonDry, Struers, Rødovre, Denmark). After samples were ultrasonically washed, rinsed with Milli-Q water, and completely dried, otoliths were carbon-coated (36 A and 25 s) with a vacuum coating machine (JEE-420, Electronics Co., Ltd., Tokyo, Japan). Elemental line and surface distributions of Sr on the otoliths were detected using an electron probe micro-analyzer (EPMA, JXA-8100, JEOL Co., Ltd., Tokyo, Japan) following similar analytical methods to Liu et al. [33]. The accelerating voltage and beam current were 15 kV and 2×10^{-8} A, respectively. The electron beam was focused on a point that was 2 μm in diameter, with measurements spaced at intervals of 4 μm. The Sr and Ca concentrations on the longest axis from the core to the edge of each otolith sample were measured. X-ray intensity maps of Sr contents were developed for each representative otolith with the EPMA, using an accelerating voltage of 15 kV, beam current of 5×10^{-7} A, counting time of 30 mS, and pixel size of 6×6 μm. The electron beam was focused on a point with a diameter of 5 μm. Measurement quality was validated using calcium carbonate ($CaCO_3$) and strontium titanate ($SrTiO_3$) standards purchased from the Chinese Academy of Geological Sciences, Beijing, China. The backscattered electron (BSE) images of otoliths were taken by a Phenom Pure Desktop Scanning Electron Microscope (Pure-SED, Thermo Fisher Scientific, Eindhoven, The Netherlands). The accelerating voltage was 5 kV.

2.3. Data Processing

Statistical analysis of data was performed using Microsoft Excel 2016. The Sr/Ca ratios of different areas on the otoliths are expressed as the mean ± standard deviation. A Mann–Whitney U-test (SPSS 26.0, IBM SPSS Statistics Inc., Chicago, IL, USA) was used to compare the Sr/Ca ratios in the different otoliths in the marked and control groups after post-marking culture, as well as the otoliths in different stages. The significance level of the difference was set at $p < 0.05$. As the Sr content in otoliths is much lower than the Ca content, the Sr/Ca ratio refers to a standardized ratio, which is Sr/Ca $\times$ 10^3. Different colors in X-ray intensity maps from blue (lowest) to green, yellow, and red (highest) indicate values corresponding to Sr concentrations.

3. Results

3.1. Effect of Sr Marking on the Survival and Growth of Juvenile Crucian Carp

During the 5-d exposure of juvenile crucian carp to Sr, there were no deaths in the marked or control group, indicating that Sr marking had no acute toxicity in the experimental fish. After the post-marking culture for 20 d, there were also no deaths in both groups. There was also no significant difference in the average total length (n = 5) and wet mass (n = 5) between the control group and the marked group ($p > 0.05$), indicating that Sr marking had no chronic effects on the experimental fish (Figure 2). In summary, Sr marking had no significant effect on the survival and growth of experimental fish within the dose range used in this experiment.

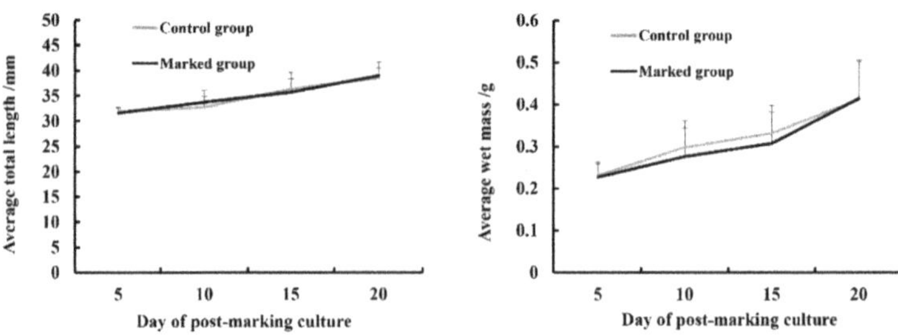

Figure 2. Changes in the means and corresponding standard deviations of wet mass and total length of crucian carp in the marked and control groups at different sampling days (n = 5).

3.2. Otolith Marking

3.2.1. Inter-Otolith Variation in Sr/Ca Ratio

Quantitative line analyses demonstrated that after being soaked in 60 mg/L $SrCl_2{\cdot}6H_2O$ for 5 d, Sr-marked peaks were formed on the sagittae, lapilli, and asterisci otoliths of the juvenile crucian carp (Figure 3). However, among the three types of otoliths, there were differences in the results of line distribution. On the fifth day after the resumption of feeding, the Sr/Ca ratio increased in the sagittae, lapilli, and asterisci. During this time, the Sr/Ca ratios of the lapilli and sagittae were far greater than those of the asterisci. By the 10th day after post-marking culture, the Sr/Ca ratio marked peak of the asterisci differed from that of the lapilli and sagittae, with a gradual decline over time. By the 15th day after post-marking culture, the peak value of Sr marking had decreased significantly, gradually approaching the value of the Sr/Ca ratio before the immersion. On the 20th day of post-marking culture, the Sr/Ca ratio of the three types of otoliths returned to values before the Sr immersion.

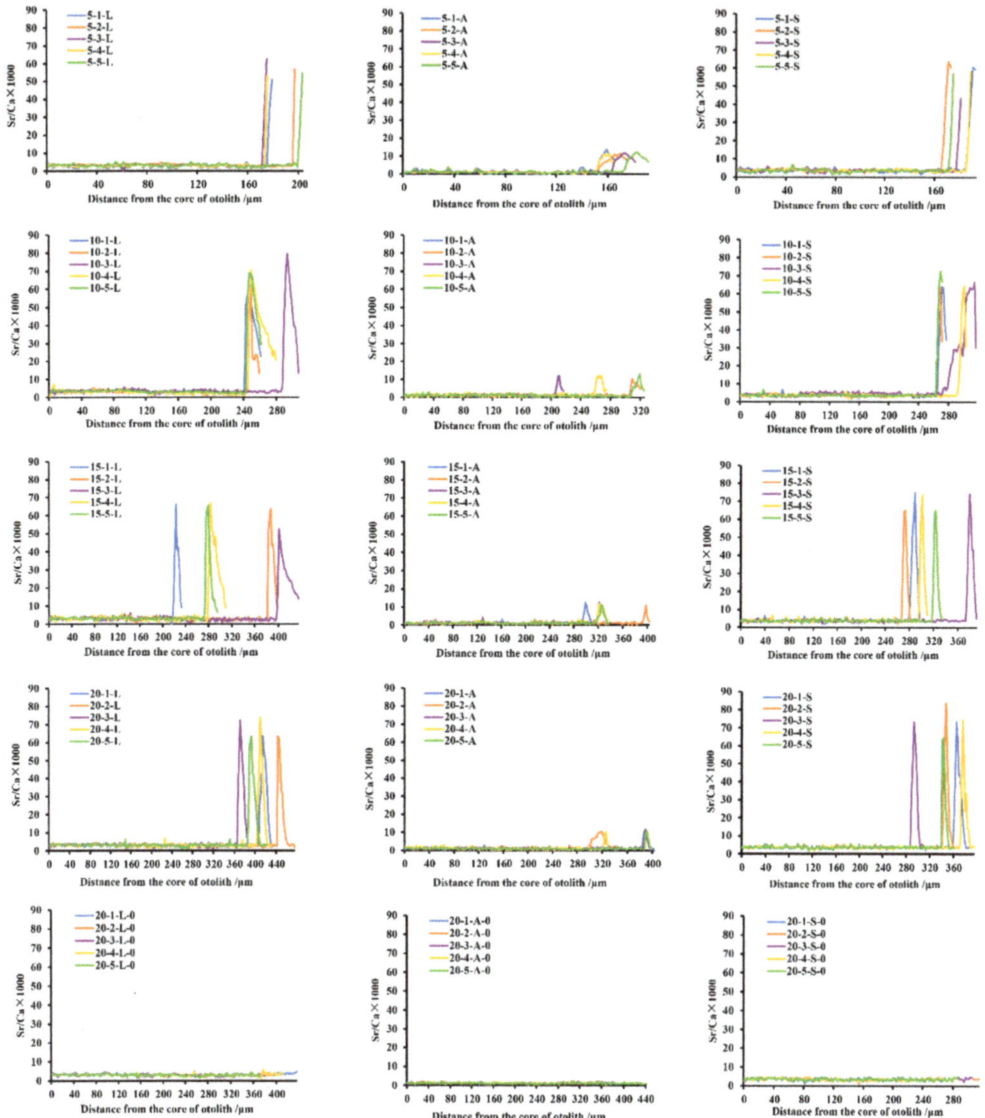

Figure 3. Dynamics of Sr/Ca ratios in the three types of otoliths (lapilli, asterisci, and sagittae) of juvenile crucian carp during 20 d of post-marking culture. Note: in the label for each line, the first numbers 5, 10, 15, and 20 represent that the fish was extracted and euthanized for analysis on the 5th, 10th, 15th, and 20th day after the immersion marking was completed; the second numbers of 1, 2, 3, 4, and 5 represent the code of each sampled fish; and the third character, "L" represents lapilli, "S" represents sagittae, and "A" represents asterisci. The last digit "0" means the control group.

After 20 d of post-marking culture, the otoliths showed complete dynamic changes from the core to the edge. The change in Sr/Ca was divided into three phases (Figure 3 and Table 1). In each of the three phases, the Sr/Ca ratio of the lapilli and sagittae was significantly different from that of the asterisci ($p < 0.05$). For the lapilli and sagittae, there was no significant difference in the second phases ($p > 0.05$) but there was a significant difference in the first and third phases ($p < 0.05$). In the lapilli and sagittae of every sample, the ratios of Sr/Ca in the first and third phases were significantly different from those in the second phases ($p < 0.05$), and there was no significant difference between the first and third phases ($p > 0.05$). However, on the second and third samples of the asterisci, the Sr/Ca ratios in the three phases were significantly different from each other ($p < 0.05$), whereas the first and third stages, were not significantly different for most samples ($p > 0.05$). In addition, the Sr/Ca ratio in the three types of otoliths in the control group remained stable during the post-marking culture for 20 d (Figure 3 and Table 2). The Sr/Ca ratio of the lapilli and sagittae differed significantly from each other ($p < 0.05$), and both differed significantly from that of the asterisci ($p < 0.05$).

Table 1. Inter-otolith microchemical phases of the Sr/Ca ratio for marked juvenile crucian carps at 20 d of post-marking culture.

Sample Cord	First Phase		Second Phase		Third Phase	
	Distance from the Core (μm)	Sr/Ca (Mean ± SD)	Distance from the Core (μm)	Sr/Ca (Mean ± SD)	Distance from the Core (μm)	Sr/Ca (Mean ± SD)
20-1-L	0–406	3.03 ± 0.59^{b}	408-426	37.65 ± 16.30^{a}	428–440	2.92 ± 0.31^{b}
20-2-L	0–440	3.09 ± 0.56^{b}	442–458	32.12 ± 22.61^{a}	460–474	3.18 ± 0.64^{b}
20-3-L	0–364	3.01 ± 0.67^{b}	366–384	38.74 ± 21.75^{a}	386–396	2.71 ± 0.54^{b}
20-4-L	0–402	3.04 ± 0.67^{b}	404–422	38.46 ± 21.89^{a}	424–432	3.23 ± 0.55^{b}
20-5-L	0–384	3.09 ± 0.62^{b}	386–408	32.26 ± 21.06^{a}	410–422	3.01 ± 0.35^{b}
Average	-	3.05 ± 0.04^{B}	-	35.85 ± 3.36^{A}	-	3.01 ± 0.21^{B}
20-1-S	0–358	3.78 ± 0.57^{b}	360–378	38.23 ± 21.54^{a}	380–386	3.66 ± 0.23^{b}
20-2-S	0–340	3.80 ± 0.52^{b}	342–356	39.06 ± 27.51^{a}	358–364	4.01 ± 0.46^{b}
20-3-S	0–286	3.82 ± 0.57^{b}	288–302	41.64 ± 23.74^{a}	304–312	3.89 ± 0.93^{b}
20-4-S	0–370	3.78 ± 0.54^{b}	372–386	33.02 ± 23.17^{a}	388–394	4.01 ± 0.36^{b}
20-5-S	0–338	3.79 ± 0.62^{b}	340–350	36.17 ± 25.03^{a}	352–360	3.40 ± 0.22^{b}
Average	-	3.79 ± 0.02^{A}	-	37.62 ± 3.24^{A}	-	3.79 ± 0.06^{A}
20-1-A	0–382	1.01 ± 0.41^{b}	384–392	8.22 ± 2.72^{a}	394–402	1.00 ± 0.79^{b}
20-2-A	0–296	1.09 ± 0.59^{b}	298–326	6.86 ± 2.66^{a}	328–346	1.61 ± 0.64^{c}
20-3-A	0–384	1.08 ± 0.43^{b}	386–392	9.04 ± 3.05^{a}	394–396	1.79 ± 0.04^{c}
20-4-A	0–320	1.06 ± 0.58^{b}	322–326	7.89 ± 2.81^{a}	328–330	1.67 ± 0.51^{b}
20-5-A	0–386	1.04 ± 0.28^{b}	388–394	9.07 ± 1.89^{a}	396–398	1.94 ± 1.63^{b}
Average	-	1.06 ± 0.03^{C}	-	7.95 ± 0.79^{B}	-	1.42 ± 0.38^{C}

Note: The Sr/Ca ratio refers to a standardized ratio, which is Sr/Ca × 10^3. In the sample code, the number "20" represents 20 d of post-marking culture was completed; the second number 1, 2, 3, 4, or 5 represents the order of the sample fish; and the third character "L" represents lapilli, "S" represents sagittae, and "A" represents asterisci. In the superscript symbols, lowercase letters "a, b, and c" indicate significant differences among the three phases of the same sample, and uppercase letters "A, B, and C" indicate a comparison of significant differences between different samples at the same stage. Values that do not share a common letter are significantly different.

3.2.2. Inter-Otolith Changing of Sr Map Patterns

The surface distribution of the Sr content in the otoliths of juvenile crucian carp was consistent with the line distribution. The different colors presented in the Sr maps corresponded to different ranges of Sr/Ca ratios. The otoliths of fish in the marked group had a low Sr central area surrounded by a high Sr ring; the otoliths of all fish in the control group were always in blue color with a low Sr content from the core to the edge (Figure 4). Therefore, it can be directly observed from the otolith surface distribution that a "Sr-marked area" was produced on the otoliths by the 20th day of post-marking culture, and the blue color, consistent with the formation of the otolith before immersion, was observed outside the marked area. However, the color of the lapilli and sagittae differed from that of the asterisci. Red and yellow-green colors in the marked area can be clearly observed against the background color in the lapilli and sagittae, whereas the dark blue in the marked area of the asterisci was relatively inconspicuous.

Table 2. The otolith Sr/Ca ratio of control juvenile crucian carps measured on the same day as marked fish on the 20th day of post-marking culture.

Sample Cord	Distance from the Core (μm)	Sr/Ca (Mean ± SD)
20-1-L-0	0–434	3.18 ± 0.53
20-2-L-0	0–398	3.10 ± 0.50
20-3-L-0	0–386	3.05 ± 0.57
20-4-L-0	0–410	3.12 ± 0.55
20-5-L-0	0–370	3.06 ± 0.50
Average	-	3.10 ± 0.05 [B]
20-1-S-0	0–268	3.61 ± 0.71
20-2-S-0	0–314	3.57 ± 0.69
20-3-S-0	0–304	3.61 ± 0.52
20-4-S-0	0–278	3.53 ± 0.52
20-5-S-0	0–286	3.61 ± 0.57
Average	-	3.59 ± 0.04 [A]
20-1-A-0	0–398	1.09 ± 0.32
20-2-A-0	0–370	1.12 ± 0.45
20-3-A-0	0–380	1.11 ± 0.37
20-4-A-0	0–356	1.10 ± 0.30
20-5-A-0	0–440	1.09 ± 0.33
Average	-	1.10 ± 0.01 [C]

Note: The Sr/Ca ratio refers to a standardized ratio, which is Sr/Ca $\times 10^3$. In the sample code, the number "20" represents that 20 d of post-marking culture was completed; the second number "1, 2, 3, 4, or 5" represents the order of the sample fish; and the third character "L" represents lapilli, "S" represents sagittae, and "A" represents asterisci. The last digit "0" represents the control group. In the superscript symbols, uppercase letters "A, B, and C" indicate a comparison of significant differences between different samples. Values that do not share a common letter are significantly different.

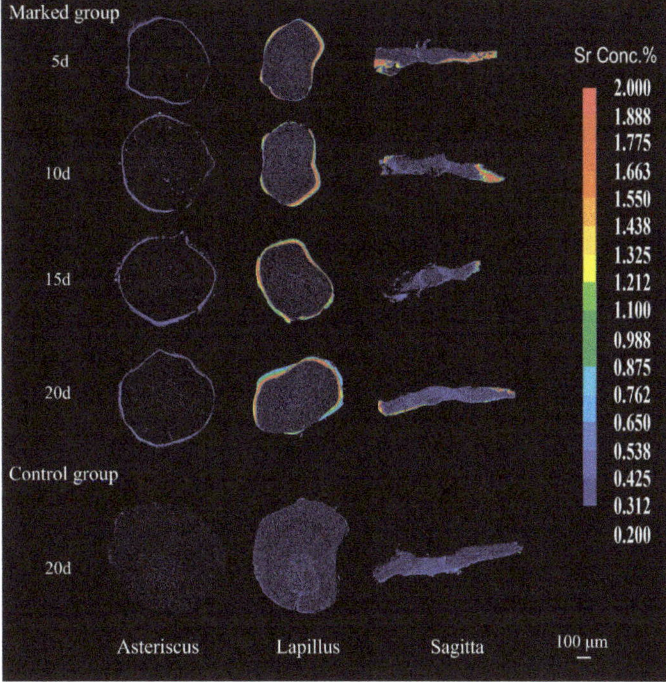

Figure 4. Typical Sr content maps in three types of otoliths (lapilli, asterisci, and sagittae) of control and marked groups of juvenile crucian carp. Note: The color from blue to red indicates a gradual increase in Sr content of otoliths.

3.2.3. Inter-Otolith Time Delay for Sr Marking

By comparing the EPMA maps of Sr and BSE images that included daily increments for the three types of otoliths (Figure 5), where the daily increments were calculated at the edge of the otolith (Figure 6), there was a specific time delay between the beginning of the immersion and the generation of the marking ring in the three types of otoliths; this time delay also existed between the end of immersion and the termination of the marking ring. On the lapilli and sagittae, the Sr signal appeared on the third day of the marking phase and was terminated on the 17th day of post-marking culture. On the asterisci, the Sr signal appeared on the second day of the marking phase and was terminated on the 13th day of post-marking culture. Therefore, in both the lapilli and sagittae, there was a time delay (i.e., time lag) of 2 d between the start of Sr marking and the appearance of the marking signal, whereas there was a 16-d time delay between the end of marking and the termination of the signal. For asterisci, there was a time delay of 1 d between the start of Sr marking and the appearance of the Sr marking signal, whereas there was a 12-d time delay between the end of immersion and the termination of the signal.

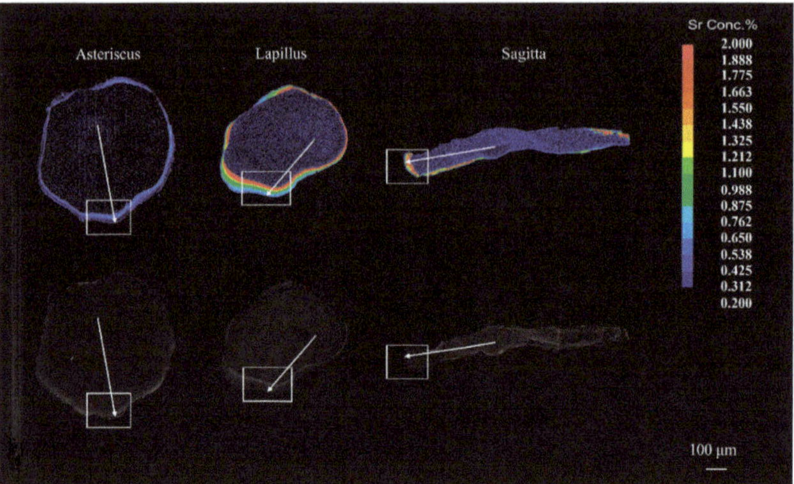

Figure 5. Typical Sr content maps and backscattered electron image of the three types of otoliths (lapilli, asterisci, and sagittae) of juvenile crucian carp on the 20th day of post-marking culture. Note: the color from blue to red indicates a gradual increase in the Sr/Ca ratio of otoliths, the white arrows indicate the region from the core to the edge of the otolith, and the white box contains the growth area of the fish otoliths during the post-marking culture phase.

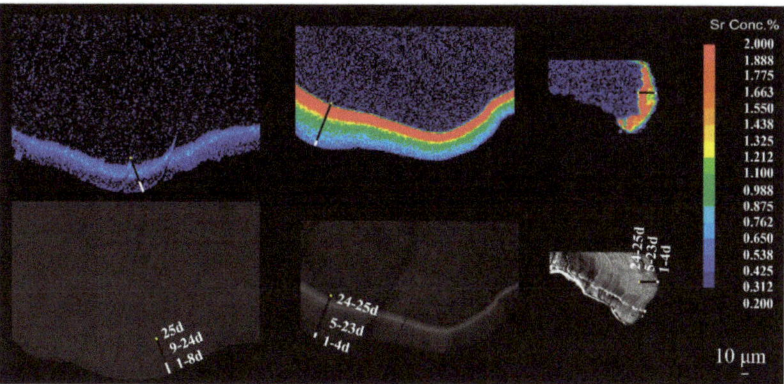

Figure 6. Time delay between the appearance and termination of Sr marks in three types of otoliths (lapilli, asterisci, and sagittae) of juvenile crucian carp. Note: the color from blue to red indicates a gradual increase in the Sr/Ca ratio of otoliths. The length of the yellow line segment indicates that the Sr/Ca ratio was still in the normal level of the otolith growth range after the start of the Sr-immersed marking. The length of the black line indicates the growth range of otoliths when the Sr/Ca ratio was higher than the normal level. The length of the white line in the figure indicates the growth range of otoliths when the Sr/Ca ratio returned to a normal level after a high value.

4. Discussion

4.1. Selection of Otolith Type for Observation of Sr Marking

When selecting the best otolith type for observation, not only should the marking effect be adequate, but the otolith sample should also be easy to obtain and process. The results of the elemental line and surface mapping of the Sr marks (Figures 3 and 4) showed that a successful marking effect could be achieved in the sagittae, lapilli, and asterisci. However, owing to the morphological characteristics of the sagittae during the growth process, both ends of the otoliths grow forward, and one end becomes extremely slender, similar to those of *Opsaridium microcephalum* [34]; furthermore, they are fragile when picked, and it is difficult to obtain the entire "horizontal plane" during the polishing process. This leads to difficulties in obtaining a complete "marking ring" owing to the damage to the otolith edge during surface distribution mapping. Compared with the asterisci, the color shown by the concentration of Sr in the marked area on the sagittae is attractive, as with the lapilli.

The peak value of the Sr mark for the asterisci was far lower than that for the sagittae and the lapilli in the line distribution (Figure 3), and the Sr concentration in the marked area appeared to be a dark-blue color, which is different from the color in the sagittae and lapilli. The width of the "Sr mark areas" was narrow; thus, the marked area was more indistinguishable than that of the lapilli and sagittae in EPMA maps. However, the asterisci of cyprinid fish are large and easy to remove, and the shape of the growth increment of the otolith is relatively irregular [34].

From both the line and mapping distribution effects of Sr marking (Figures 2 and 3), the marking effect was more obvious for the lapilli than for the asterisci and sagittae. Lapilli are moderate in size and are oval, with regular shapes; hence, they are easy to process [34]. In summary, it would be most suitable and feasible to choose lapilli as the material to observe the Sr marking of juvenile crucian carp in future experiments. Moreover, in previous studies on *Hemigrammopetersius barnardi* and *O. microcephalum*, lapilli were concluded to be the most appropriate for the analysis of daily increments [34,35].

4.2. *Variations in Sr-Marked Effects among the Three Types of Otoliths*

In freshwater, Sr can replace Ca in a $CaCO_3$ matrix, which is similar to Ba, and shows stable characteristics over time [7]. According to the data shown in Table 1, the three pairs of otoliths showed significant increases in Sr, after Sr immersion, even though the level of increase varied among them. Different pairs of otoliths are formed by different $CaCO_3$ polymorph types, namely aragonite, vaterite, and calcite [36], and different polymorph types have varied distribution coefficients. Vaterite typically has lower elemental concentrations than aragonite [37], which might explain why the asterisci (composed mainly of vaterite [7,38]) had lower Sr levels than the sagittae and lapilli (aragonite [38,39]). Similar results were obtained in the determination of strontium element concentrations in the lapilli and asterisci of goldfish (*Carassius auratus*) [40,41].

In the samples obtained on the 15th day of post-marking culture, when the Sr mark peaks were observed, the sagittae and lapilli were still in the Sr uptake phase. The Sr mark of the asterisci reached its peak and started to gradually decline and return to normal levels. This is probably because otolith biomineralization has been found to be tightly coupled to the metabolic rate [42,43], and it is also possible that different types of otoliths grow at different speeds at certain stages. For example, during the juvenile period, the growth rate of the lapilli and sagittae is lower than that of the asterisci [41]. To fully explain this observation, an in-depth exploration of the crystalline structure and mechanism of element formation of the three types of otoliths in juvenile crucian carp is required.

4.3. *Significance of Time Delay to Sr Markings of Otoliths*

The deposition of elements into otoliths from the environment is a multi-stage process. The elements are transported from the plasma into the endolymph of the inner ear through ion transport and finally settle on the surface of the otoliths [7]. In this process, there will be a certain time delay from an element entering the fish body to its deposition in the otoliths. Generally, specific time delays occur between the beginning of the immersion and the generation of the marking signals of the three types of otoliths, as well as between the end of immersion and the termination of the marking signals, i.e., the times (e.g., date) are not equal to each other for the former and latter. Brown and Harris used exogenous Sr to immerse golden perch and found that the Sr marking could not be detected until the 20th day of post-marking culture [44]. In the Sr marking of Japanese eels, the changes in the Sr/Ca ratio in the otoliths were visible after 10 d, whereas the complete deposition in the otolith required at least 30–60 d [45].

From the findings of previous studies and this study, it can be concluded that there is a time delay in the deposition of environmental elements into the otoliths, which differs among fish species, growth stages, and water environments. For a better understanding of the marking process and to objectively and accurately assess marking effectiveness, otolith sampling should avoid the beginning and end of immersion, because otolith Sr marking may not have started or finished during these periods. The mechanism of this time delay must be studied further in terms of Sr deposition and the otolith structure of different fish species, [38] and combined with basic characteristics of the water environment, such as salinity [46] and temperature [40]. Other biological factors, such as fish genetics, development stage, growth rate, food, and physiological conditions, can also affect the time lag of Sr deposition in otoliths [47]. Future research on the mechanisms of Sr deposition in otoliths among different species will be useful to improve Sr marking methods.

5. Conclusions

This is the first time that 60 mg/L $SrCl_2{\cdot}6H_2O$ was used for the immersion of juvenile crucian carp to evaluate the Sr marking among otoliths. Otoliths were collected every 5 d during the 20-d period of post-marking culture for EPMA analysis. Both quantitative line analysis of otoliths and surface distribution detection obtained ideal marking results for the three types of otoliths. Considering the influence of otolith element content and otolith morphology on marker observations, the lapilli provide the best option for detecting

Sr-marked juvenile crucian carp. There is a time delay (or time lag) between the timing of the appearance or termination of the Sr marker and the beginning and end of Sr immersion for the three types of otoliths. The experimental protocol can be optimized to determine the optimal immersion concentration and time and elucidate the mechanisms of Sr deposition in a future study.

Author Contributions: Conceptualization, J.Y., Y.Z., T.J. and Q.P.; methodology, Y.Z., T.J., X.C., H.L. and J.Y.; investigation, Y.Z., T.J., X.C., H.L. and J.Y.; resources, Y.Z., T.J. and X.C.; funding acquisition, J.Y.; writing—original draft, Y.Z., T.J., X.C., H.L. and Q.P.; writing—review and editing, J.Y.; supervision, J.Y. All authors have read and agreed to the published version of the manuscript.

Funding: This work was supported by the China Central Governmental Research Institutional Basic Special Research Project from Public Welfare Funds [grant numbers 2021GH08 and 2019JBFM06] and the Agricultural Finance Special Project [grant number CJDC-2017-22].

Institutional Review Board Statement: The study was conducted according to the guidelines of the Declaration of Helsinki and was approved by the Ethics Committee of Animal Care and Use Committee of the Freshwater Fisheries Research Center at the Chinese Academy of Fishery Sciences. The analysis was carried out following the Guidelines for the Care and Use of Laboratory Animals set by the Animal Care and Use Committee of the Freshwater Fisheries Research Center (2003WXEP61). All operations were carried out with field permit no. 20181AC1128.

Data Availability Statement: Data that support the findings of this study are available from the corresponding author upon reasonable request (Y.J.: jiany@ffrc.cn).

Acknowledgments: We thank Junren Xue, Yuhai Hu, and Jing Tang for their technical support in the aquaculture of crucian carp.

Conflicts of Interest: The authors declare no conflict of interest.

References

1. Warren-Myers, F.; Dempster, T.; Swearer, S.E. Otolith mass marking techniques for aquaculture and restocking: Benefits and limitations. *Rev. Fish. Biol. Fish.* **2018**, *28*, 485–501. [CrossRef]
2. Morita, K.; Saito, T.; Miyakoshi, Y.; Fukuwaka, M.-A.; Nagasawa, T.; Kaeriyama, M. A review of Pacific salmon hatchery programmes on Hokkaido Island, Japan. *ICES J. Mar. Sci.* **2006**, *63*, 1353–1363. [CrossRef]
3. Avigliano, E.; Martinez, C.F.R.; Volpedo, A.V. Combined use of otolith microchemistry and morphometry as indicators of the habitat of the silverside (*Odontesthes bonariensis*) in a freshwater–estuarine environment. *Fish. Res.* **2014**, *149*, 55–60. [CrossRef]
4. Fowler, A.M.; Smith, S.M.; Booth, D.J.; Stewart, J. Partial migration of grey mullet (*Mugil cephalus*) on Australia's east coast revealed by otolith chemistry. *Mar. Environ. Res.* **2016**, *119*, 238–244. [CrossRef] [PubMed]
5. Hogan, J.D.; Kozdon, R.; Blum, M.J.; Gilliam, J.F.; Valley, J.W.; Mcintyre, P.B. Reconstructing larval growth and habitat use in an amphidromous goby using otolith increments and microchemistry. *J. Fish Biol.* **2017**, *90*, 1338–1355. [CrossRef] [PubMed]
6. Taddese, F.; Reid, M.R.; Closs, G.P. Direct relationship between water and otolith chemistry in juvenile estuarine triplefin *Forsterygion nigripenne*. *Fish. Res.* **2019**, *211*, 32–39. [CrossRef]
7. Campana, S.E. Chemistry and composition of fish otoliths: Pathways, mechanisms and applications. *Mar. Ecol. Prog. Ser.* **1999**, *188*, 263–297. [CrossRef]
8. Elsdon, T.S.; Gillanders, B.M. Fish otolith chemistry influenced by exposure to multiple environmental variables. *J. Exp. Mar. Biol. Ecol.* **2004**, *313*, 269–284. [CrossRef]
9. Bath Martin, G.; Thorrold, S.R. Temperature and salinity effects on magnesium, manganese, and barium incorporation in otoliths of larval and early juvenile spot *Leiostomus xanthurus*. *Mar. Ecol.-Prog. Ser.* **2005**, *293*, 223–232. [CrossRef]
10. Izzo, C.; Reis-Santos, P.; Gillanders, B.M. Otolith chemistry does not just reflect environmental conditions: A meta-analytic evaluation. *Fish. Fish.* **2018**, *19*, 441–454. [CrossRef]
11. Davoren, G.K.; Halden, N.M. Connectivity of capelin (*Mallotus villosus*) between regions and spawning habitats in Newfoundland inferred from otolith chemistry. *Fish. Res.* **2014**, *159*, 95–104. [CrossRef]
12. Crook, D.A.; Gillanders, B.M. Use of otolith chemical signatures to estimate carp recruitment sources in the mid-Murray River, Australia. *River Res. Appl.* **2006**, *22*, 871–879. [CrossRef]
13. Avigliano, E.; Carvalho, B.M.D.; Leisen, M.; Romero, R.; Velasco, G.; Vianna, M.; Barra, F.; Volpedo, A.V. Otolith edge fingerprints as approach for stock identification of *Genidens barbus*. *Estuar. Coast Shelf Sci.* **2017**, *194*, 92–96. [CrossRef]
14. Campana, S.E.; Thorrold, S.R. Otoliths, increments, and elements: Keys to a comprehensive understanding of fish populations? *Can. J. Fish. Aquat. Sci.* **2001**, *58*, 30–38. [CrossRef]
15. Hayden, T.A.; Limburg, K.E.; Pine, W.E. Using otolith chemistry tags and growth patterns to distinguish movements and provenance of native fish in the Grand Canyon. *River Res. Appl.* **2013**, *29*, 1318–1329. [CrossRef]

16. Wickström, H.; Sjöberg, N.B. Traceability of stocked eels—The Swedish approach. *Ecol. Freshw. Fish* **2014**, *23*, 33–39. [CrossRef]
17. Stötera, S.; Degen-Smyrek, A.K.; Krumme, U.; Stepputtis, D.; Bauer, R.; Limmer, B.; Hammer, C. Marking otoliths of Baltic cod (*Gadus morhua* Linnaeus, 1758) with tetracycline and strontium chloride. *J. Appl. Ichthyol.* **2019**, *35*, 427–435. [CrossRef]
18. Duffy, W.J.; McBride, R.S.; Hendricks, M.L.; Oliveira, K. Otolith age validation and growth estimation from oxytetracycline-marked and recaptured American shad. *Trans. Am. Fish. Soc.* **2012**, *141*, 1664–1671. [CrossRef]
19. Frenkel, V.; Kindschi, G.; Zohar, Y. Noninvasive, mass marking of fish by immersion in calcein: Evaluation of fish size and ultrasound exposure on mark endurance. *Aquaculture* **2002**, *214*, 169–183. [CrossRef]
20. Cottingham, A.; Hall, N.G.; Loneragan, N.R.; Jenkins, G.I.; Potter, I.C. Efficacy of restocking an estuarine-resident species demonstrated by long-term monitoring of cultured fish with alizarin complexone-stained otoliths. A case study. *Fish. Res.* **2020**, *227*, 105556. [CrossRef]
21. Caraguel, J.-M.; Charrier, F.; Mazel, V.; Feunteun, E. Mass marking of stocked European glass eels (*Anguilla anguilla*) with alizarin red S. Ecol. *Freshw. Fish* **2015**, *24*, 435–442. [CrossRef]
22. Simon, J.; Wickström, H. Long-term retention of alizarin red S marks and coded wire tags in European eels. *Fish. Res.* **2020**, *224*, 105453. [CrossRef]
23. Braux, E.D.; Warren-Myers, F.; Dempster, T.; Fjelldal, P.G.; Hansen, T.; Swearer, S.E. Osmotic induction improves batch marking of larval fish otoliths with enriched stable isotopes. *ICES J. Mar. Sci.* **2014**, *71*, 2530–2538. [CrossRef]
24. Woodcock, S.H.; Gillanders, B.M.; Munro, A.R.; Crook, D.A.; Sanger, A.C. Determining mark success of 15 combinations of enriched stable isotopes for the batch marking of larval otoliths. *N. Am. J. Fish. Manag.* **2011**, *31*, 843–851. [CrossRef]
25. Ennevor, B.C.; Beames, R.M. Use of lanthanide elements to mass mark juvenile salmonids. *Can. J. Fish. Aquat. Sci.* **1993**, *50*, 1039–1044. [CrossRef]
26. Brown, P.; Green, C.; Sivakumaran, K.P.; Stoessel, D.; Giles, A. Validating otolith annuli for annual age determination of common carp. *Trans. Am. Fish. Soc.* **2004**, *133*, 190–196. [CrossRef]
27. Swanson, R.G.; Gagnon, J.E.; Miller, L.M.; Dauphinais, J.D.; Sorensen, P.W. Otolith microchemistry of common carp reflects capture location and differentiates nurseries in an interconnected lake system of the North American Midwest. *N. Am. J. Fish. Manag.* **2020**, *40*, 1100–1118. [CrossRef]
28. Smith, B.b.; Walker, K.F. Validation of the aging of 0+ carp (*Cyprinus carpio* L.). *Mar. Freshw. Res.* **2003**, *54*, 1005–1008. [CrossRef]
29. Pan, X.D.; Ye, Z.J.; Xu, B.D.; Jiang, T.; Yang, J.; Tian, Y.J. Population connectivity in a highly migratory fish, Japanese Spanish mackerel (*Scomberomorus niphonius*), along the Chinese coast, implications from otolith chemistry. *Fish. Res.* **2020**, *231*, 105690. [CrossRef]
30. Santos, R.O.; Schinbeckler, R.; Viadero, N.; Larkin, M.F.; Rennert, J.J.; Shenker, J.M.; Rehage, J.S. Linking bonefish (*Albula vulpes*) populations to nearshore estuarine habitats using an otolith microchemistry approach. *Environ. Biol. Fishes* **2019**, *102*, 267–283. [CrossRef]
31. Yang, J.; Arai, T.; Liu, H.; Miyazaki, N.; Tsukamoto, K. Reconstructing habitat use of *Coilia mystus* and *Coilia ectenes* of the Yangtze River estuary, and of *Coilia ectenes* of Taihu Lake, based on otolith strontium and calcium. *J. Fish Biol.* **2006**, *69*, 1120–1135. [CrossRef]
32. Bostanci, D. Otolith biometry-body length relationships in four fish species (chub, pikeperch, crucian carp, and common carp). *J. Freshw. Ecol.* **2009**, *24*, 619–624. [CrossRef]
33. Liu, H.; Jiang, T.; Yang, J. Unravelling habitat use of *Coilia nasus* from the Rokkaku River and Chikugo River estuaries of Japan by otolith strontium and calcium. *Acta Oceanol. Sin.* **2018**, *37*, 52–60. [CrossRef]
34. Morioka, S.; Matsumoto, S. Otolith features and utility of lapillus for daily increment analysis in *Opsaridium microcephalum* (Cyprinidae) juveniles collected from Lake Malawi. *Ichthyol. Res.* **2003**, *50*, 82–85. [CrossRef]
35. Morioka, S.; Matsumoto, S.; Kaunda, E. Otolith features and growth of Malawian characid *Hemigrammopetersius barnardi* from the southwestern coast of Lake Malawi. *Ichthyol. Res.* **2006**, *53*, 143–147. [CrossRef]
36. Oliveira, A.M.; Farina, M.; Ludka, I.P.; Kachar, B. Vaterite, calcite, and aragonite in the otoliths of three species of piranha. *Naturwissenschaften* **1996**, *83*, 133–135. [CrossRef]
37. Schulz-Mirbach, T.; Ladich, F.; Plath, M.; Heß, M. Enigmatic ear stones: What we know about the functional role and evolution of fish otoliths. *Biol. Rev. Camb. Philos. Soc.* **2019**, *94*, 457–482. [CrossRef]
38. Ren, D.; Meyers, M.A.; Zhou, B.; Feng, Q. Comparative study of carp otolith hardness: Lapillus and Asteriscus. *Mater. Sci. Eng. C Mater. Biol. Appl.* **2013**, *33*, 1876–1881. [CrossRef]
39. Thomas, O.R.B.; Swearer, S.E.; Kapp, E.A.; Peng, P.; Tonkin-Hill, G.Q.; Papenfuss, A.; Roberts, A.; Bernard, P.; Roberts, B.R. The inner ear proteome of fish. *FEBS J.* **2019**, *286*, 66–81. [CrossRef]
40. Mugiya, Y.; Tanaka, S. Incorporation of waterborne strontium into otoliths and its turnover in the goldfish *Carassius auratus*: Effects of strontium concentrations, temperature, and 17β-estradiol. *Fish. Sci.* **1995**, *61*, 29–35. [CrossRef]
41. Mugiya, Y.; Satoh, C. Strontium accumulation in slow-growing otoliths in the goldfish *Carassius auratus*. *Fish. Sci.* **1997**, *63*, 361–364. [CrossRef]
42. Hüssy, K.; Mosegaard, H. Atlantic cod (*Gadus morhua*) Growth and otolith accretion characteristics modelled in a bioenergetics context. *Can. J. Fish. Aquat. Sci.* **2004**, *61*, 1021–1031. [CrossRef]
43. Wright, P.J.; Fallon-Cousins, P.; Armstrong, J.D. The relationship between otolith accretion and resting metabolic rate in juvenile Atlantic salmon during a change in temperature. *J. Fish Biol.* **2001**, *59*, 657–666. [CrossRef]

44. Brown, P.; Harris, J.H. Strontium batch marking of golden perch (*Macquaria ambigua* Richardson) (Percichthyidae) and trout cod (*Maccullochella macquariensis*) (Cuvier). In *Recent Developments in Fish Otolith Research*; Secor, D.H., Dean, J.M., Campana, S.E., Eds.; University of South Carolina Press: Columbia, SC, USA, 1995; pp. 693–702.
45. Yokouchi, K.; Fukuda, N.; Shirai, K.; Aoyama, J.; Daverat, F.; Tsukamoto, K. Time lag of the response on the otolith strontium/calcium ratios of the Japanese eel, *Anguilla japonica* to changes in strontium/calcium ratios of ambient water. *Environ. Biol. Fish.* **2011**, *92*, 469–478. [CrossRef]
46. Panfili, J.; Darnaude, A.M.; Vigliola, L.; Jacquart, A.; Labonne, M.; Gilles, S. Experimental evidence of complex relationships between the ambient salinity and the strontium signature of fish otoliths. *J. Exp. Mar. Biol. Ecol.* **2015**, *467*, 65–70. [CrossRef]
47. Sturrock, A.M.; Hunter, E.; Milton, J.A.; Johnson, R.C.; Waring, C.P.; Trueman, C.N. Quantifying physiological influences on otolith microchemistry. *Methods Ecol. Evol.* **2015**, *6*, 806–816. [CrossRef]

MDPI

Article

Fluctuating Asymmetry in *Asteriscii* Otoliths of Common Carp (*Cyprinus carpio*) Collected from Three Localities in Iraqi Rivers Linked to Environmental Factors

Laith Jawad [1] and Kélig Mahé [2,*]

[1] School of Environmental and Animal Health Sciences, Unitec Institute of Technology, 139 Carrington Road, Mt Albert, Auckland 1025, New Zealand; laith_jawad@hotmail.com

[2] IFREMER, Fisheries Laboratory, 150 Quai Gambetta, BP 699, 62321 Boulogne-sur-Mer, France

* Correspondence: kelig.mahe@ifremer.fr; Tel.: +33-321-995-602

Abstract: Otoliths, calcified structures in the inner ears, are used to estimate fish age, and their shape is an efficient fish stock identification tool. Otoliths are thus very important for the management and assessment of commercial stocks. However, most studies have used left or right otoliths, chosen arbitrarily without evaluation of the difference between these otoliths. In this study, the *asteriscii* otoliths from 263 common carp (*Cyprinus carpio*; Linnaeus, 1758) were sampled in three Iraqi rivers to test the potential asymmetry and the geographical effect on otolith growth from three measurements (length, width and weight), and on shape from two shape indices (ellipticity and form-factor). Among all *asteriscii* otolith features, there was significant fluctuating asymmetry between fish length and every otolith descriptor. At one fish length, the size and/or the shape of otoliths could be different between two individuals and/or between left and right *asteriscii* otoliths for the same individual. Moreover, the relationship between fish length and otolith shape/growth was significantly dependent on the studied geographical area and, more especially, the environmental effects as the water temperature and pH. Finally, the relationships between fish length and otolith shape indices showed that the otolith evolves into the elliptical shape during the life of the fish. To use the otolith shape, it is essential to take into account the developmental stage of individuals to integrate the ontogenetic effect. Our results highlight the importance of verifying potential otolith asymmetry, especially for the *asteriscii* otoliths (lagenar otoliths) before their use in fisheries research.

Keywords: *asteriscus*; side effect; growth; otolith shape; geographical effect; temperature effect; pH

Citation: Jawad, L.; Mahé, K. Fluctuating Asymmetry in *Asteriscii* Otoliths of Common Carp (*Cyprinus carpio*) Collected from Three Localities in Iraqi Rivers Linked to Environmental Factors. *Fishes* **2022**, *7*, 91. https://doi.org/10.3390/fishes7020091

Academic Editor: Josipa Ferri

Received: 7 March 2022
Accepted: 8 April 2022
Published: 15 April 2022

1. Introduction

Otoliths, calcified structures in the inner ears [1,2], are used to identify species in taxonomic or phylogenic studies and to collect age data for management and assessment of stocks. The otoliths grow throughout the fish life and are metabolically inert [3], and their shape (i.e., their outline resulting from genetic, environmental and ontogenetic effects) is used as an efficient tool to recognize the species at the interspecific level and to identify the fish stocks at the intraspecific level (to see the Stock Identification Methods Working Group of which identifies all studies each year; SIMWG). Each head side (i.e., left and right inner ear), however, has three pairs of otoliths (*sagittae*, *asteriscii* and *lapilli*) [4]. Among the three different otoliths, the sagittal otolith is usually used due to its larger size and easier removal [5]. The *asteriscii* (lagenar otoliths) are the smallest otolith in marine species and so rarely used. Conversely, *asteriscus* otoliths are the most frequently used otoliths in Cypriniforms species such as common carp [6]. For these species, *asteriscii* are the largest of the three otolith pairs [6]. Several growth studies focused on the common carp have used *asteriscii* otoliths [6–9]; however, no study has focused on the *astericus* shape for this species.

Since 1990, when the first report on the asymmetry in fishes was published on fishes of Iraq [10], few scientific publications have appeared on the phenomenon of asymmetry in freshwater fish species. In total, there are four publications concerning the Iraqi waters reporting asymmetry in four fish species (*Heteropneustes fossilis*, Bloch, 1794; *Mystus pelusius*, Solander, 1794; *Planiliza abu*, Heckel, 1843; and *Parasilurus triostegus*, Heckel, 1843) related to different fish characters [10–14]. There are similarly few reports for the marine fish species of Iraq: the first publication on the asymmetry of this group was in 2020, with six reports studying otolith asymmetry in six marine species. The present study focuses on potential asymmetry and the geographical effects on otoliths growth and shape in common carp *Cyprinus carpio*, a freshwater species of Iraq. A recent study using the otolith shape to identify the stock structure of the bogue (*Boops boops*; Linnaeus, 1758) in the Mediterranean Sea showed that the significant asymmetry could modify the boundaries of stocks according to the use of the left or right otolith [15]. Moreover, this study showed that this significant asymmetry could be due to environmental differences. Consequently, the aims of this study are to compare the fluctuations in asymmetry of the *asteriscus* in the carp, *Cyprinus carpio*, and to identify if this asymmetry could be a possible response to the environmental variables in several rivers in Iraq.

2. Materials and Methods

2.1. Sample Collection

A total of 263 individual carp were collected in May–July 2021 from three different locations on the Euphrates (Nasiriya City; 31°2′28″ N, 46°14′45″ E; n = 90), Tigris (Amarah City; 31°51′25″ N, 47°8′15″ E; n = 89) and Shatt al-Arab (Basrah City; 30°33′41″ N, 47°47′41″ E; n = 84) rivers (Figure 1; Supplementary Table S1). All individuals were from commercial catches and macroscopic observation of the gonads did not identify sex with a good accuracy.

2.2. Study Areas

The Euphrates is the longest river in Western Asia (around 3000 km; [16–19]). At Nasiriya City, where the common carp specimens were collected for this study, the water of the Euphrates river is seasonally variable in temperature, salinity and hydrogen ion concentration. Water temperature ranged from 11 °C in January to 34 °C in July and pH values fluctuated from 7.12 in August to 8.43 in December [16].

The Tigris is 1750 km long, rising in the Taurus Mountains of eastern Turkey, about 25 km southeast of Elazig city and about 30 km from the headwaters of the Euphrates [20]. The water temperature of the Tigris river in Mysan Province varied between 15 °C and 25 °C and pH values from 7.05 to 7.8 [21].

The Shatt al-Arab river, around 200 km long, is formed at the confluence of the Euphrates and Tigris rivers in al-Qurnah City in southern Iraq [22]. Currently, the river depends mainly on freshwater flow from the Tigris river [23,24]. The water flow in the Shatt Al-Arab river is affected by the tidal phenomenon of the Arabian Gulf, which has a semidiurnal pattern [25,26]. The recorded water temperature of the Shatt al-Arab river at Basrah City varied between 19 and 32 °C and the pH values between 7.86 and 8.07 [27,28].

2.3. Morphometrical Analysis

The commercial catches were sampled directly at the laboratory to limit the storage bias on the fish and otolith data. The first step was to measure the total length (TL ± 0.1 cm) of fish, then extract and clean both *asteriscii* otoliths. To describe the otolith shape, only univariate data as size parameters and shape factors were used. *Asteriscii* Otolith weight (O_{WEIGHT}) was measured using a digital balance to the nearest 0.0001 g (Sartorius Precision Balance Entris; Model BCE6231-1CFR, Sartorius Lab Instruments GmbH & Co. KG Otto-Brenner-Strasse 20 37079 Goettingen, Germany). Otolith images were captured using a camera (Efix 07-45x, 13MP, Alibaba.com, China (accessed on 28 March 2022)) with a stereomicroscope. *Asteriscii* Otolith length (O_{LENGTH}, mm) and width (O_{WIDTH}, mm) were taken using image processing systems (detailed descriptions in Figure 2). Size parameters (otolith length, width and weight) are measurements linked directly to otolith size, and are linked to otolith growth. Conversely, shape indices are dimensionless (and thus independent from otolith size) measures of otolith morphology similarity (i.e., otolith shape), compared with ideal geometric shapes calculated using size features. Two shape indices were used: ellipticity (to quantify similarity to an ellipse) and aspect-ratio (to compare to a rectangle) [29].

Figure 1. Sampling locations where common carp (*Cyprinus carpio*) were collected: the Shatt al-Arab (1), Euphrates (2) and Tigris (3) rivers.

$$\text{Ellipticity} = \frac{\text{OLENGTH} - \text{OWIDTH}}{\text{OLENGTH} + \text{OWIDTH}} \quad (1)$$

$$\text{Aspect} - \text{Ratio} = \frac{\text{OLENGTH}}{\text{OWIDTH}} \quad (2)$$

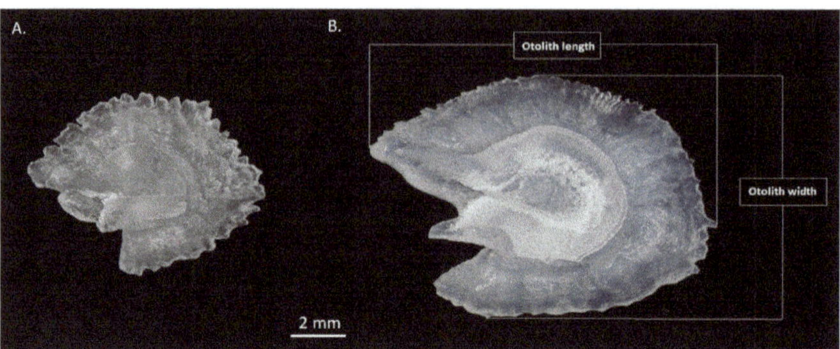

Figure 2. *Asteriscii* otoliths of *Cyprinus carpio* illustrating different features of the otolith measurements for two life stages ((**A**): small individual, TL = 155 mm; (**B**): large individual, TL = 380 mm).

The relationship between fish total length and each *asteriscii* otolith feature (otolith length, width, weight, ellipticity and aspect-ratio) according to the geographical position (River) and head side (Side) was modelled according to Equation (3):

$$\text{TL} \sim \text{Otolith feature} + \text{Otolith feature:River} + \text{Otolith feature:Side} + \text{Otolith feature:River:Side} \quad (3)$$

One generalized linear model was performed by each otolith feature (otolith length, width, weight, ellipticity and aspect-ratio). The geographical (*Asteriscii* Otolith feature:River) and side (*Asteriscii* Otolith feature:Side) effects and their interaction (Otolith feature:River:Side) were tested by the relationships between fish length and otolith feature. A second model was applied to evaluate the environmental effects (water temperature: T °C and acidification: pH) on the relationship between fish size and all otolith features according to the head side (Equation (4)). The environmental data were extracted from the literature for each location [16,21,27,28]. To test the environmental parameters, we used three values by factor with the annual mean value, the minimum and the maximum values registered by area.

$$\text{TL} \sim \text{Otolith feature:T °C:Side} + \text{Otolith feature:pH:Side} \quad (4)$$

The normality and the homoscedasticity of the residuals were assessed by visual inspection of diagnostic plots. The significance of explanatory variables was tested by likelihood ratio tests between nested models, while respecting the marginality of the effects (type 2 tests [29]) that are supposed to follow an χ^2 distribution under the null hypothesis, while correcting for multiple comparisons using a Bonferroni procedure. Statistical analyses were performed in the R statistical environment [30], using car [31], sp. [32], HH [33], vegan [34] and ggplot2 [35] packages.

3. Results

There were some differences between left and right *asteriscii* otoliths from the same individuals with the right otolith being bigger (i.e., otolith length and width) than the left otolith (Figure 3). Analysis of the relationships between fish length and each otolith feature (three size parameters and two shape indices) showed that there was a significant relationship between only two otolith size descriptors (length and width) with the total length of fish (column "TL", Table 1). For three other otoliths features, the relationship with the total length was not significant (Figure 4).

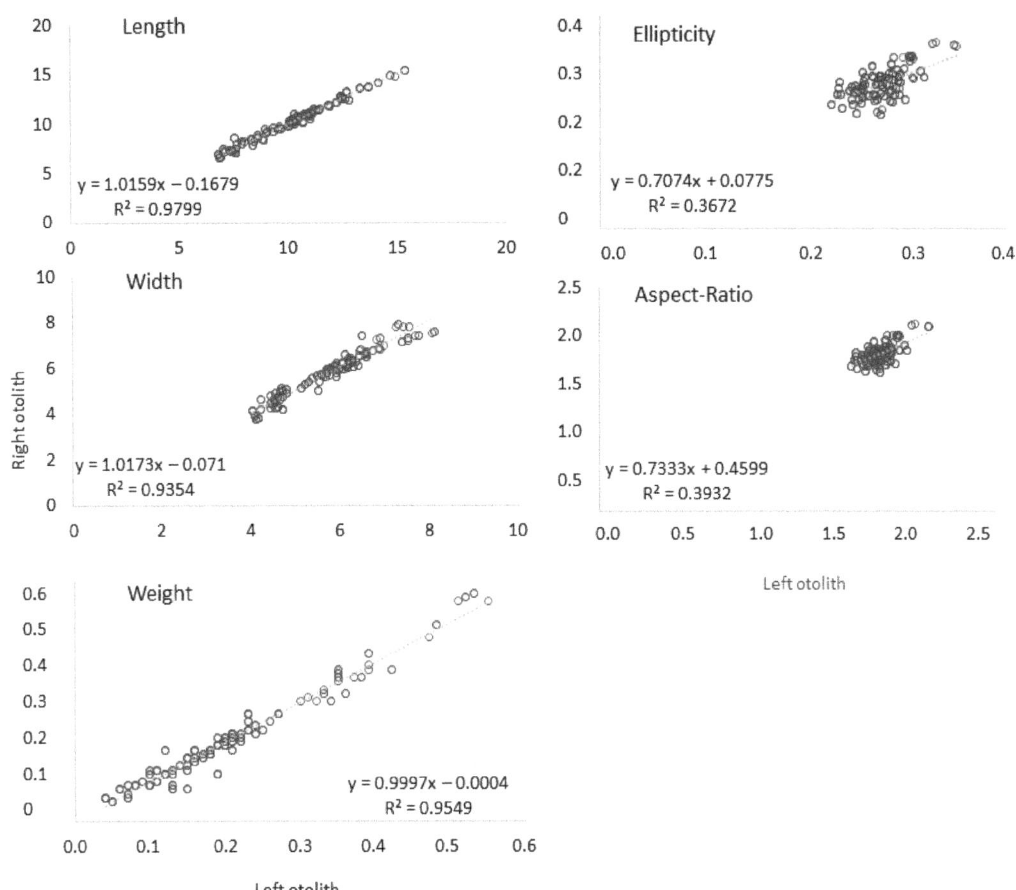

Figure 3. Difference between left and right otoliths of *Cyprinus carpio* for each otolith measurement.

Table 1. *p*-values of generalized linear models for the relationship between fish length and each otolith variable for *Cyprinus carpio* collected from three localities in Iraqi rivers with by geographical and Side effects (observed from the interaction between Otolith feature and river or side in the GLM: Otolith feature: River/Side). The environmental effects on the relationship between fish length and each otolith variable (Otolith feature: T °C/Side and Otolith feature: pH/Side) were tested with three different values (mean, minimal and maximum). Grey cases show significant effects ($p < 0.05$).

Otolith Descriptor	TL	Geographical Effect	Side Effect	Geographical Effect/Side Effect	T°C Effect/Side Effect			pH Effect/Side Effect		
					mean	min	max	mean	min	max
O$_{LENGTH}$	0.004	0.213	0.005	0.331	0.011	0.053	0.434	0.004	0.005	0.222
O$_{WEIGHT}$	0.051	0.278	<0.001	0.187	0.059	0.212	0.529	0.029	0.048	0.307
O$_{WIDTH}$	0.005	0.064	0.027	0.058	0.007	0.042	0.416	0.003	0.010	0.283
Ellipticity	0.326	0.008	0.014	0.029	<0.001	0.009	0.265	<0.001	0.003	0.151
AR	0.912	0.002	0.032	0.043	<0.001	0.005	0.236	<0.001	0.006	0.154

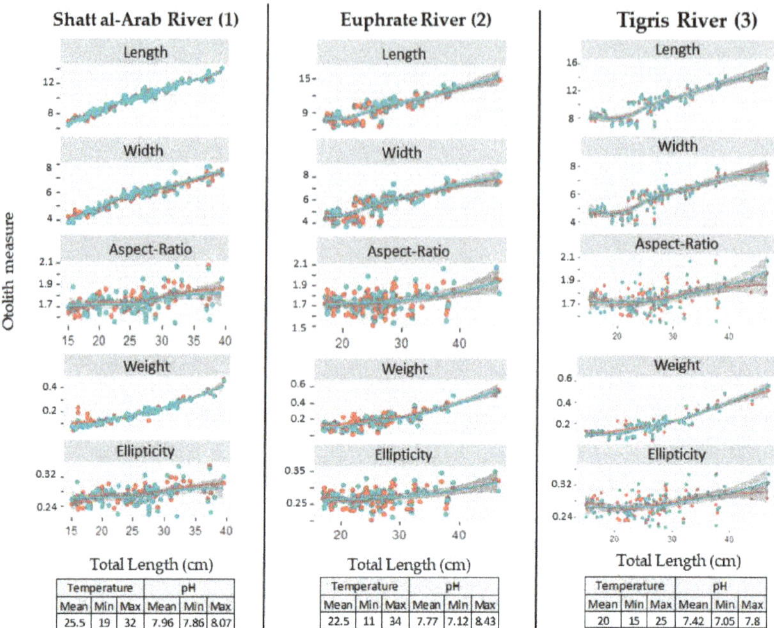

Figure 4. Relationships between fish length and *asteriscii* otolith morphological features (red points and line = left otolith, and green points and line = right otolith) according to the head side of *Cyprinus carpio* for each Iraqi River (for each river, the environmental data are presented).

Among the five *asteriscii* otolith features (size parameters or shape indices), there is always a significant effect of the head side on the relationship between fish length and each otolith descriptor (observed from the interaction between otolith feature and area or side in the GLM: Otolith feature: River/Side; Table 1). The relationship between fish length and otolith shape (i.e., two shape indices) was significantly dependent on the studied geographical area (i.e., river). The data collected in Tigris river presented the highest level of the fluctuating asymmetry in *asteriscii* otoliths of common carp collected from three localities in Iraqi Rivers. The relationship between fish and otolith growth was not directly linked to the location of sampling (Table 1). Finally, this generalized linear model showed that the relationship between fish length and otolith shape described by ellipticity and aspect-ratio is significantly different between the right and left otoliths, and this result was also linked to the location of sampling. Moreover, two mainly environmental factors were tested on the side effect applied to the relationship between fish length and each otolith descriptor. The results showed the interaction between both water temperature and pH and head side on the relationship between fish length and each otolith descriptor, especially with the minimum and mean values of these factors (Table 1). The Tigris river, which presented the lowest value of water pH and temperature, was the location with the highest level of the fluctuating asymmetry in *asteriscii* otoliths.

The significant correlation between fish total length and otolith growth showed that the following factors of otolith size also increased with an increase in total length: length, width and weight (Figure 4). The relationship between body length and otolith shape indices showed that the otolith shape evolves into a rectangular or elliptical shape during the life of the fish. The difference between the shape of the left and right otoliths in the same fish increased mainly in large individuals over 40 cm. This asymmetry is explained mainly by the width of the *asteriscii* otolith (Figure 4).

4. Discussion

The morphogenesis of the otolith is the result of a combination of exogenous (environment), endogenous (genetics) and developmental (ontogeny) factors [36,37]. Thus, the variability of these different factors and the biological scale at which they are expressed will determine the variation in the growth and shape of the otolith. Environmental factors that influence the morphogenesis of the otolith are separated into two groups: firstly, there are abiotic factors with temperature [38,39], pH [40], the depth of the water in which the fish live [41], the nature of the substrate [42] and salinity [43]. The growth and the morphogenesis of otoliths during the early life stages of fish are due to ontogenetic effects, but they also temperature-dependent [38]. In the same way, water acidification can also alter otolith shape [40]. For several species, individuals exposed to high pCO_2 had a larger otolith area and maximum length compared with controls; the increases were larger than could be explained by an increase in $CaCO_3$ precipitation in the otoliths driven by the modification of the pH regulation in the endolymph [44–47]. Our study corroborated that the water temperature and pH modified the relationship between fish length and each otolith measure (i.e., otolith length and width) and shape (i.e., ellipticity and aspect-ratio). Secondly, biotic factors such as the quantity [48] and specific composition [49] of the food available to the fish also affect the morphogenesis of the otolith. Genetic variability is also a strong factor influencing this morphogenesis process [50–55]. Ontogeny is also a factor generating shape variability, which can be reflected in effects of sex, age [53], body size [48] or the stage of sexual maturity [42,50,53] of the individual. The development of the otolith is spatially heterogeneous. The otoliths are paired structures present in the right and left inner ears, and an asymmetry between the two otoliths, or fluctuating asymmetry (FA), can be measured. This fluctuating asymmetry is related to the developmental trajectory of the otoliths, which is itself guided by developmental regulatory processes, such as the evolutionary canalization [54]. Fluctuating asymmetry may be related to developmental instability and thus provide an indicator measure of stress or micro-environmental variability [55–58]. In particular, Lemberget and McCormick [56] identify FA as an indicator of fish health as this type of asymmetry can directly affect the balance and hearing performance of the fish. This asymmetry can also be a functional adaptation to the environment. Our study showed the significant interaction between the level of asymmetry and the environmental factors (water temperature and pH). The acidification and the lowest temperature (i.e., lowest value of pH and temperature of water) observed in the Tigris river showed the highest level of asymmetry registered in Iraqi Rivers. As the trends of pH and temperature between the three locations were the same direction, it was not possible to dissociate the acidification and the temperature effects. The asymmetry can be measured in the otolith growth from the morphological descriptors (i.e., the relationship between fish length and otolith descriptors such as weight, length and width) and/or the otolith shape from univariate (i.e., Shape index) or multivariate (i.e., Elliptic Fourier Descriptors) measures. Otolith mass asymmetry is the most commonly used measure to compare the growth between left and right otoliths. Our results showed that there was a significant asymmetry in otolith growth of common carp (*Cyprinus carpio*) observed from all otolith descriptors (i.e., length, width and weight). This asymmetry corroborates that observed in previous studies on other marine and freshwater species [49,58–67]. In contrast, one study focused on Cyprinid fishes from Turkish inland waters observed no difference in the relationship between otolith descriptors and fish length according to the side effect [68]. However, the previous study on common carp (*Cyprinus carpio*) estimated that this species showed fluctuating asymmetry in otolith weight [69].

Otolith mass asymmetry has been used as an indicator to evaluate condition between several habitats [70], in particular, as a consequence of environmental stress, human activities, genetic disposition and a combination of these factors [71]. Our results showed that there was a significant asymmetry in otolith shape of common carp (*Cyprinus carpio*) observed from two shape indices (i.e., ellipticity and form factor) with increased asymmetry with size. During the life of the fish, the ontogenetic effect on the shape of the otolith decreased, unlike the environmental effects (temperature, food, etc. [36,39,42]). Consequently, this observed trend showed that asymmetry could be the result of the environmental conditions in each river. The water temperature and pH showed that the fluctuating asymmetry could be linked significantly to these two environmental factors. Our results corroborate those of previous studies on other marine and freshwater species [54,72–75].

5. Conclusions

This study showed that the observed asymmetry level was significantly linked to the geographical area, with the otolith shape being linked only to the environmental effect (i.e., water temperature and pH). In the climate change context, the temperature and pH are mainly environmental factors, which will be modified. Consequently, in the future, the asymmetry level may change and alter the balance and hearing performance of the fish. Sagittal otoliths, such as *asteriscii*, may also show significant asymmetry. Moreover, the experimental approach should be used to better understand the factors controlling otolith shape, integrating the ontogenetic effect, as has been undertaken for other species. Finally, this is the first study showing that it is necessary to estimate the level of asymmetry between otolith pairs to limit the potential bias due to a difference between left and right sides when using otoliths, and that the asymmetry could be linked directly to environment factors such as temperature and pH.

Supplementary Materials: The following are available online at https://www.mdpi.com/article/10.3390/fishes7020091/s1, Table S1: Size ranges of *Cyprinus carpio* collected from three localities in Iraq Rivers by head side (sample size, characteristics of fish length and each otolith variable).

Author Contributions: L.J. and K.M. designed the research; L.J. realized the sampling. K.M. performed the statistical analyses. All authors provided input for the results and discussion and wrote the paper. All authors provided critical comments and were involved in the writing of the manuscript. All authors accepted the final version of the manuscript. All authors have read and agreed to the published version of the manuscript.

Funding: This research received no external funding.

Institutional Review Board Statement: Ethical review and approval were waived for this study, because the fish were obtained from fisheries and were already dead when the otoliths were extracted.

Data Availability Statement: Data Availability Statements in section "MDPI Research Data Policies" at https://www.mdpi.com/ethics (accessed on 28 March 2022).

Acknowledgments: We would like to express our gratitude to all people involved in the collection of samples required in this study. We would especially thank Kirsteen MacKenzie for her valuable help in editing this manuscript.

Conflicts of Interest: The authors declare no conflict of interest.

References

1. Popper, A.N.; Lu, Z. Structure-function relationships in fish otolith organs. *Fish. Res.* **2000**, *46*, 15–25. [CrossRef]
2. D'Iglio, C.; Natale, S.; Albano, M.; Savoca, S.; Famulari, S.; Gervasi, C.; Lanteri, G.; Panarello, G.; Spanò, N.; Capillo, G. Otolith Analyses Highlight Morpho-Functional Differences of Three Species of Mullet (Mugilidae) from Transitional Water. *Sustainability* **2022**, *14*, 398. [CrossRef]
3. Casselman, J.M. Determination of age and growth. In *The Biology of Fish Growth*; Weatherley, A.H., Gill, H.S., Eds.; Academic Press: New York, NY, USA, 1987; pp. 209–242.
4. Wright, P.J.; Panfili, J.; Morales-Nin, B.; Geffen, A.J. Types of calcified structures. A. Otoliths. In *Manual of Fish Sclerochronology*; Panfili, J., de Pontual, H., Troadec, H., Wright, P.J., Eds.; Ifremer-IRD Coédition: Brest, France, 2002; pp. 31–57.
5. Panfili, J.; de Pontual, H.; Troadec, H.; Wright, P.J. *Manual of Fish Sclerochronology*; Coédition Ifremer-IRD: Brest, France, 2002.
6. Phelps, Q.E.; Edwards, K.R.; Willis, D.W. Precision of five structures for estimating age of common carp. *N. Am. J. Fish. Manag.* **2007**, *27*, 103–105. [CrossRef]
7. Vilizzi, L.; Walker, K.F. Age and growth of common carp, *Cyprinus carpio*, in the River Murray, Australia: Validation, consistency of age interpretation and growth models. *Env. Biol. Fish.* **1999**, *54*, 77–106. [CrossRef]
8. Brown, P.; Green, C.; Sivakumaran, K.P.; Stoessel, D.; Giles, A. Validating otolith annuli for annual age determination of commoncarp. *Trans. Am. Fish. Soc.* **2004**, *133*, 190–196. [CrossRef]

9. Winker, H.; Weyl, O.L.; Booth, A.J.; Ellender, B.R. Life history and population dynamics of invasive common carp, *Cyprinus carpio*, within a large turbid African impoundment. *Mar. Freshw. Res.* **2011**, *62*, 1270–1280. [CrossRef]
10. Al-Hassan, L.A.J.; Al-Dubaikel, A.Y.; Wahab, N.K.; Al-Daham, N.K. Asymmetry analysis in the catfish, *Heteropneustes fossilis* collected from Shatt Al-Arab River, Basrah, Iraq. *Riv. Idrobiol.* **1990**, *29*, 775–780.
11. Al-Hassan, L.A.J.; Hassan, S.S. Asymmetry study in *Mystus pelusius* collected from Shatt al-Arab River, Basrah, Iraq. *Pak. J. Zool.* **1994**, *26*, 276–278.
12. Jawad, L.A. Asymmetry analysis in the mullet, *Liza abu* collected from Shatt al-Arab River, Basrah, Iraq. *Bol. Mu. Reg. Sci. Nat. Torino* **2004**, *21*, 145–150.
13. Jawad, L.A.; Al-Janabi, M.I.; Rutkayová, J. Directional fluctuating asymmetry in certain morphological characters as a pollution indicator: Tigris catfish collected from the Euphrates, Tigris, and Shatt al-Arab Rivers in Iraq. *Fish. Aquat. Life* **2020**, *28*, 18–32. [CrossRef]
14. Jawad, L.A.; Abed, J.M. Morphological asymmetry in the greater lizardfish *Saurida tumbil* (Bloch, 1795) collected from the marine waters of Iraq. *Mar. Pollut. Bull.* **2020**, *159*, 111523. [CrossRef] [PubMed]
15. Mahé, K.; Ider, D.; Massaro, A.; Hamed, O.; Jurado-Ruzafa, A.; Goncalves, P.; Anastasopoulou, A.; Jadaud, A.; Mytilineou, C.; Elleboode, R.; et al. Directional bilateral asymmetry in otolith morphology may affect fish stock discrimination based on otolith shape analysis. *ICES J. Mar. Sci.* **2019**, *76*, 232–243. [CrossRef]
16. Abdullah, A.D. *Modelling Approaches to Understand Salinity Variations in a Highly Dynamic Tidal River, the Case of the Shatt Al-Arab River*; dissertation of Delft University of Technology and of the Academic Board of the UNESCO-IHE; CRC Press: Boca Raton, FL, USA, 2016.
17. IME. *Iraqi Ministries of Environment, Water Resources and Municipalities and Public Works, Volume I: Overview of Present Conditions and Current Use of the Water in the Marshlands Area/Book 1: Water Resources, New Eden Master Plan for Integrated Water Resources Management in the Marshlands Areas*; New Eden Group: Brescia, Italy, 2006.
18. Daoudy, M. Le Partage des Eaux Entre la Syrie, l'Irak et la Turquie. Négociation, Sécurité et Asymétrie des Pouvoirs, Moyen-Orient (in French), Paris. 2005. Available online: http://books.openedition.org/editionscnrs/2449 (accessed on 28 March 2022).
19. Zarins, J. Euphrates. In *The Oxford Encyclopedia of Archaeology in the Ancient Near East*; Meyers, E.M., Ed.; Oxford University Press: New York, NY, USA, 2005.
20. Isaev, V.A.; Mikhailova, M.V. The hydrology, evolution, and hydrological regime of the mouth area of the Shatt al-Arab River. *Wat. Res.* **2009**, *36*, 380–395.
21. Lazem, I.I.; Al-Naqeeb, N.A. Measuring pollution based on total petroleum hydrocarbons and total organic carbon in Tigris River, Maysan Province, Southern Iraq. *Casp. J. Environ. Sci.* **2021**, *19*, 535–545.
22. Al-Asadi, S.A.; Abdullah, S.S.; Al-Mahmood, H.K. Estimation of minimum amount of the net discharge in the Shatt Al-Arab River (south of Iraq). *J. Adab Al-Basrah* **2015**, *2*, 285–314.
23. Al-Asadi, S.A. The future of freshwater in Shatt Al-Arab River (Southern Iraq). *J. Geogr. Geol.* **2017**, *9*, 24–38. [CrossRef]
24. Al-Tememi, M.K.; Hussein, M.A.; Khaleefa, U.Q.; Ghalib, H.B.; AL-Mayah, A.M.; Ruhmah, A.J. The salts diffusion between East Hammar marsh area and Shatt Al-Arab River Northern Basra City. *Marsh. Bull.* **2015**, *10*, 36–45.
25. Al-Ramadhan, B.; Pastour, M. Tidal characteristics of Shatt Al-Arab River. *Mesopotamian J. Mar. Sci.* **1987**, *2*, 15–28.
26. Abdullah, A.H.J. Evaluation of fish assemblages' composition in the Euphrates River, southern Thiqar province, Iraq. *Mesopot. J. Mar. Sci.* **2020**, *35*, 83–96.
27. Al-Asadi, S.A.; Al Hawash, A.B.; Alkhlifa, N.H.A.; Ghalib, H.B. Factors affecting the levels of toxic metals in the Shatt Al-Arab River, Southern Iraq. *Earth Syst. Environ.* **2019**, *3*, 313–325. [CrossRef]
28. Gatea, M.H. Study of water quality changes of Shatt Al-Arab River, south of Iraq. *J. Univ. Babylon Eng. Sci.* **2018**, *26*, 228–241.
29. Russ, J. *Computer-Assisted Microscopy: The Measurement and Analysis of Image*; Plenum Press Corporation: New York, NY, USA, 1990.
30. R Core Team. *R: A Language and Environment for Statistical Computing*; R Foundation for Statistical Computing: Vienna, Austria, 2021.
31. Fox, J.; Weisberg, S. *Multivariate Linear Models in R: An R Companion to Applied Regression*; SAGE Publications: Thousand Oaks, LA, USA, 2011.
32. Bivand, R.S.; Pebesma, E.; Gomez-Rubio, V. *Applied Spatial Data Analysis with R*, 2nd ed.; Springer: Berlin/Heidelberg, Germany, 2013.
33. Heiberger, R.M.; Holland, B. *Statistical Analysis and Data Display: An Intermediate Course with Examples in R*, 2nd ed.; Springer: New York, NY, USA, 2015.
34. Dixon, P. VEGAN, a package of R functions for community ecology. *J. Veg. Sci.* **2003**, *14*, 927–930. [CrossRef]
35. Wickham, H. *Ggplot2: Elegant Graphics for Data Analysis*, 2nd ed.; Springer International Publishing: Cham, Switzerland, 2016.
36. Mille, T. Sources de Variation Intra-Populationelle de la Morphologie des Otolithes: Asymétrie Directionnelle et Régime Alimentaire. Ph.D. Thesis, Université de Lille 1—Sciences et Technologies, Lille, France, 2015.
37. Vignon, M. Disentangling and quantifying sources of otolith shape variation across multiple scales using a new hierarchical partitioning approach. *Mar. Ecol. Prog. Ser.* **2015**, *534*, 163–177. [CrossRef]
38. Mahé, K.; Gourtay, C.; Bled Defruit, G.; Chantre, C.; de Pontual, H.; Amara, R.; Claireaux, G.; Audet, C.; Zambonino-Infante, J.L.; Ernande, B. Do environmental conditions (temperature and food composition) affect otolith shape during fish early-juvenile phase? An experimental approach applied to European Seabass (*Dicentrarchus labrax*). *J. Exp. Mar. Biol. Ecol.* **2019**, *521*, e151239. [CrossRef]

39. Bolles, K.L.; Begg, G.A. Distinction between silver hake (*Merluccius bilineariz*) stocks in U.S. waters of the northwest Atlantic using whole otolith morphometric. *Fish. Bull.* **2000**, *98*, 451–462.
40. Holmberg, R.J.; Wilcox-Freeburg, E.; Rhyne, A.L.; Tlusty, M.F.; Stebbins, A.; Nye, S.W., Jr.; Honig, A.; Johnston, A.E.; San Antonio, C.M.; Bourque, B.; et al. Ocean acidification alters morphology of all otolith types in Clark's anemonefish (*Amphiprion clarkii*). *PeerJ* **2019**, *7*, e6152. [CrossRef]
41. Lombarte, A.; Lleonart, J. Otolith size changes related with body growth, habitat depth and temperature. *Environ. Biol. Fish.* **1993**, *37*, 297–306. [CrossRef]
42. Mille, T.; Mahé, K.; Villanueva, M.C.; De Pontual, H.; Ernande, B. Sagittal otolith morphogenesis asymmetry in marine fishes. *J. Fish Biol.* **2015**, *87*, 646–663. [CrossRef]
43. Capoccioni, F.; Costa, C.; Aguzzi, J.; Menesatti, P.; Lombarte, A.; Ciccotti, E. Ontogenetic and environmental effects on otolith shape variability in three Mediterranean European eel (*Anguilla anguilla*, L.) populations. *J. Exp. Mar. Biol. Ecol.* **2011**, *397*, 1–7. [CrossRef]
44. Checkley, D.M.; Dickson, A.G.; Takahashi, M.; Radich, J.A.; Eisenkolb, N.; Asch, R. Elevated CO_2 Enhances Otolith Growth in Young Fish. *Science* **2009**, *324*, 1683. [CrossRef]
45. Munday, P.L.; Gagliano, M.; Donelson, J.M.; Dixson, D.L.; Thorrold, S.R. Ocean acidification does not affect the early life history development of a tropical marine fish. *Mar. Ecol. Prog. Ser.* **2011**, *423*, 211–221. [CrossRef]
46. Réveillac, E.; Lacoue-Labarthe, T.; Oberhänsli, F.; Teyssié, J.L.; Jeffree, R.; Gattuso, J.P.; Martin, S. Ocean acidification reshapes the otolith-body allometry of growth in juvenile sea bream. *J. Exp. Mar. Bio. Ecol.* **2015**, *463*, 87–94. [CrossRef]
47. Coll-Lladó, C.; Giebichenstein, J.; Webb, P.B.; Bridges, C.R.; Garcia de la Serrana, D. Ocean acidification promotes otolith growth and calcite deposition in gilthead sea bream (*Sparus aurata*) larvae. *Sci. Rep.* **2018**, *8*, 8384. [CrossRef] [PubMed]
48. Hüssy, K. Otolith shape in juvenile cod (*Gadus morhua*): Ontogenetic and environmental effects. *J. Exp. Mar. Biol. Ecol.* **2008**, *364*, 35–41. [CrossRef]
49. Mérigot, B.; Letourneur, Y.; Lecomte-Finiger, R. Characterization of local populations of the common sole Solea solea (Pisces, Soleidae) in the NW Mediterranean through otolith morphometrics and shape analysis. *Mar. Biol.* **2007**, *151*, 997–1008. [CrossRef]
50. Cardinale, M.; Doerin-Arjes, P.; Kastowsky, M.; Mosegaard, H. Effects of sex, stock, and environment on the shape of known-age Atlantic cod (*Gadus morhua*) otoliths. *Can. J. Fish. Aquat. Sci.* **2004**, *61*, 158–167. [CrossRef]
51. Cadrin, S.X.; Friedland, K.D. The utility of image processing techniques for morphometric analysis and stock identification. *Fish. Res.* **1999**, *43*, 129–139. [CrossRef]
52. Castonguay, M.; Simard, P.; Gagnon, P. Usefulness of Fourier analysis of otolith shape for Atlantic mackerel (*Scomber scombrus*) stock discrimination. *Can. J. Fish. Aquat. Sci.* **1999**, *48*, 296–302. [CrossRef]
53. Lleonart, J.; Salat, J.; Torres, G.J. Removing allometric effects of body size in morphological analysis. *J. Theor. Biol.* **2000**, *205*, 85–93. [CrossRef]
54. Mahé, K. Sources de Variation de la Forme des Otolithes: Implications Pour la Discrimination des Stocks de Poissons. Ph.D. Thesis, Université du Littoral Côte d'Opale, Boulogne-sur-mer, France, 2019.
55. Dowhower, J.F.; Blumer, L.S.; Lejeune, P.; Gaudin, P.; Marconato, A.; Bisazza, A. Otolith asymmetry in Cottus bairdi and Cottus gobio. *Pol. Arch. Hydrobiol.* **1990**, *37*, 209–220.
56. Lemberget, T.; Mccormick, M.I. Replenishment success linked to fluctuating asymmetry in larval fish. *Oecologia* **2009**, *159*, 83–93. [CrossRef] [PubMed]
57. Díaz-Gil, C.; Palmer, M.; Catalán, I.A.; Alós, J.; Fuiman, L.A.; García, E.; del Mar Gil, M.; Grau, A.; Kang, A.; Maneja, R.H.; et al. Otolith fluctuating asymmetry: A misconception of its biological relevance? *ICES J. Mar. Sci.* **2015**, *72*, 2079–2089. [CrossRef]
58. Hilbig, R.; Anken, R.; Rahmann, H. On the origin of susceptibility to kinetotic swimming behaviour in fish: A parabolic aircraft flight study. *J. Vest. Res.* **2003**, *12*, 185–189. [CrossRef]
59. Lychakov, D.V. Behavioural lateralization and otolith asymmetry. *J. Evol. Biochem. Physiol.* **2003**, *49*, 441–456. [CrossRef]
60. Lychakov, D.V.; Rebane, Y.T.; Lombarte, A.; Demestre, M.; Fuiman, L.A. Saccular otolith mass asymmetry in adult flatfishes. *J. Fish Biol.* **2008**, *72*, 2579–2594. [CrossRef]
61. Jawad, L.A.; Al-Mamry, J.; Al-Busaidi, H. Relationship between fish length and otolith length and width in the lutjanid fish, Lutjanus bengalensis (Lutjanidae) collected from muscat city coast on the sea of oman. *J. Black Sea/Mediterr. Environ.* **2011**, *17*, 116–126.
62. Jawad, L.A.; Sadighzadeh, Z. Otolith mass asymmetry in the mugilid fish, Liza klunzingeri (Day, 1888) collected from Persian Gulf near Bandar Abbas. *Anal. Biol.* **2013**, *35*, 105–107. [CrossRef]
63. Jawad, L.A. Otolith Mass Asymmetry in Carangoides caerulepinnatus (Rüppell, 1830) (Family: *Carangidae*) Collected from the Sea of Oman. *Croat. J. Fish.* **2013**, *71*, 37–41. [CrossRef]
64. Jawad, L.A.; Park, J.M.; Kwak, S.N.; Ligas, A. Study of the relationship between fish size and otolith size in four demersal species from the south-eastern yellow sea. *Cah. Biol. Mar.* **2017**, *58*, 9–15.
65. Jawad, L.A.; Qasim, A.M.; Al-Faiz, N.A. Bilateral asymmetry in size of otolith of Otolithes ruber (Bloch & Schneider, 1801) collected from the marine waters of Iraq. *Mar. Poll. Bull.* **2021**, *165*, 112110.
66. Yedier, S.; Bostanci, D.; Kontaş, S.; Kurucu, G.; Polat, N. Comparison of Otolith Mass Asymmetry in Two Different Solea solea Populations in Mediterranean Sea. *J. Sci. Tech.* **2018**, *8*, 125–133.

67. Osman, Y.A.A.; Mahé, K.; El-Mahdy, S.M.; Mohammad, A.S.; Mehanna, S.F. Relationship between Body and Otolith Morphological Characteristics of Sabre Squirrelfish (*Sargocentron spiniferum*) from the Southern Red Sea: Difference between Right and Left Otoliths. *Oceans* **2021**, *2*, 624–633. [CrossRef]
68. Emre, N. Biometric Relation between Asteriscus Otolith Size and Fish Total Length of Seven Cyprinid Fish Species from Inland Waters of Turkey. *Turk. J. Fish. Aquat. Sci.* **2019**, *20*, 171–175.
69. Takabayashi, A.; Ohmura-Iwasaki, T. Functional asymmetry estimated by measurements of otolith in fish. *Biol. Sci. Space.* **2003**, *17*, 293–297. [CrossRef] [PubMed]
70. Gronkjaer, P.; Sand, M.K. Fluctuating asymmetry and nutritional condition of Baltic cod (*Gadus morhua*) larvae. *Mar. Biol.* **2003**, *143*, 191–197. [CrossRef]
71. Palmer, A.R. What determines direction of asymmetry: Genes, environment or chance? *Philos. Trans. R. Soc. B Biol. Sci.* **2016**, *371*, 20150417. [CrossRef]
72. Trojette, M.; Ben Faleh, A.; Fatnassi, M.; Marsaoui, B.; Mahouachi, N.H.; Chalh, A.; Quignard, J.-P.; Trabelsi, M. Stock discrimination of two insular populations of *Diplodus annularis* (Actinopterygii: Perciformes: Sparidae) along the coast of Tunisia by analysis of otolith shape. *Acta Ichthyol. Piscat.* **2015**, *45*, 363–372. [CrossRef]
73. Bostanci, D.; Yilmaz, M.; Yedier, S.; Kurucu, G.; Kontas, S.; Darçin, M.; Polat, N. Sagittal otolith morphology of sharpsnout seabream *Diplodus puntazzo* (Walbaum, 1792) in the Aegean sea. *Int. J. Morphol.* **2016**, *34*, 484–488. [CrossRef]
74. Zhang, C.; Ye, Z.; Li, Z.; Wan, R.; Ren, Y.; Dou, S. Population structure of Japanese Spanish mackerel Scomberomorus niphonius in the Bohai Sea, the Yellow Sea and the East China Sea: Evidence from retom forests based on otolith features. *Fish. Sci.* **2016**, *82*, 251–256. [CrossRef]
75. Rebaya, M.; Ben Faleh, A.R.; Allaya, H.; Khedher, M.; Trojette, M.; Marsaoui, B.; Fatnassi, M.; Chalh, A.; Quignard, J.P.; Trabelsi, M. Otolith shape discrimination of *Liza ramada* (Actinopterygii: Mugiliformes: Mugilidae) from marine et estuarine populations in Tunisia. *Acta Ichthyol. Piscat.* **2017**, *47*, 13–21. [CrossRef]

Article

Assessing the Speciation of *Lutjanus campechanus* and *Lutjanus purpureus* through Otolith Shape and Genetic Analyses

Angel Marval-Rodríguez [1], Ximena Renán [2,*], Gabriela Galindo-Cortes [1], Saraí Acuña-Ramírez [1], María de Lourdes Jiménez-Badillo [1], Hectorina Rodulfo [3], Jorge L. Montero-Muñoz [2], Thierry Brulé [2] and Marcos De Donato [3,*]

[1] Instituto de Ciencias Marinas y Pesquerías, Universidad Veracruzana, Boca del Rio 94290, Mexico; avgelo7@gmail.com (A.M.-R.); gagalindo@uv.mx (G.G.-C.); acua.sarai@gmail.com (S.A.-R.); ljimenez@uv.mx (M.d.L.J.-B.)
[2] Departamento de Recursos del Mar, Centro de Investigación y de Estudios Avanzados del Instituto Politécnico Nacional, Mérida 97205, Mexico; jorge.montero@cinvestav.mx (J.L.M.-M.); tbrule@cinvestav.mx (T.B.)
[3] Tecnológico de Monterrey, Escuela de Ingeniería y Ciencias, Querétaro 76130, Mexico; herodulfo@tec.mx
* Correspondence: ximena.renan@cinvestav.mx (X.R.); mdedonate@tec.mx (M.D.D.); Tel.: +52-999-242-1587 (X.R.); +52-442-231-2927 (M.D.D.)

Abstract: Based on their morphological and genetic similarity, several studies have proposed that *Lutjanus campechanus* and *Lutjanus purpureus* are the same species, but there is no confirmed consensus yet. A population-based study concerning otolith shape and genetic analyses was used to evaluate if *L. campechanus* and *L. purpureus* are the same species. Samples were collected from populations in the southwestern Gulf of Mexico and the Venezuelan Caribbean. Otolith shape was evaluated by traditional and outline-based geometric morphometrics. Genetic characterization was performed by sequencing the mtDNA control region and intron 8 of the nuclear gene FASD2. The otolith shape analysis did not indicate differences between species. A nested PERMANOVA identified differences in otolith shape for the nested population factor (fishing area) in morphometrics and shape indexes ($p = 0.001$) and otolith contour (WLT4 anterior zone, $p = 0.005$ and WLT4 posterodorsal zone, $p = 0.002$). An AMOVA found the genetic variation between geographic regions to be 10%, while intrapopulation variation was 90%. Network analysis identified an important connection between haplotypes from different regions. A phylogenetic analysis identified a monophyletic group formed by *L. campechanus* and *L. purpureus*, suggesting insufficient evolutionary distances between them. Both otolith shape and molecular analyses identified differences, not between the *L. campechanus* and *L. purpureus* species, but among their populations, suggesting that western Atlantic red snappers are experiencing a speciation process.

Keywords: red snapper; sagittal otolith; wavelet; genetic diversity; population structure

Citation: Marval-Rodríguez, A.; Renán, X.; Galindo-Cortes, G.; Acuña-Ramírez, S.; Jiménez-Badillo, M.d.L.; Rodulfo, H.; Montero-Muñoz, J.L.; Brulé, T.; De Donato, M. Assessing the Speciation of *Lutjanus campechanus* and *Lutjanus purpureus* through Otolith Shape and Genetic Analyses. *Fishes* **2022**, *7*, 85. https://doi.org/10.3390/fishes7020085

Academic Editor: Josipa Ferri

Received: 21 January 2022
Accepted: 23 March 2022
Published: 7 April 2022

1. Introduction

Correct species identification and data on population spatial distribution are vital to improving fishery resource assessment and management [1–3]. Although it is practically impossible to analyze all the variables (morphometrics and/or genetics) characteristic of a species, variation in the individual phenotypes of organisms or their genotypes can be examined to identify species and delimit populations [4,5]. Traditionally, fish are identified using molecular genetics, morphometric measurements and meristic characters, and otolith shape analysis, among other techniques [6]. Sequencing of conserved genes (a molecular biology technique) is the most common tool currently used for species identification, in addition to the characterization of population structure and gene flow [7]. Additionally, otolith shape analysis has been in use for more than 20 years as an objective method for species identification, fish stock discrimination, systemic and taxonomic studies, aging of fish, fish auditory neuroscience studies, and the study of the ecomorphological patterns of fish [8–16].

Particularly for the Lutjanidae family, otolith shape analysis has been used to identify environmental and genetic influences on otolith morphology, to age juvenile red snappers, discriminate between stocks, identify closely-related species, and to analyze the morphometric relationships between two species [17–21].

The Lutjanidae family comprises a large group of species distributed in tropical and subtropical marine ecosystems in the Atlantic, Pacific, and Indian Oceans [22]. In the western Atlantic, 6 genera and 18 species have been identified, of which 12 species belong to the genus *Lutjanus* [23]. *Lutjanus campechanus* and *L. purpureus* are the most important species captured in the western Atlantic, and fetch high market prices [24]. Populations of *L. campechanus* are distributed throughout the Gulf of Mexico, from the Yucatan Peninsula to Key West, and along the Atlantic coast of the United States to Massachusetts [25]. *Lutjanus purpureus* are distributed from the southern coast of Cuba and the Yucatan Peninsula throughout the Caribbean Sea, and from the north and northeast of South America to Pernambuco in Brazil, approximately [26].

Lutjanus campechanus and *L. purpureus* are remarkably similar in their life cycle, population parameters, and morphology [27]. Taxonomic identification of these species is difficult because of the similarities in their external morphology and the overlap in the characteristics commonly used to identify them, such as spines, hard and soft rays of the pectoral, dorsal and anal fins, lateral line scales, and gill rakers [28,29]. Based on these morphological similarities, Cervigón et al. [30] hypothesized the existence of a single species of red snapper in the western Atlantic Ocean; that is, that *L. campechanus* and *L. purpureus* are actually the same species, with morphological differences between populations in the western Atlantic. Based on genetic analysis, lack of phylogeographic structure, and intense intermingling between individuals, they suggested the probable existence of just one red snapper species throughout the western Atlantic [29,31]. However, a recent study [32] used molecular delimitation to discriminate *L. campechanus* and *L. purpureus* as distinct evolutionary units, although the groups did share a significant number of haplotypes, suggesting important gene flow between them. The objective of the present study was to elucidate the species-specific boundaries between *L. campechanus* and *L. purpureus* for the western Atlantic by combining, for the first time, otolith morphometrics and genetic analyses.

2. Materials and Methods

2.1. Sample Collection

Biological samples of *L. campechanus* and *L. purpureus* were collected between 2015 and 2017 from dead individuals caught by a commercial multi-species artisanal fleet (Veracruz and Tabasco State) and an industrial shrimp trawl fleet in the southwestern Gulf of Mexico (Campeche State) and the Venezuelan Caribbean (Nueva Esparta and Sucre State) (Figure 1). The individuals comprised 108 *L. campechanus* (72–452 mm total length (TL)) and 24 *L. purpureus* (214–460 mm TL). Sagittal otoliths were extracted through the gill arch, washed with distilled water and stored in labeled plastic containers. Collected muscle tissue was stored in 96% ethanol and kept frozen until laboratory processing.

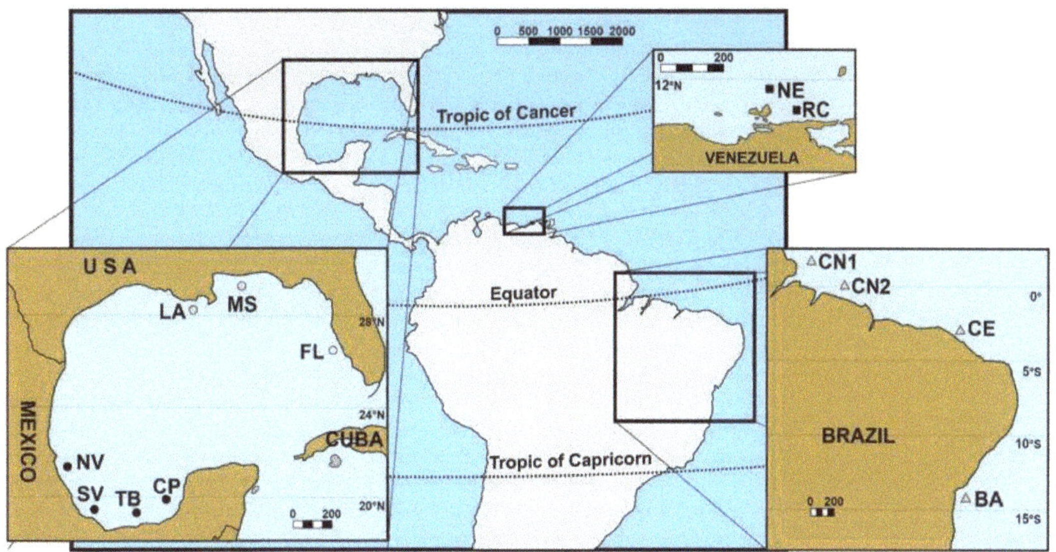

Figure 1. Map of *Lutjanus campechanus* and *Lutjanus purpureus* spatial distribution. Black point: fishing areas used in this study. Gray point: sequences from GenBank of samples collected in the USA and Brazil. (NV = North Veracruz; SV = South Veracruz; TB = Tabasco; CP = Campeche; NE = Nueva Esparta; RC = Rio Caribe; LA = Louisiana; MS = Mississippi, FL = Florida; CN1 = Coast North Brazil (46° and 50° W); CN2 = Coast North Brazil (43° and 45° W); CE = Ceará; BA = Bahia).

2.2. Otolith Analysis

2.2.1. Otolith Morphological Description

The morphology of the left otoliths from *L. campechanus* and *L. purpureus* were described based on high-definition photographs taken with a scanning electron microscope (FESEM-JOEL-7600F). Morphological description was based on thirteen characteristics considering general otolith shape, edges, sulcus acusticus, ostium, cauda, rostrum and antirostrum, and variations in shape, position and orientation [33,34].

2.2.2. Otolith Shape Analyses

Digital photographs of the left otoliths were taken using a stereoscopic microscope (Leica-EZ4E), with the sulcus acusticus facing downwards, under reflected light on a black background. Based on the morphometrics classification established by Pavlov [35], we used traditional morphometrics and outline-based geometric morphometrics to analyze otolith shape. Traditional otolith morphometrics, which describe shape and length measurements or indices between vectors passing through certain points, were automatically generated using the Image Pro Plus v.7 software (Media Cybernetics Inc., USA). Four otolith morphometrics were measured (area (A); perimeter (P); maximum diameter (MaxD); and minimum diameter (MinD)), as well as five shape indices (aspect (AS); ellipticity (E); rectangularity (RE); roundness (RD) and fractal dimension index (FI)).

The outline-based geometric otolith morphometrics analysis was performed by extracting discrete Wavelet Transforms (DWLT) using the "Shape" module in the Age & Shape program (Infaimon, Spain). This program reconstructs otolith contour by tracing equally sampled angles (radii) from the otolith geometrical center (mean x and y polar coordinates) to 512 equidistant points in the edge. It then automatically generates ten multi-scale decompositions from the finest (DWLT1) to the coarsest (DWLT10) [36–38]. Following Tuset et al. [34], the differential characteristics of each otolith, by subsections (anterior, ventral, posterodorsal and anterodorsal), were defined by the graphic representation of wavelet scale mean and standard deviation.

2.2.3. Otolith Statistical Analysis

Fish size effects were removed from the magnitude of otolith morphometrics and shape indices, as recommended [39,40]. To remove amplitude effects, all DWLT multi-scale decompositions were standardized by dividing each radius by the mean radius length [41]. A principal component analysis (PCA) was applied to reduce the dimensions without loss of information, selecting the DWLT scale with the strongest correlation to component 1 [42]. This analysis indicated that the DWLT4 scale explained 94% of data variability; it was used in the subsequent analyses.

Six steps were applied in the statistical analysis to estimate differences in otolith shape in relation to the population factor and region factor. First, morphometrics, shape indices and DWLT4 data were transformed using the Hellinger distance. Second, a Euclidean triangular matrix was calculated using the Euclidean distance function applied to the Hellinger-transformed data [43]. Third, the resulting Hellinger distance matrix was used in a multivariate permutational analysis of variance (PERMANOVA) [44] to assess the region factor and the nested factor of population. Fourth, for the significant factor in the PERMANOVA ($\alpha = 0.05$), pairwise tests were used to estimate differences in otolith shape between factor levels. Fifth, the significant factor was plotted using the multivariate of the metric dimensional scale (MDS) analysis. Sixth and final, an average of bootstrap samples (95% confidence bootstrap region) was generated for the MDS to compare factor levels. The minimum dimension value of the MDS metric was 0.99 Pearson's correlation. The triangular Hellinger distance matrix, PERMANOVA and MDS analyses were run with PERMANOVA + for PRIMER (FIRST-E: Plymouth, UK) [45].

2.3. *Molecular Analysis*

2.3.1. Sample Processing, Fragment Amplification and Sequencing

DNA was extracted from alcohol-preserved muscle tissue using the Wizard® Genomic DNA Purification Kit (Promega, Madison, USA), following the manufacturer protocol. DNA concentration and quality were measured using a NanoDrop One (Thermo Scientific, Waltham, USA) and by visual inspection of the DNA by electrophoresis. DNA extracts were stored at −20 °C until use.

Genetic characterization of the samples was performed by sequencing the D-loop or control region of the mitochondrial DNA (mtDNA-CR) and the intron 8 of the nuclear gene for the enzyme fatty acid desaturase 2 (FADS2), according to previously published studies [46,47]. The fragments were sequenced using both primers and the BigDye Terminator system at the National Biodiversity Genomic Laboratory (Laboratorio Nacional de Genómica para la Biodiversidad—Langebio), Irapuato, Mexico.

2.3.2. Population Analysis

Evaluation of population structure was performed by calculating the number of haplotypes (h) and nucleotide diversity (π) within and between the populations (fishing areas) using the Arlequin v.3.5 software [48]. Sequence population structure was assessed using the fixation index (FST), and an analysis of molecular variance (AMOVA). Haplotype relationships were reconstructed using the Network ver. 10.1 software (Fluxus Technology, Ltd., Santa Clara, USA), and calculated with the median joining algorithm [49] using default settings (weight: 10 and ε: 0). A test to identify the correlation between genetic and geographic distances between populations was performed with a Mantel test using the geographic distance calculated from Google Earth (i.e., straight lines between sampling areas) and pairwise FST (obtained from the AMOVA). This was run with the GenAlEx v.6.51b2 software [50].

2.3.3. Phylogenetic Analysis

The generated sequences were viewed using the GeneStudio v.2.2.0 program (http://genestudio.com, accessed on 14 Mach 2022) to assess sequence quality and identify differences between them. They were compared to sequences in GenBank using BLAST.

Phylogenetic analyses were run using a Bayesian phylogenetic analysis performed with the Mr.Bayes v.3.2.1 program [51], implementing the general time-reversible (GTR) model using the rate at each site as a random variable. A discrete gamma distribution was applied to model evolutionary rate differences between sites (G) and a proportion of invariable sites. Markov chain Monte Carlo (MCMC) chains were run for 1,000,000 generations.

The sequences generated in the present study were deposited in GenBank (Table S1). Sequences from GenBank from samples collected in the USA and Brazil for the mtDNA-CR region, and from Brazil for the FADS2 gene, were used in the present study (Table S2). To evaluate the phylogeographic attributes of *L. campechanus* and *L. purpureus*, phylogenetic analysis consensus sequences were used for three *Lutjanus* species (Table S3) distributed along the Pacific and Atlantic coasts of the Americas: *L. peru*, *L. synagris* and *L. guttatus*.

3. Results

3.1. Otolith Analysis

3.1.1. Otolith Morphological Description

A total of 132 otoliths were analyzed (108 from *L. campechanus* and 24 from *L. purpureus*). All exhibited a generally pentagonal-like shape with a concave-convex profile. The sulcus acusticus was heterosucoidal, ostial with a middle position and downward orientation. Both species exhibited a developed rostrum and a moderately curved cauda (Figure 2). Despite their morphological similarities, the otoliths did vary, primarily in terms of the anterior and posterodorsal regions, ostium shape and margin type. Those from individuals from the Gulf of Mexico had an angled anterior region, an oblique posterodorsal region, and a poorly developed antirostrum. Also present were a sinuous ventral edge and angular dorsal edge with a funnel-liked ostium. In contrast, otoliths from individuals from the Venezuelan Caribbean exhibited a rounded anterior region and angular posterodorsal region, with crenate dorsal and ventral margins, a developed antirostrum and a rectangular ostium.

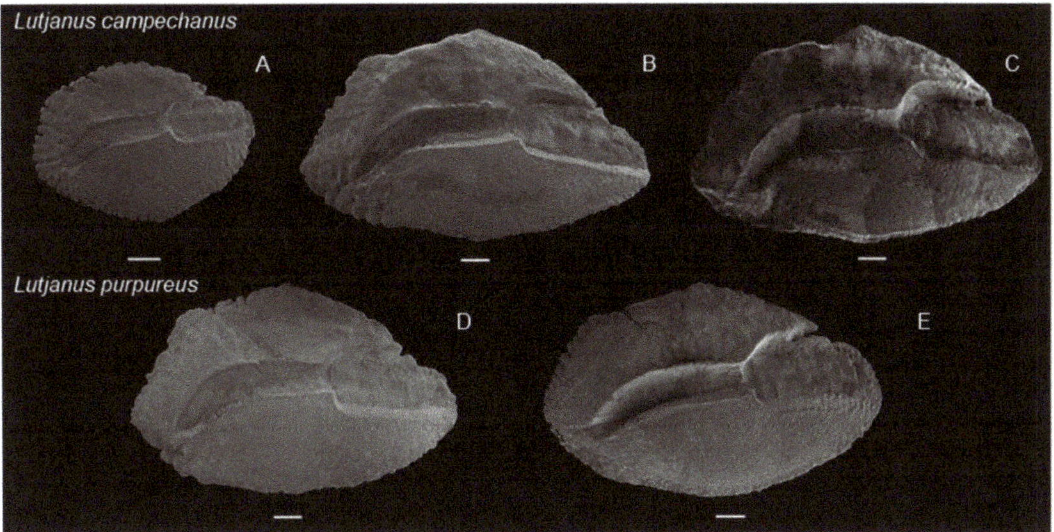

Figure 2. Red snapper sagittae otoliths: Gulf of Mexico (**A** = Campeche: 114 mm total length (TL); **B** = Tabasco: 360 mm TL; **C** = Veracruz: 378 mm TL) and Venezuelan Caribbean (**D** = Sucre: 285 mm TL; **E** = Nueva Esparta: 299 mm TL). Scale = 1 mm.

3.1.2. Inter-region Factor Variation

The nested PERMANOVA results using the data from the two regions (Gulf of Mexico and Caribbean basin) exhibited no differences in the morphometrics and shape indices ($p = 0.116$), DWLT4 anterior zone ($p = 0.967$), or DWLT4 posterodorsal zone ($p = 0.475$).

3.1.3. Intrapopulation Factor Variation

The nested PERMANOVA analysis identified differences between the otoliths from organisms caught in the different populations: morphometrics and shape indices ($p = 0.001$), DWLT4 anterior zone ($p = 0.005$), and DWLT4 posterodorsal zone ($p = 0.002$). The MDS analyses found that, although there was high variability in otolith shape between the individuals from different populations, only the individuals from Campeche (Gulf of Mexico) could be discriminated from other populations from the Gulf of Mexico and the Caribbean basins (Figure 3).

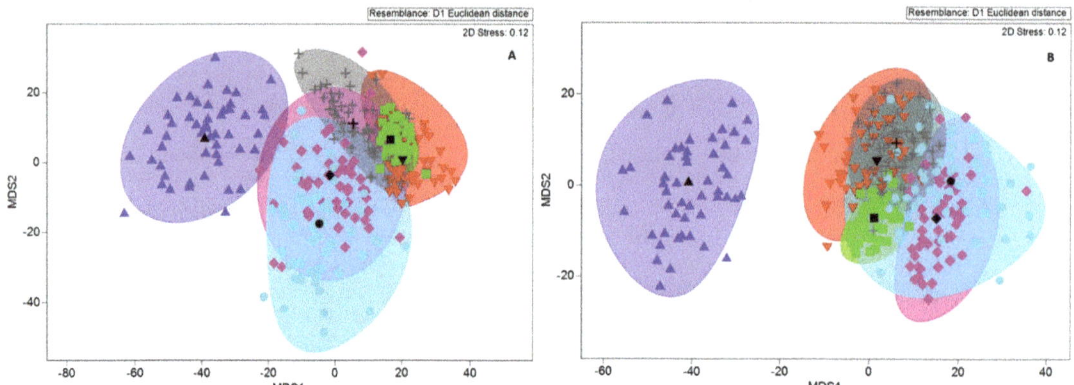

Figure 3. MDS arrangement diagram of Euclidean distances from the nested factor population in PERMANOVA of the DWLT4 anterior region (**A**) and DWLT4 posterior region (**B**) of sagittae otoliths from western Atlantic red snappers. (Blue = Campeche; Red = South Veracruz; Green = North Veracruz; Pink = Tabasco; Gray = Rio Caribe; Light blue = Nueva Esparta).

A pairwise comparison found differences between otoliths from individuals from Campeche vs. Tabasco ($p = 0.005$) and North Veracruz ($p = 0.010$) in the morphometrics/shape indices and anterior subsection. There were also differences in the DWLT4 posterodorsal subsection in otoliths from Campeche and Tabasco versus all other Gulf of Mexico populations ($p = 0.001$). Differences were also present in North Veracruz vs. Nueva Esparta ($p = 0.025$) for the morphometrics/shape indices and DWLT4 anterior subsection ($p = 0.023$), and in the DWLT4 posterodorsal otolith subsection in Campeche and Tabasco vs. Nueva Esparta and Rio Caribe ($p = 0.001$).

3.2. Molecular Analysis

3.2.1. Intrapopulation Factor Variation

A total of 1363 nucleotide positions were analyzed in the final dataset, 798 nucleotides for the mtDNA-CR and 565 nucleotides for the nuclear gene FADS2. For the mtDNA-CR region, significant intrapopulation values were found for nucleotide diversity, ranging from 0.013–0.029 substitutions per site. The populations from Venezuela and Brazil exhibited the highest nucleotide diversity values. The Tajima D test for neutrality for the mtDNA-CR region revealed that the changes were significantly different from random changes (non-neutral, $p < 0.05$) among regions of the USA and Mexico, but not within any of their populations. In Brazil, there were non-neutral changes between populations, and between North Coast individuals (Table S4).

Intrapopulational nucleotide diversity for the FADS2 gene was significantly lower, ranging from 0.002–0.012 substitutions per site. The Tajima test revealed that changes found within South Veracruz were non-neutral. It also identified non-neutral changes between the Venezuela populations and between Rio Caribe and Nueva Esparta individuals (Table S5), suggesting that selection or other forces are directly driving variability.

The AMOVA found that for mtDNA-CR, 53% of variation was intrapopulational and 47.9% was attributed to variation among geographical regions. For intron 8 (FADS2), 90% of variation was intrapopulational, with only 18% between regions (Table 1).

Table 1. Analysis of molecular variance (AMOVA) between regions (Gulf of Mexico and Caribbean basins) and populations (fishing areas) of red snapper using mitochondrial DNA based on the D-loop region and nuclear DNA based on the intron 8 (FADS2) sequences. (DF = degrees of freedom; SS = sum of squares; VC = variance components).

Gene	Source of Variation	DF	SS	VC	% Variation
mtDNA D-loop	Inter-regional	4	1231.22	8.59	47.86
	Interpopulational within regions	8	57.24	−0.17	−0.94
	Intrapopulational	199	1896.17	9.53	53.08
	Total	211	3184.62	17.95	
FADS2, intron 8	Inter-regional	2	25.60	0.49	18.21
	Interpopulational within regions	4	3.09	−0.23	−8.44
	Intrapopulational	99	241.61	2.44	90.23
	Total	105	270.30	2.70	

3.2.2. Interpopulation Factor Differentiation

The inter-region pairwise FST values calculated using the mtDNA-CR (Table 2), were significant between the Gulf of Mexico and the Caribbean, as well as between the Gulf of Mexico and Brazil, but not between the USA and Mexico, nor Venezuela and Brazil. Average calculated inter-region FST values were five times lower than the average values between any of the regions and *Lutjanus peru*.

Table 2. Pairwise F_{ST} values (above the diagonal) and the associated p value (below the diagonal) from the inter-regional comparison using the mtDNA D-loop sequences. * Statistically significant.

Regions	USA	Mexico	Venezuela	Brazil	*L. peru*
USA	_	0.050	0.255	0.144	0.751
Mexico	0.121	_	0.273	0.155	0.765
Venezuela	<0.001 *	<0.001 *	_	0.003	0.716
Brazil	<0.001 *	<0.001 *	0.387	_	0.659
L. peru	<0.001 *	<0.001 *	<0.001 *	<0.001 *	_

Inter-population pairwise FST values (Table S6) were significant between some of the USA and Mexican populations, and between all the Gulf of Mexico populations and the Venezuela and Brazil populations, but not between populations from the same region nor between the Venezuela and Brazil populations.

In the case of the FADS2 gene, significant FST values were found only between Mexico and Brazil, but not between these two regions and Venezuela. Similar results were observed in the inter-population FST values, in which only populations from Mexico and Brazil differed (Table S7).

The Mantel tests revealed a significant correlation between geographic and genetic distance between populations (R = 0.571; $p = 0.010$; Figure 4), suggesting minimal inter-regional gene flow.

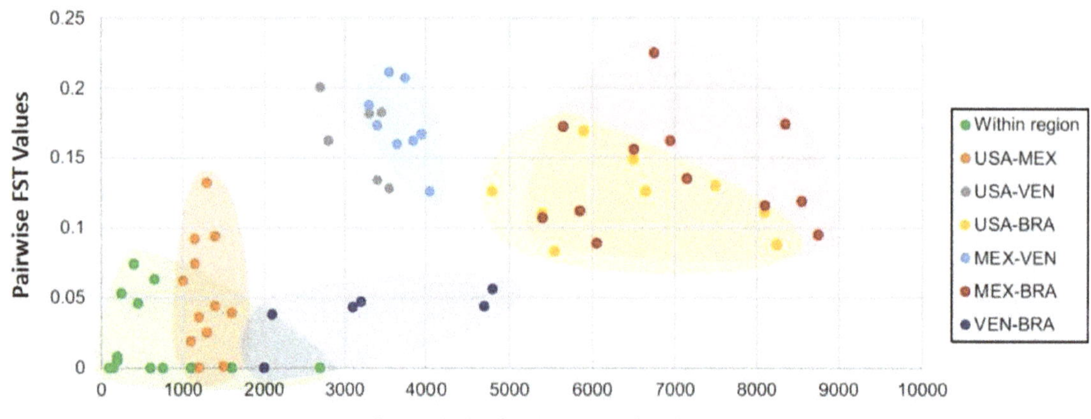

Figure 4. Relationship between geographic distance and linearized FST for western Atlantic red snapper.

3.2.3. Haplotype Network and Phylogenetic Analysis

The network analysis of the mtDNA-CR haplotypes (Figure 5A) showed wide variation in haplotypes. However, there was also substantial connection between haplotypes from different regions, with the most frequent haplotype being found in the USA, Mexico and Brazil. In many cases, the inter-regional difference between haplotypes was 1–6 mutations, as shown in the central area of the network. The haplotype network for the FADS2 gene (Figure 5B) exhibited a significantly lower level of variation compared to the mtDNA-CR region results, and the most frequent haplotypes were detected in all seven studied locations. The highest level of variation occurred in Brazil, and was most likely related to the total number of haplotypes from that region.

Phylogenetic analyses were performed using consensus sequences of all the sampling sites and the published mtDNA-CR region consensus sequences from *L. synagris* (same distribution as *L. campechanus*/*L. purpureus*), *L. guttatus* (Pacific coast of North and South America) and *L. peru* (Pacific coast of North and South America), used as an outgroup, demonstrated monophyly for individuals sampled in the Gulf of Mexico and for those sampled in the Caribbean and Brazil (Figure 6). Even though the statistical support was lower within each *L. campechanus* and *L. purpureus* population, and the relative evolutionary distance between them was small, statistical support still showed 100% incipient separation between these two groups. Of course, the distance between them was relatively much lower than the distances between them and the other *Lutjanus* species in the Americas.

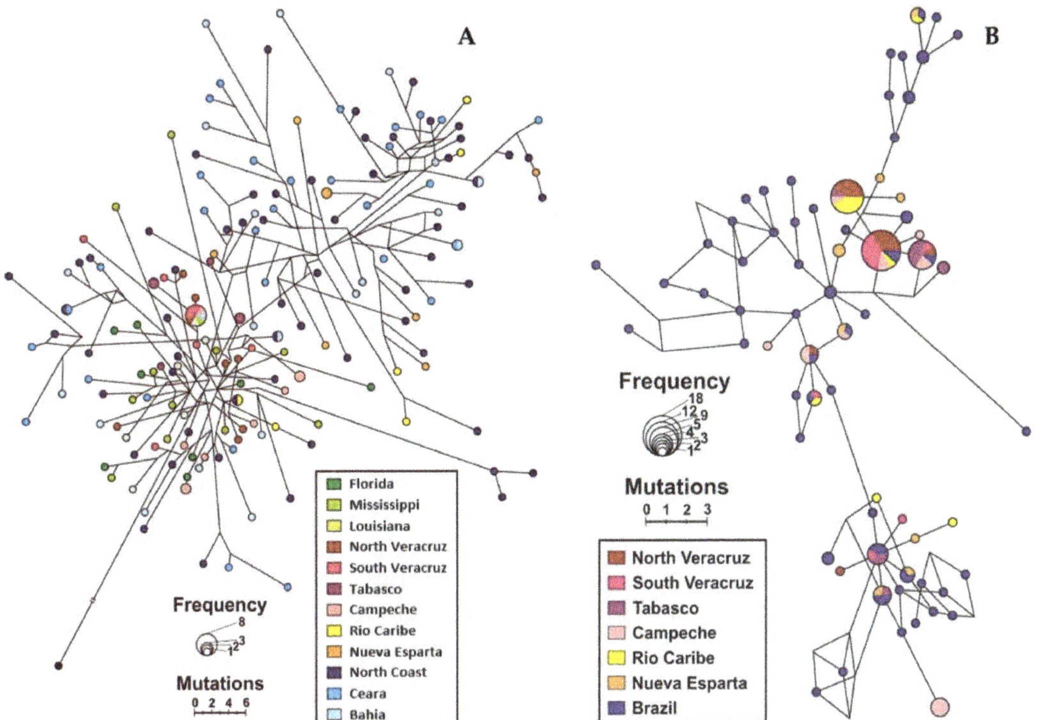

Figure 5. Median-joining network analysis for mtDNA-CR haplotypes (**A**) and FADS2 haplotypes (**B**) for *Lutjanus campechanus* and *Lutjanus purpureus* from the western Atlantic. Median vectors are not shown for clarity. The circles represent each haplotype, and circle size is proportional to haplotype frequency. Each color corresponds to a specific population and each circle shows a proportion of individuals in the haplotypes. Branch lengths are proportional to the number of S substitutions per nucleotide site. Mutational steps between haplotypes are represented by dashes.

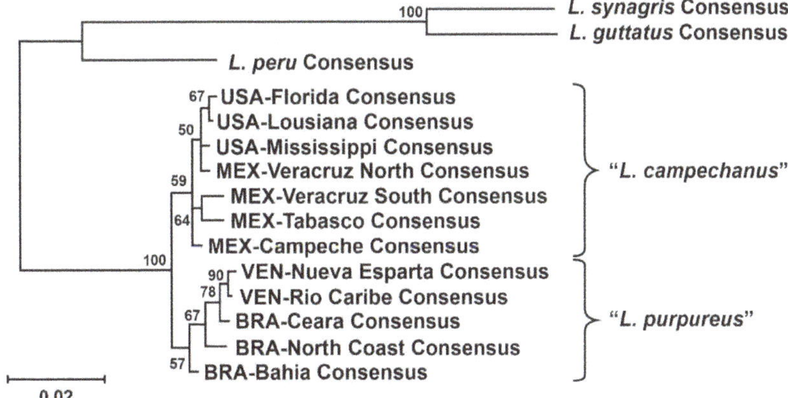

Figure 6. Phylogenetic analysis of the consensus sequences of the mtDNA region for *Lutjanus campechanus* and *Lutjanus purpureus* from the western Atlantic, and consensus sequences for other *Lutjanus* species from the Atlantic and Pacific coasts of the Americas.

4. Discussion

4.1. Otolith Analysis

Otolith morphologies in *L. campechanus* and *L. purpureus* were similar to that described for other species of the Lutjanidae family [25]. Only one description of otolith morphology in *L. campechanus* has been published to date [52], meaning the present study greatly expands the descriptions available for *L. campechanus*. It is also the first description of *L. purpureus* otoliths. The studied otoliths varied primarily in terms of the anterior and posterodorsal regions, ostium shape, and margin type. This variability has been noted in many types of fishes, especially in terms of otolith growth [15,20]. Several studies have verified that the anterior and posterodorsal regions and margin type are the most important in defining overall otolith shape, which is linked to ecological traits [15].

Otolith morphometric analysis cannot identify the main sources of variability. However, environmental conditions and genetic factors, or a combination thereof [53,54], can produce changes in fish growth rate and therefore in otolith shape, following an allometric increase in dimensions [55]. Interpopulation variation in otolith shape was noted in the otolith morphometrics/shape indices and anterior and posterior otolith contour subsections, which are all partially linked to different environmental factors for each area, and/or ontogenic factors [52]. The intrapopulation differences observed in otolith shape in individuals captured in Campeche is partly due to a combination of fish size and ontogenetic development. During early life stages, otoliths are still small, with a relatively high accretion rate, and can therefore be strongly influenced by environmental factors [56]. While otolith shape is genetically constrained, growth patterns in calcified structures can be affected by a wide range of exogenous factors, producing variation among conspecific individuals that have experienced contrasting life histories [57].

4.2. Genetic Differentiation

The molecular results from the mtDNA-CR and FADS2 intron showed that most of the genetic variation was intrapopulational. Nucleotide diversity values were low between the studied populations, indicating low genetic differentiation among red snapper populations from regions in the western Atlantic. The results also indicated an excess of polymorphisms, which is consistent with increasing population size and thus suggests that the studied populations are currently expanding. Gomes et al. [31], reported similar results when comparing the population structure of *L. campechanus* and *L. purpureus* in the western Atlantic. Considering the combined values of haplotype diversity and nucleotide diversity, Grant and Bowen [58] classified marine fish into four categories. Red snapper falls into category 1, that is, populations with low nucleotide diversity. This suggests that a population is expanding after a period of low effective population size, with rapid population growth based on one or a few lineages. This nucleotide diversity behavior is caused by a lack of physical barriers (terrestrial) and/or soft interregional barriers (non-terrestrial), facilitating the migration of adults and larvae and egg dispersal which promotes population growth or the establishment of new populations [59].

The present results indicate a generalized distribution of many haplotypes among the studied western Atlantic red snapper populations. The mixture of current and historical haplotypes by gene flow between different locations probably contributed to this result. Grant and Bowen [58] suggest that high levels of haplotype diversity indicate a long, stable evolutionary history or secondary contact between differentiated lineages. Genetic connectivity via ocean currents has been reported in different species of lutjanids [60] and serranids [61] in the western Atlantic.

The connection between haplotypes from populations in different regions of the western Atlantic shows the presence of gene flow (past and/or present), and suggests that small amounts of gene flow may be enough to homogenize red snapper populations, even in the face of demographic discontinuity. Although the analyzed populations are spatially separated, they may have had sufficient contact in the recent past to allow enough gene flow for haplotypes to spread into different geographical areas [58]. The phylogeny inferred from mtDNA-CR in the studied western Atlantic red snapper populations clearly showed a shared common evolutionary history, while the gene flow reflected an incomplete separation of lineages with the retention of an ancestral polymorphism. This would explain the absence of simple monophyly between the two species and, given that they still share many mitochondrial haplotypes, may indicate that the cladogenetic event that gave rise to the two groups was relatively recent. A second alternative is the possibility of hybridization or introgression resulting from the generation of a fertile hybrid. This would then reproduce with members of one or both original species creating gene flow, as suggested by Pedraza-Marron et al. [62].

Gomes et al. [29] were the first to use molecular data in an attempt to differentiate between the two species by studying mtDNA control regions in populations in the USA and Brazil. Their phylogenetic and population genetic analyses showed high similarity between the two species, which is compatible with the single species hypothesis. Their group later studied a larger set of samples (414 individuals) using the same mtDNA region [31]. The resulting phylogenetic tree and haplotype network did not indicate phylogeographic substructuring between the two species, but rather intense haplotype sharing. In further studies, they expanded the number of samples and added nuclear genes to the studied genetic regions [32]. They found that, in Brazil, *L. purpureus* had high levels of genetic diversity distributed homogeneously throughout the analyzed geographic region, implying high effective population size and a large dispersal of individuals.

In a very recent study interrogating nuclear and mtDNA regions, da Silva et al. [47] found significant numbers of haplotypes shared between the two species, particularly in the analyzed nuclear regions. The molecular delimitation of the species supported limited discrimination between *L. purpureus* and *L. campechanus* as distinct evolutionary units. However, it did identify a substantial north–south unidirectional gene flow, suggesting that introgression was responsible for the presence of shared haplotypes. In a more comprehensive study, Pedraza-Marron et al. [62] studied samples from different populations in the USA, Mexico, Caribbean and Brazil, interrogating mtDNA regions and thousands of nuclear SNPs genotyped by RADseq. They found that the mtDNA regions failed to delimit the nominal species as distinct haplo-groups, which agrees with previous studies [29,31]. On the contrary, even though they did find evidence of introgression in neighboring populations in northern South America, they suggested that the genomic analyses strongly support the isolation and differentiation of these species, and that the northern and southern red snapper populations should be treated as distinct taxonomic entities.

The apparent contradiction between mtDNA gene-based studies and the present genomic data-based study, as well as the sharing of haplotypes among populations separated by large geographic distances, can be understood as evidence supporting strong gene flow between the two species. We believe this is evidence that the two species are going through a process of recent speciation caused by geographic isolation and environmental adaptation, and that reproductive barriers have not yet been established. The fact that the Tajima test indicated non-neutral changes in some of the populations suggests that the studied populations are subject to natural forces that are driving greater genetic variation and differentiation, which may explain previous findings based on nuclear data [62].

Generally, the high degree of mixing between the northern and southern red snapper populations in the western Atlantic is due to egg and larvae introgression. This is at the mercy of the ocean currents, suggesting that gene flow patterns among the studied populations are influenced by oceanic currents that flow from Brazil towards the Caribbean, and from the Caribbean into the Gulf of Mexico [61,63]. For example, a virtual larval tracking model for *L. analis* suggested that the marine areas of the Mesoamerican Reef are closely connected to the Gulf of Mexico through a south-to-north ocean current [64]. Nonetheless, apparent north-to-south gene flow between *L. campechanus* and *L. purpureus* has also been reported [62], which would explain the broad distribution and low genetic differentiation between red snapper populations in the western Atlantic.

The greater genetic differentiation observed between the Gulf of Mexico and Brazil populations is probably caused by an isolation-by-distance effect. A common observation is that genetic divergence between populations increases as geographic distance increases; this is the expected isolation-by-distance pattern if gene flow and genetic drift are roughly in balance [65]. Therefore, the diversification pattern observed between the Gulf of Mexico and Brazil populations may be manifesting an incipient speciation process.

5. Conclusions

Investigating speciation among marine organisms is complex. The combination of otolith morphometrics and genetic analyses used in the present study provided salient insights into the speciation processes. The new data generated here confirm that, in the western Atlantic, the two studied red snapper taxonomic entities *L. campechanus* and *L. purpureus* exhibit some otolith shape and genetic differentiation between populations in the Gulf of Mexico, the Caribbean, and the southwestern Atlantic, but not enough to consider them as two distinct species. It is more probable that they are in a recent speciation process generated by isolation-by-distance and adaptation to different environmental conditions.

Supplementary Materials: The following are available online at https://www.mdpi.com/article/10.3390/fishes7020085/s1. Table S1: Accession numbers of the sequences of the red snapper Lutjanus purpureus/Lutjanus campechanus generated in this study. Table S2: Accession numbers of the sequences for red snapper Lutjanus purpureus and Lutjanus campechanus published in GenBank used in the analyses. Table S3: Accession numbers of the sequences of Lutjanus peru, Lutjanus synagris and Lutjanus guttatus used to calculate the consensus sequences for the phylogenetic analysis. Table S4: Measures of mitochondrial DNA based on the D-loop region from red snapper captured in the western Atlantic. Table S5: Measures of nuclear DNA based on intron 8 (FADS2) from red snapper captured in the western Atlantic. Table S6: Pairwise FST values (above the diagonal) and the associated p value (below the diagonal) from the comparison between populations using the mtDNA D-loop sequences. Table S7: Pairwise FST values (above the diagonal) and the associated *p* value (below the diagonal) from the comparison between populations of red snapper using intron 8 (FADS2) sequences.

Author Contributions: A.M.-R., conceptualization, methodology, software, data analysis, writing—original draft; X.R., conceptualization, data analysis, writing—review and editing; G.G.-C., conceptualization, data analysis, review and editing; S.A.-R., sampling, data analysis; M.d.L.J.-B., data analysis, review and editing; J.L.M.-M.: data curation, methodology, review and editing; H.R., methodology, validation, review and editing; T.B., data analysis, review and editing; M.D.D., conceptualization, data analysis, writing—original draft, review and editing. All authors have read and agreed to the published version of the manuscript.

Funding: This research was funded by CONACyT postgraduate scholarship No. 730163; the Universidad Veracruzana supported the publication process.

Institutional Review Board Statement: All samples were obtained from dead individuals caught by the commercial multi-species artisanal fleet and following the recommendations of the Ethics Committee of the Academic Group: Management and Conservation of Aquatic Resources of the Instituto de Ciencias Marinas y Pesquerias of the Universidad Veracruzana, Boca del Rio, Veracruz, Mexico.

Data Availability Statement: All the supporting data can be obtained by request from the corresponding author. Sequence data was submitted to GenBank and can be accessed directly in the NCBI web page. See Table S1.

Acknowledgments: The authors thank the Laboratorio de Evaluación de Recursos Pesqueros (Venezuela) for technical support, Project CONACyT LAB-2009-123913 for acquisition of MEB otolith images, Universidad Veracruzana for manuscript translation and ICIMAP and Centro de Bioingenierías del Tecnológico de Monterrey for genetic analysis.

Conflicts of Interest: This manuscript was read and approved by all the authors listed and has not been submitted for publication, in whole or in part, to any other journal. The authors declare no conflict of interest.

References

1. Bryant, D.; Bouckaert, R.; Felsenstein, J.; Rosenberg, N.A.; Roychoudhury, A. Inferring species trees directly from biallelic genetic markers: Bypassing gene trees in a full coalescent analysis. *Mol. Biol. Evol.* **2012**, *29*, 1917–1932. [CrossRef] [PubMed]
2. Leaché, A.; Fujita, M.; Minin, V.; Bouckaert, R. Species delimitation using genome-wide SNP data. *Syst. Biol.* **2014**, *63*, 534–542. [CrossRef] [PubMed]
3. Cowen, R.K.; Paris, C.B.; Srinivasan, A. Scaling of connectivity in marine populations. *Science* **2006**, *311*, 522–527. [CrossRef] [PubMed]
4. Sturrock, A.M.; Trueman, C.N.; Darnaude, A.M.; Hunter, E. Can otolith elemental chemistry retrospectively track migrations in fully marine fishes? *J. Fish Biol.* **2012**, *81*, 766–795. [CrossRef]
5. Corrigan, S.; Maisano, P.; Eddy, C.; Duffy, C.; Yang, L.; Li, C.; Bazinet, A.L.; Mona, S.; Naylor, G.J. Historical introgression drives pervasive mitochondrial admixture between two species of pelagic sharks. *Mol. Phylogenet. Evol.* **2017**, *110*, 122–126. [CrossRef]
6. Vignon, M.; Morat, F. Environmental and genetic determinant of otolith shape revealed by a non-indigenous tropical fish. *Mar. Ecol. Prog. Ser.* **2010**, *411*, 231–241. [CrossRef]
7. Waples, R.S.; Do, C. LDNE: A program for estimating effective population size from data on linkage disequilibrium. *Mol. Ecol. Resour.* **2017**, *8*, 753–756. [CrossRef]
8. Tuset, V.M.; Lozano, I.J.; González, J.A.; Pertusa, J.F.; García-Díaz, M.M. Shape indices to identify regional differences in otolith morphology of comber *Serranus cabrilla* (L., 1758). *J. Appl. Ichthyol.* **2003**, *19*, 88–93. [CrossRef]
9. Tuset, V.; Rosin, P.; Lombarte, A. Sagittal otolith shape used in the identification of fishes of the genus *Serranus*. *Fish. Res.* **2006**, *81*, 316–325. [CrossRef]
10. Tuset, V.M.; Imondi, R.; Aguado, G.; Otero-Ferrer, J.L.; Santschi, L.; Lombarte, A.; Love, M. Otolith Patterns of Rockfishes from the Northeastern Pacific. *J. Morphol.* **2015**, *276*, 458–469. [CrossRef]
11. DeVries, D.A.; Grimes, C.B.; Prager, M.H. Using otolith shape analysis to distinguish eastern Gulf of Mexico and Atlantic Ocean stocks of king mackerel. *Fish. Res.* **2002**, *57*, 51–62. [CrossRef]
12. Volpedo, A.V.; Echevarría, D.D. Ecomorphological patterns of the sagitta in fish on the continental shelf off Argentine. *Fish. Res.* **2003**, *60*, 551–560. [CrossRef]
13. Galley, E.A.; Wright, P.J.; Gibb, F.M. Combined methods of otolith shape analysis improve identification of spawning areas of Atlantic cod. *ICES J. Mar. Sci.* **2006**, *63*, 1710–1717. [CrossRef]
14. Avigliano, E.; Martínez-Riaños, F.; Volpedo, A.V. Combined use of otolith microchemistry and morphometry as indicators of the habitat of the silverside (*Odontesthes bonariensis*) in a freshwater-estuarine environment. *Fish. Res.* **2014**, *149*, 55–60. [CrossRef]
15. Callicó Fortunato, R.; Benedito Durà, V.; González-Castro, M.; Volpedo, A. Morphological and morphometric changes of sagittae otoliths related to fish growth in three Mugilidae species. *J. Appl. Ichthyol* **2017**, *33*, 1137–1145. [CrossRef]
16. Da Silva Santos, R.; Costa de Azevedo, M.C.; Queiroz de Albuquerque, C.; Gerson Araújo, F. Different sagitta otolith morphotypes for the whitemouth croaker *Micropogonias furnieri* in the Southwestern Atlantic coast. *Fish. Res.* **2017**, *195*, 222–229. [CrossRef]
17. Vignon, M.; Morat, F.; Galzin, R.; Sasal, P. Evidence for spatial limitation of the bluestripe snapper *Lutjanus kasmira* in French Polynesia from parasite and otolith shape analysis. *J. Fish Biol.* **2008**, *73*, 2305–2320. [CrossRef]
18. Beyer, S.G.; Szedlmayer, S.T. The use of otolith shape analysis for ageing juvenile red snapper, *Lutjanus campechanus*. *Environ. Biol. Fishes* **2010**, *89*, 333–340. [CrossRef]
19. Sadighzadeh, Z.; Tuset, V.M.; Valinassab, T.; Dadpour, M.R.; Lombarte, A. Comparison of different otolith shape descriptors and morphometrics for the identification of closely related species of Lutjanus spp. from the Persian Gulf. *Mar. Biol. Res.* **2012**, *8*, 802–814. [CrossRef]
20. Sadighzadeh, Z.; Valinassab, T.; Vosugi, G.; Motallebi, A.A.; Fatemi, M.R.; Lombarte, A.; Tuset, V.M. Use of otolith shape for stock identification of John's Snapper, *Lutjanus johnii* (Pisces: Lutjanidae), from the Persian Gulf and the Oman Sea. *Fish. Res.* **2014**, *155*, 59–63. [CrossRef]
21. Puentes-Granada, V.; Rojas, P.; Pavolini, G.; Gutiérrez, C.F.; Villa, A.A. Morphology and morphometric relationships for sagitta otoliths in *Lutjanus argentiventris* (Pisces: Lutjanidae) and *Hyporthodus acanthistius* (Pisces: Serranidae) from the Colombian Pacific Ocean. *Univ. Sci.* **2019**, *24*, 337–361. [CrossRef]
22. Stevens, M.M. *Commercially Important Gulf of Mexico/South Atlantic Snappers*; Monterey Bay Aquarium: Monterey, CA, USA, 2009; pp. 2–4.
23. Nelson, J.S.; Grande, T.C.; Wilson, M.V. *Fishes of the World*; John Wiley & Sons: Hoboken, NJ, USA, 2016; pp. 366–367.
24. FAO: Yearbook of Fishery Statistics. Available online: http://www.fao.org/fi/statist/statist.asp (accessed on 23 June 2021).
25. Anderson, W.D. Lutjanidae. In *The living Marine Resources of the Western Central Atlantic, FAO Species Identification Guide for Fishery Purposes*; Carpenter, K.E., Ed.; Food and Agriculture Organization of the United Nations: Rome, Italy, 2003; pp. 1479–1504.
26. Moura, R.L.; Lindeman, K.C. A new species of snapper (Perciformes: Lutjanidae) from Brazil, with comments on the distribution of *Lutjanus griseus* and *L. apodus*. *Zootaxa* **2007**, *1422*, 31–43. [CrossRef]
27. Monroy-García, C.; Garduño-Andrade, M.; Espinosa, J.C. Análisis de la pesquería de huachinango (*Lutjanus campechanus*) en el Banco de Campeche. *Proc. Gulf Caribb. Fish. Inst.* **2002**, *53*, 507–515.
28. Drass, D.M.; Bootes, K.L.; Lyczkowski-Shultz, J.; Comyns, B.H.; Holt, G.H.; Riley, C.M.; Phelps, R.P. Larval development of red snapper, *Lutjanus campechanus*, and comparisons with co-occurring snapper species. *Fish. Bull.* **2000**, *98*, 507–527.

29. Gomes, G.; Schneider, H.; Vallinoto, M.; Santos, S.; Ortil, G.; Sampaio, I. Can *Lutjanus purpureus* (South red snapper) be "legally" considered a red snapper (*Lutjanus campechanus*)? *Genet. Mol. Biol.* **2008**, *31*, 372–376. [CrossRef]
30. Cervigón, F.; Cipriani, R.; Fischer, W.; Garibaldi, L.; Hendrickx, M.; Lemus, A.; Márquez, R.; Poutiers, J.; Robaina, G.; Rodríguez, B. Fichas FAO de identificación de especies para los fines de la pesca. In *Guía de Campo de las Especies Comerciales Marinas y de Aguas Salobres de la Costa Septentrional de Sur América*; Food and Agriculture Organization of the United Nations: Rome, Italy, 1992; pp. 353–355.
31. Gomes, G.; Sampaio, I.; Schneider, H. Population Structure of *Lutjanus purpureus* (Lutjanidae-Perciformes) on the Brazilian coast: Further existence evidence of a single species of red snapper in the western Atlantic. *An. Acad. Bras. Ciênc.* **2012**, *84*, 979–999. [CrossRef]
32. da Silva, R.; Pedraza-Marrón, C.R.; Sampaio, I.; Betancur-R, R.; Gomes, G.; Schneider, H. New insights about species delimitation in red snappers (*Lutjanus purpureus* and *L. campechanus*) using multilocus data. *Mol. Phylogenet. Evol.* **2020**, *147*, 106780. [CrossRef]
33. Assis, C.A. *Guia Para a Identificação de Algumas Famílias de Peixes Ósseos de Portugal Continental, Através da Morfologia dos Seus Otólitos Sagitta*; Câmara Municipal de Cascais: Cascais, Portugal, 2004.
34. Tuset, V.M.; Lombarte, A.; Assis, C.A. Otolith atlas for the western Mediterranean, north and central eastern Atlantic. *Sci. Mar.* **2008**, *72*, 7–198. [CrossRef]
35. Pavlov, D.A. Differentiation of three species of Genus Upeneus (Mullidae) based on otolith shape analysis. *J. Ichthyol.* **2016**, *56*, 37–51. [CrossRef]
36. Parisi-Baradad, V.; Lombarte, A.; García-Ladona, E.; Cabestany, J.; Piera, J.; Chic, O. Otolith shape contour analysis using affine transformation invariant wavelet transforms and curvature scale space representation. *Mar. Freshw. Res.* **2005**, *56*, 795–804. [CrossRef]
37. Parisi-Baradad, V.; Manjabacas, A.; Lombarte, A.; Olivella, R.; Chic, Ò.; Piera, J.; García-Ladona, E. Automatic taxon identification of teleost fishes in an otolith online database. *Fish. Res.* **2010**, *105*, 13–20. [CrossRef]
38. Renán, X.; Montero-Muñoz, J.; Garza-Pérez, J.R.; Brulé, T. Age and stock analysis using otolith shape in gags from the southern Gulf of Mexico. *Trans. Am. Fish. Soc.* **2016**, *145*, 1252–1265. [CrossRef]
39. Lleonart, J.; Salat, J.; Torres, G.J. Removing allometric effects of body size in morphological analysis. *J. Theor. Biol.* **2000**, *205*, 85–93. [CrossRef] [PubMed]
40. Cardinale, M.; Doering-Arjes, P.; Kastowsky, M.; Mosegaard, H. Effects of sex, stock, and environment on the shape of known-age Atlantic Cod (*Gadus morhua*) otoliths. *Can. J. Fish. Aquat. Sci.* **2004**, *61*, 158–167. [CrossRef]
41. Watkinson, D.A.; Gills, D.M. Stock discrimination of Lake Winnipeg walleye based on Fourier and wavelet description of scale outline signals. *Fish. Res.* **2005**, *72*, 193–203. [CrossRef]
42. Quinn, G.P.; Keough, M.J. *Experimental Design and Data Analysis for Biologists*, 2nd ed.; Cambridge University Press: Cambridge, UK, 2003.
43. Borcard, D.; Gillet, F.; Legendre, P. *Numerical Ecology with R*, 2nd ed.; Springer International Publishing: New York, NY, USA, 2018.
44. Anderson, M.J. A new method for non-parametric multivariate analysis of variance. *Austral Ecol.* **2001**, *26*, 32–46.
45. Anderson, M.J.; Gorley, R.N.; Clarke, K.R. *PERMANOVA+ for PRIMER: Guide to Software and Statistical Methods*; PRIMER-E: Plymouth, MA, USA, 2008.
46. Lee, W.; Coroy, J.; Howell, W.H.; Koocher, T.D. Structure and evolution of teleost mitochondrial control regions. *J. Mol. Evol.* **1995**, *41*, 54–66. [CrossRef]
47. da Silva, R.; Sampaio, I.; Schneider, H.; Gomes, G. Lack of spatial subdivision for the snapper *Lutjanus purpureus* (Lutjanidae-Perciformes) from southwest Atlantic based on multi-locus analyses. *PLoS ONE* **2016**, *11*, e0161617. [CrossRef]
48. Excoffier, L.; Lischer, H.E. Arlequin suite ver 3.5: A new series of programs to perform population genetics analyses under Linux and Windows. *Mol. Ecol. Resour.* **2010**, *10*, 564–567. [CrossRef]
49. Bandelt, H.J.; Forster, P.; Rohl, A. Median-joining networks for inferring intraspecific phylogenies. *Mol. Biol. Evol.* **1999**, *16*, 37–48. [CrossRef]
50. Smouse, P.E.; Banks, S.C.; Peakall, R. Converting quadratic entropy to diversity: Both animals and alleles are diverse, but some are more diverse than others. *PLoS ONE* **2017**, *12*, e0185499. [CrossRef] [PubMed]
51. Ronquist, F.; Teslenko, M.; van der Mark, P.; Ayres, D.L.; Darling, A.; Hohna, S.; Larget, B.; Liu, L.; Suchard, M.A.; Huelsenbeck, J.P. MrBayes 3.2: Efficient Bayesian phylogenetic inference and model choice across a large model space. *Syst. Biol.* **2012**, *61*, 539–542. [CrossRef] [PubMed]
52. Mier-Uco, L.A. *Catálogo de otolitos de Peces Capturados en el Parque Nacional Sistema Arrecifal Veracruzano*; Instituto Tecnológico de Boca del Río: Veracruz, México, 2011; 119p.
53. Capoccioni, F.; Costa, C.; Aguzzi, J.; Menesatti, P.; Lombarte, A.; Ciccotti, E. Ontogenetic and environmental effects on otolith shape variability in three Mediterranean European eel (*Anguilla anguilla*) local stocks. *J. Exp. Mar. Biol. Ecol.* **2011**, *397*, 1–7. [CrossRef]
54. Mille, T.; Mahé, K.; Cahcera, M.; Villanueva, M.C.; de Pontual, H.; Ernande, B. Diet is correlated with otolith shape in marine fish. *Mar. Ecol. Prog. Ser.* **2016**, *555*, 167–184. [CrossRef]
55. Begg, G.A.; Brown, R.W. Stock identification of haddock *Melanogrammus aeglefinus* on Georges Bank based on otolith shape analysis. *Trans. Am. Fish. Soc.* **2001**, *129*, 935–945. [CrossRef]

56. Hüssy, K. Otolith shape in juvenile cod (*Gadus morhua*): Ontogenetic and environmental effects. *J. Exp. Mar. Biol. Ecol.* **2008**, *364*, 35–41. [CrossRef]
57. Vignon, M. Ontogenetic trajectories of otolith shape during shift in habitat use: Interaction between otolith growth and environment. *J. Exp. Mar. Biol. Ecol.* **2012**, *420–421*, 26–32. [CrossRef]
58. Grant, W.S.; Bowen, B.W. Shallow population histories in deep evolutionary lineages of marine fishes: Insights from sardines and anchovies and lessons for conservation. *Genetics* **1998**, *89*, 415–426. [CrossRef]
59. Floeter, S.R.; Rocha, L.A.; Robertson, D.R.; Joyeux, J.C.; Smith-Vaniz, W.F.; Wirtz, P.; Edwards, A.J.; Barreiros, J.P.; Ferreira, C.E.; Gasparini, J.L.; et al. Atlantic reef fish biogeography and evolution. *J. Biogeogr.* **2008**, *35*, 22–47. [CrossRef]
60. Rosado-Nic, O.J.; Hogan, J.D.; Lara-Arenas, J.H.; Rosas-Luis, R.; Carrillo, L.; Villegas-Sánchez, C.A. Gene flow between subpopulations of gray snapper (*Lutjanus griseus*) from the Caribbean and Gulf of Mexico. *PeerJ* **2020**, *8*, e8485. [CrossRef]
61. Jue, N.K.; Brulé, T.; Coleman, F.C.; Koenig, C.C. From Shelf to Shelf: Assessing Historical and Contemporary Genetic Differentiation and Connectivity across the Gulf of Mexico in Gag, *Mycteroperca microlepis*. *PLoS ONE* **2015**, *10*, e0120676. [CrossRef] [PubMed]
62. Pedraza-Marrón, C.D.R.; Silva, R.; Deeds, J.; Van Belleghem, S.M.; Mastretta-Yanes, A.; Domínguez-Domínguez, O.; Rivero-Vega, R.A.; Lutackas, L.; Murie, D.; Parkyn, D.; et al. Genomics overrules mitochondrial DNA, siding with morphology on a controversial case of species delimitation. *Proc. Biol. Sci.* **2019**, *286*, 20182924. [CrossRef] [PubMed]
63. Rocha, L.A.; Rocha, C.R.; Robertson, D.R.; Bowen, B.W. Comparative phylogeography of Atlantic reef fishes indicates both origin and accumulation of diversity in the Caribbean. *Evol. Biol.* **2008**, *8*, 157. [CrossRef] [PubMed]
64. Martínez, S.; Carrillo, L.; Marinone, S.G. Potential connectivity between marine protected areas in the Mesoamerican Reef for two species of virtual fish larvae: *Lutjanus analis* and *Epinephelus striatus*. *Ecol. Ind.* **2019**, *102*, 10–20. [CrossRef]
65. Silva, D.; Martins, K.; Oliveira, J.; da Silva, R.; Sampaio, I.; Schneider, H.; Gomes, G. Genetic differentiation in populations of lane snapper (*Lutjanus synagris*-Lutjanidae) from Western Atlantic as revealed by multilocus analysis. *Fish. Res.* **2018**, *198*, 138–149. [CrossRef]

Article

Unravelling Stock Spatial Structure of Silverside *Odontesthes argentinensis* (Valenciennes, 1835) from the North Argentinian Coast by Otoliths Shape Analysis

Santiago Morawicki [1,2], Patricio J. Solimano [1] and Alejandra V. Volpedo [3,4,*]

1 Centro de Investigaciones y Transferencia de Río Negro, Rotonda Cooperación y Ruta Provincial N° 1, Universidad Nacional de Río Negro, Viedma C.P 8500, Río Negro, Argentina; smorawicki@gmail.com (S.M.); psolimano@unrn.edu.ar (P.J.S.)

2 Consejo Nacional de Investigaciones Científicas y Técnicas (CONICET), Centro de Investigaciones y Transferencia de Río Negro, Rotonda Cooperación y Ruta Provincial N° 1, Viedma C.P 8500, Río Negro, Argentina

3 Instituto de Investigaciones en Producción Animal (INPA), Facultad de Ciencias Veterinarias, CONICET—Universidad de Buenos Aires, Av. Chorroarín 280, Buenos Aires C1427CWO, Argentina

4 Centro de Estudios Transdisciplinarios del Agua (CETA), Buenos Aires C1427CWO, Argentina

* Correspondence: avolpedo@fvet.uba.ar

Abstract: The marine silverside (*Odontesthes argentinensis*) is an euryhaline species, distributed along the southwest coast of the Atlantic Ocean, present in estuaries, brackish coastal lagoons and shallow marine waters. It is a significant economic resource for local fisheries in southern Brazil, Uruguay and Argentina. The aim of this work was to contribute to knowledge on the stock spatial structure of the silverside, using otolith shape analysis, based on samples from nine locations in the Argentinian Sea, covering a large distribution range of the species. A combination of elliptic Fourier descriptors, Wavelet coefficients and otolith Shape indices were explored by multivariate statistical methods. The application of wavelet and combined wavelet, Fourier and Shape Indices were the most effective variables to discriminate between sampling sites (7.42 total error). PERMANOVA analysis of otolith shape revealed multivariate significant differences between north versus south locations ($p < 0.0001$). The results obtained show that the spatial structure of *O. argentinensis* presents a North–South gradient with marked differences between the extreme localities of the north (Mar del Plata, Quequén) with more elliptical shapes than those in the south (San Blas, San Antonio Este) and an isolated group conformed by Puerto Lobos.

Keywords: morphometry; otolith; stock spatial structure; atherinids

Citation: Morawicki, S.; Solimano, P.J.; Volpedo, A.V. Unravelling Stock Spatial Structure of Silverside *Odontesthes argentinensis* (Valenciennes, 1835) from the North Argentinian Coast by Otoliths Shape Analysis. *Fishes* **2022**, *7*, 155. https://doi.org/10.3390/fishes7040155

Academic Editor: Josipa Ferri

Received: 26 April 2022
Accepted: 31 May 2022
Published: 29 June 2022

1. Introduction

Marine coastal and euryhaline species have different ecological strategies to survive in dynamic environments. These strategies include a wide range of diet, reproductive alternatives and, especially, organization at the level of stock spatial structure [1].

The term "stock" as it refers to fish has been variously defined [2,3]. In summary, fish stock mainly describes the characteristics of a population unit with genetic integrity within which some particular type of management is carried out [3]. Common stock assessment techniques and management strategies assume discrete populations [4]; this type of population structure is the exception more than the rule [5]. It is now clear that the population structure of marine species falls along a continuum with different arrangements within the range of distribution, depending on the species and the environmental characteristics [6–9]. Migration and mixing generate complex stock spatial structures and so caution is called for in seeking correct management strategies regarding spatial structure and its complexity within the species [4].

There are different methods to identify fish stocks, such as catch–recapture, population parameters, size structure, morphometric and meristic characteristics, genetic identification, otoliths, among others (e.g., Park and Moran [10]; Volpedo and Cirelli [11]). The otoliths of teleost fish are complex polycrystalline bodies composed mainly of calcium carbonate in the form of aragonite and small amounts of other minerals immersed in an organic matrix [12,13]. In recent years, the use of otolith morphometry, morphology and the geometric morphometric methods (commonly known as "otolith shape analysis") has allowed the identification of breeding areas and fish stocks, as well as the migratory routes of different commercial species [7,14–17].

Atherinopsids are an important group of coastal fishes for artisanal and/or recreational fisheries, and have a great inter-population phenotypic plasticity, presenting morphological, morphometric and meristic differences along the marine coast [18–21] and in continental environments [22–24]. Due to its complexity, this group has been studied in different aspects of its biological cycle, but very few authors have established the identification of stocks of this group [14,18,25,26], and even less so in marine atherinids.

Among the Atherinopsids, *Odontesthes argentinensis*, commonly known as silverside, is an euryhaline species, distributed along the southwest coast of the Atlantic Ocean, between Rio de Janeiro, Brazil (22° S) and Rawson, Argentina (43° S) [27,28], These fish are found in estuaries, brackish coastal lagoons and shallow marine waters [29]. It is considered one of the most ecologically and economically important species in the area. It is a significant economic resource for local fisheries in southern Brazil, Uruguay and Argentina [30–34].

In Argentina, *O. argentinensis* is exploited by recreational and artisanal fisheries along the coast of Buenos Aires province [35], and in the North of Patagonia [34].

Studies show how the otoliths of this species change during growth, using morphology and geometric morphometry [36,37]. Levy et al. [26], using the parasitic community of *O. argentinenesis* specimens caught in the Mar del Plata area (38°02′ S, 57°31′ W) and in the San Matías Gulf (40°50′ S, 64°50′ W), found that the evaluated individuals belonged to two different stocks. However, little is known about the stock spatial structure of these species in the rest of their distribution in the Argentinian Sea, where there are important fisheries for this resource; both in the coastal province of Buenos Aires and in northern Patagonia [34].

This study is guided by two central hypotheses: first, that there are at least two different stocks in the analyzed area, as proposed by Levy et al. [26], and, second, that there is a spatial structure to these stocks, and the oceanographic barrier that structures them is the area known as El Rincón. In this context, the objective of this study is to evaluate, for the first time, the stock spatial structure of *O. argentinensis* in a great portion of its distributions in the Argentinian Sea (>1000 km). The aim was to establish the amount of stock present in the area, their spatial distribution and the relations between them, taking into account the ones considered by Levy et al. [26], so as to generate management recommendations of the correct scale, with a focus on local fisheries of the area.

2. Materials and Methods

2.1. Study Area

The study area extends 1000 km over the coast of the Argentinian Sea and includes nine sampling sites approximately 150 km apart (Figure 1). The sites are located from north to south in Buenos Aires province: Mar del Plata (MDQ), Quequén (QQN), Claromecó (CLM) Monte Hermoso (MTH), La Chiquita (LCH), San Blas (SBL), and in Río Negro province: El Cóndor (ELC), San Antonio Este (SAE) and Puerto Lobos (PLB), (Figure 1).

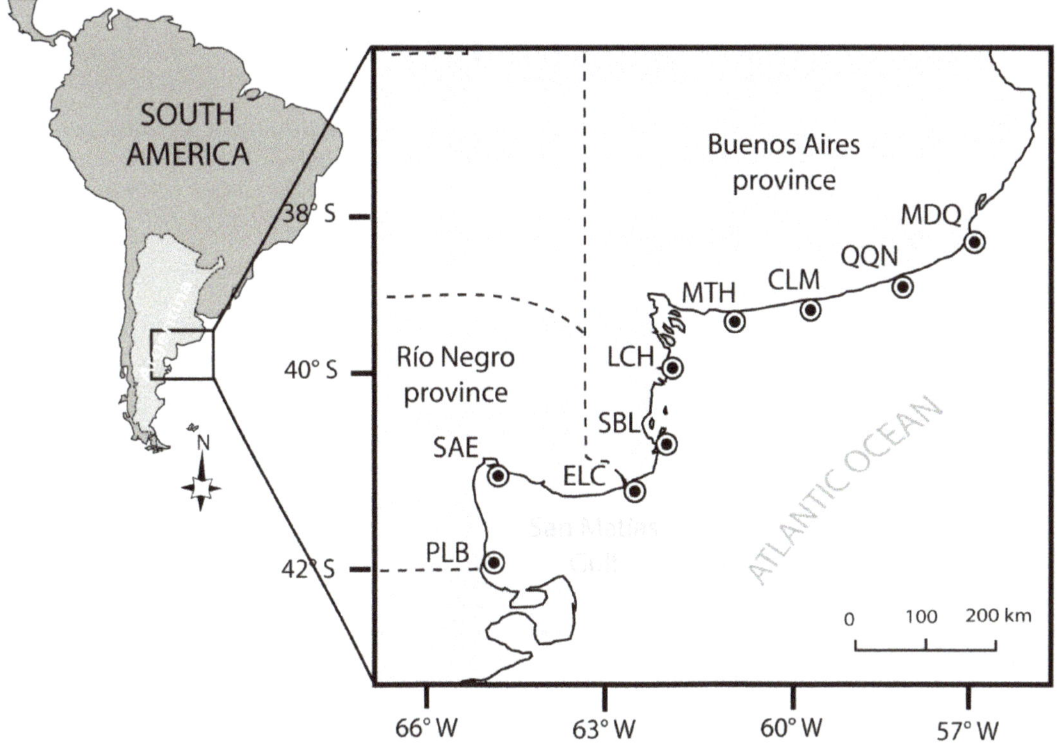

Figure 1. Sampling sites. MDQ: Mar del Plata; QQN: Quequén; CLM: Claromecó; MTH: Monte Hermoso; LCH: La Chiquita; SBL: San Blas; ELC: El Cóndor; SAE: San Antonio Este; PLB: Puerto Lobos.

The sites in Buenos Aires province are dominated primarily by extensive sandy beaches with fields of coastal dunes, which are interrupted by estuaries, such as those of the Quequén River, Colorado and the estuary of Bahía Blanca. Blanca and Anegada bays represent wide environments dominated by strong sludge deposits [38]. SBL and LCH are located within the coastal area denominated El Rincón (39°–41°30′ S), an area where the characteristics of water mass (temperature, salinity, density) differ from the rest of the coast of the province of Buenos Aires, due to the conjunction of inland waters from the Negro and Colorado rivers, high salinity waters from the San Matías Gulf and the typical shelf waters [39,40].

ELC is located on the estuary of the Negro River, at the entrance of the San Matias Gulf, and it is characterized by wide sandy beaches and large cliffs [34]. The localities PLB and SAE are located in the interior of the San Matías Gulf (Figure 1). This area has oceanographic characteristics different from the rest of the coastal area of the Argentinian Sea, because its geomorphology restricts the exchange of water with the open sea [40].

2.2. Sampling

Sampling was carried out during the months of September 2018 and March 2019 using different fishing gear (gillnets having mesh sizes of 30, 40 and 50 mm, and 25 m long, coastal trawl and with rod by the contribution of sport fishing participants). The individuals were identified according to the taxonomic keys proposed by Dyer [27] and Mancini et al. [41]. First, 354 fish were collected and frozen (−18 °C) until laboratory analysis. Each individual's weight (W) in g, total length (TL) and standard length (SL) in mm were recorded (Table 1). The fishes were adults and the TL of all the sampling sites showed high overlap sizes.

Sagittae otoliths were extracted and cleaned with Milli-Q water with a resistivity of 18.2 mΩ cm (Merck Millipore) and dried.

Table 1. Sampling locations ordered from north to south. Mean values of Total Length (TL) and Standard Length (STL) ± Standard Deviation. N: sample number.

Localities		Latitude	Longitude	N	TL (mm)	STL (mm)
Mar Del Plata	MDQ	38°02′	57°31′	57	209 ± 30	175 ± 25
Quequén	QQN	38°24′	58°40′	75	200 ± 30	168 ± 25
Claromecó	CLM	38°51′	60°04′	47	208 ± 67	178 ± 57
Monte Hermoso	MTH	38°59′	61°15′	39	226 ± 68	191 ± 60
La Chiquita	LCH	39°35′	62°05′	13	255 ± 99	214 ± 85
San Blas	SBL	40°31′	62°16′	64	269 ± 45	227 ± 36
El Cóndor	ELC	41°00′	62°48′	20	317 ± 53	267 ± 49
San Antonio Este	SAE	40°50′	64°50′	30	280 ± 46	245 ± 41
Puerto Lobos	PLB	41°51′	65°02′	19	334 ± 17	285 ± 14

The left *sagitta* otolith of each specimen was photographed on a dark background (Figure 2) with a Leica® MC170HD camera mounted on a Schönfeld stereoscopic magnifier® and the Leica Application Suite software (LAS version 4.5). The left otolith was used since both otoliths are morphologically equal [36].

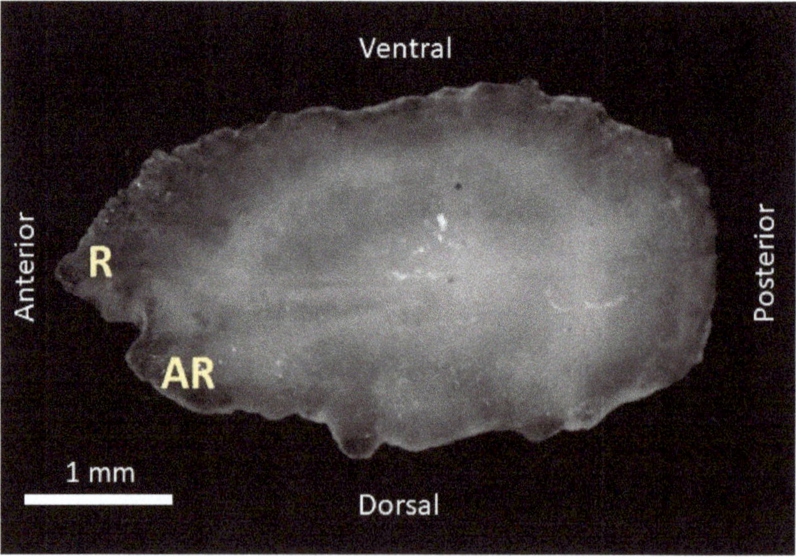

Figure 2. Left *sagitta* otolith of *Odontesthes argentinensis*. Features variables. AR: Anti rostrum. R: Rostrum.

2.3. Shape Analysis

Otolith contours were analyzed using the ShapeR package [42] of the statistical software R. The "detect. outline" function was used to extract the contour of all the otoliths (threshold = 0.15) and to eliminate the pixel noise around the contours that could impair the analyses; the "smooth out" function was used [42,43].

The Fourier descriptors (FD) and Wavelet coefficients (WC), respectively, were extracted from the digital images using the Wave thresh package [42,44]. The deviation of the reconstructed otolith's outline from the original outline was evaluated with the ShapeR package [42] to determine the number of FD and WC needed for the analysis, with 45 FD and 64 WC; an accuracy rate higher than 98.5% was obtained. The contour coefficients that showed a relationship with the length of the fish ($p < 0.05$) were discarded from the analysis [45,46]. Among the 45 FD and 64 WC integrated within the ShapeR package, only 24 FD and 44 WC met our assumptions and were, therefore, used in the following statistical analysis.

The mean otoliths reconstructed by normalized WC for each group of samples were plotted and the differences among the groups visually evaluated.

To estimate which part of the otolith outline contributed most to the difference between the potential stocks, mean shape coefficients and their standard deviation (SD) were plotted against the angle of the outline from where the coefficients were extracted, using the function plotCI from the plots package [47].

2.4. Morphometric Analysis

Morphometric variables, such as otolith length (OL mm), width (OW mm), area (OA mm^2) and perimeter (OP) in mm, were obtained from the contours of otoliths using ShapeR [42]. Four morphometric indices (MI) were calculated: (i) Circularity (OP^2/OA) [37,48,49]; (ii) rectangularity ($OA/(OW \times OL)$) [37,48,49]; (iii) aspect ratio (OW/OL), [37]; (iv) roundness $(4 \times OA)/(\pi \times OL^2)$ [13]. All morphometrical variables fitted a normal distribution and homogeneity of variance (Shapiro–Wilk, $p > 0.05$; Levene, $p > 0.05$). The morphometric indices were then analyzed by ANOVA, and Bonferroni contracts were used to evaluate the differences among sampling areas.

Indices were corrected to eliminate possible allometry effects in otolith shape related to fish body size, for a proper comparison between groups; the formula: $y' = yi \times (x0/xi)b$ was used, in which y′ is the corrected predictive variable, yi is the original value of the obtained index, x0 is a referential standard length (SL) value (SL minimum = 142 mm), xi is the original SL value, and b is the Huxley coefficient of each regression index to SL [37,49,50].

2.5. Multivariate Analysis

A Linear Discriminant Analysis (LDA) using morphometric indices, Fourier descriptors, and Wavelet coefficients separately and together was performed using the R "MASS" [51]. The overall power of the discriminant function analysis was evaluated by Wilks' λ test [52]. The success of classification among the different groups was evaluated using a Jackknifed classification matrix [52]. Comparisons among sample locations, as well as among the groups retrieved from the LDA, were conducted using a permutational multivariate analysis of variance [PERMANOVA; [53], based on the Bray–Curtis dissimilarity measure (4999 random permutations)].

Statistical analyses were performed with R (R Core Team, 2021) using ShapeR packages ShapeR [42], Vegan [54], ipred [55] and MASS [51].

3. Results

3.1. Mean Shape Features

Otolith contours of the fish from the different studied localities presented differences in mean shape (Figure 3). The morphometry of the otoliths presented modifications in the rostrum-antirostrum, the dorsal margin and the posterior end (Figure 4). These modifications were displayed in the wavelet coefficients (ICC) for these regions on the otolith outline at 170°–200°, 260°–290° and 320°–20° (Figure 4).

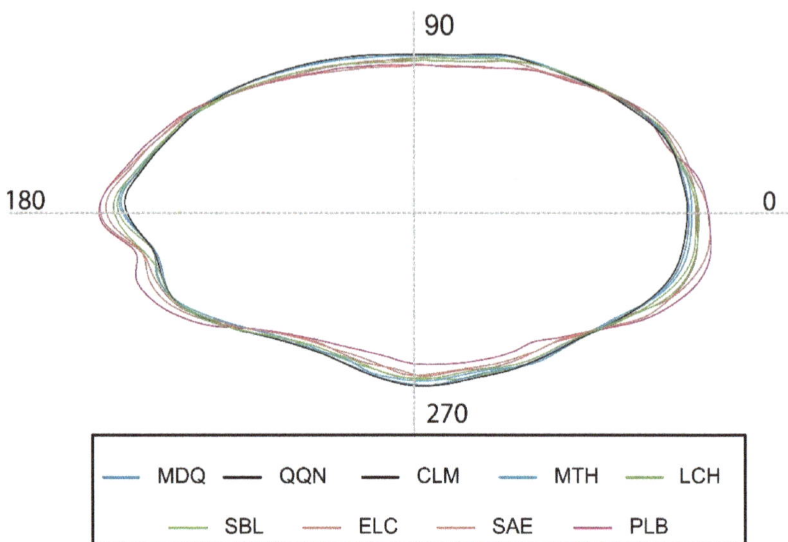

Figure 3. Mean shape of otolith shapes of silverside from nine localities. MDQ: Mar del Plata; QQN: Quequén; CLM: Claromecó; MTH: Monte Hermoso; LCH: La Chiquita; SBL: San Blas; ELC: El Cóndor; SAE: San Antonio Este; PLB: Puerto Lobos.

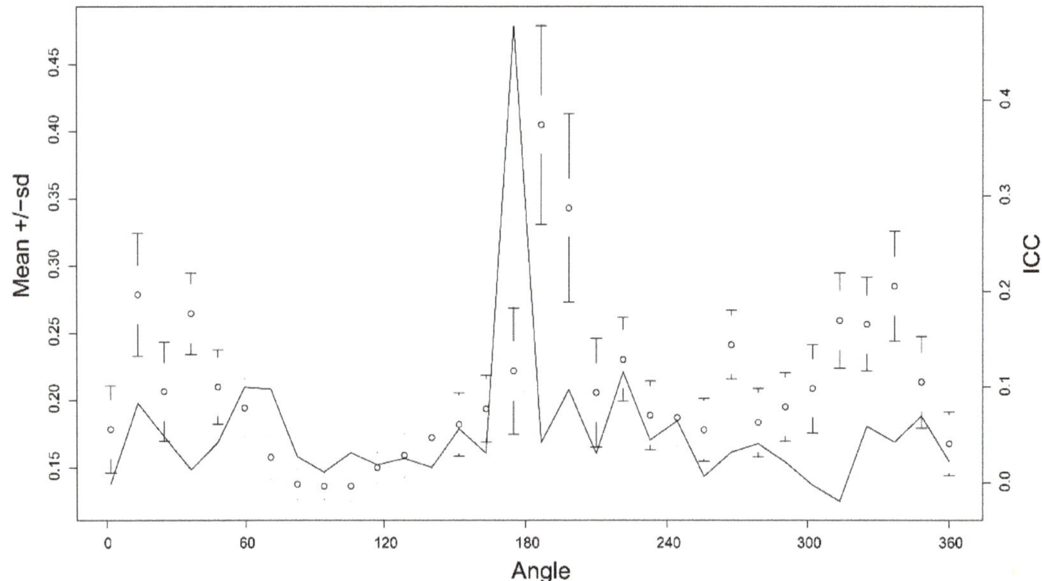

Figure 4. Mean and Standard deviation (SD) of the wavelet coefficients representing shape for all combined otoliths and the proportion of variance among silverside of different localities or interclass correlation (ICC, black solid line). The central axis shows the angle in degrees (°) based on polar coordinates where the centroid of the otolith is the central point of the polar coordinates.

The otoliths of the fish show a gradient from the northern localities (MDQ, QQN, CLM) that have a more elliptical shape, unlike the otoliths of the fish caught in the localities of the extreme south (PLB and SAE). The otoliths of the fish caught in MTH, LCH, SBL and ELC presented intermediate forms.

3.2. Shape Indices

The shape indices obtained from the variables OL, OW and OP showed higher values of *circularity* towards the south of the spatial distribution studied, while higher values of *Aspect ratio* and *Roundness* towards the North were observed. The *rectangularity* values did not show major latitudinal changes (Table 2).

Table 2. Median values ± standard deviation of shape indices by locality, (ANOVA statistical analysis: F [statistic] and p [significance]). MDQ: Mar del Plata; QQN: Quequén; CLM: Claromecó; MTH: Monte Hermoso; LCH: La Chiquita; SBL: San Blas; ELC: El Cóndor; SAE: San Antonio Este; PLB: Puerto Lobos.

Shape Indices	MDQ	QQN	CLM	MTH	LCH	SBL	ELC	SAE	PLB	F	p
Circularity	15.06 ± 0.6 ^{a,b}	14.73 ± 0.47 ^{a}	14.81 ± 1.1 ^{a,b}	15.31 ± 0.89 ^{b,c}	14.88 ± 1.02 ^{b,c,d}	15.82 ± 0.67 ^{c,d,e}	16.42 ± 1.04 ^{f}	16.19 ± 0.88 ^{d,e,f}	16.49 ± 0.6 ^{e,f}	23.69	<0.001 **
Rectangularity	0.73 ± 0.02 ^{a}	0.74 ± 0.01 ^{a}	0.74 ± 0.02 ^{a}	0.73 ± 0.02 ^{a}	0.73 ± 0.01 ^{a}	0.74 ± 0.02 ^{a}	0.72 ± 0.03 ^{a}	0.74 ± 0.02 ^{a}	0.74 ± 0.02 ^{a}	1.93	0.0544
Aspect ratio	0.63 ± 0.04 ^{a}	0.64 ± 0.03 ^{a}	0.63 ± 0.05 ^{a}	0.62 ± 0.05 ^{a}	0.65 ± 0.08 ^{a,b}	0.59 ± 0.04 ^{b}	0.56 ± 0.06 ^{b,c}	0.53 ± 0.05 ^{b}	0.53 ± 0.03 ^{c}	20.41	<0.001 **
Roundness	0.59 ± 0.04 ^{a}	0.59 ± 0.03 ^{a}	0.6 ± 0.05 ^{a}	0.58 ± 0.05 ^{a}	0.61 ± 0.08 ^{a,b}	0.55 ± 0.04 ^{b}	0.52 ± 0.05 ^{b,c}	0.53 ± 0.04 ^{b}	0.5 ± 0.03 ^{c}	20.06	<0.001 **

** Significant p value. Different letters show significant differences ($p < 0.05$).

3.3. Multivariate Analysis

The results of the Linear Discriminant Analysis showed different percentages of errors in the classification of individuals, depending on the variables used in the analysis (Table 3).

Table 3. Cross-classification table of Linear Discriminant Analysis. MDQ: Mar del Plata; QQN: Quequén; CLM: Claromecó; MTH: Monte Hermoso; LCH: La Chiquita; SBL: San Blas; ELC: El Cóndor; SAE: San Antonio Este; PLB: Puerto Lobos. ** Significant p value.

Localities	MDQ	QQN	CLM	MTH	LCH	SBL	ELC	SAE	PLB	Error (%)	Total Error (%)
(a) Morphometric Indices (MI) (Wilks' $\lambda = 0.652$, $p < 0.001$ **)											
MDQ	14	19	10	4	4	0	4	1	1	75	
QQN	9	42	18	4	0	1	1	0	0	44	
CLM	3	23	10	2	0	0	4	2	3	79	
MTH	6	7	8	6	1	1	5	3	2	85	
LCH	2	2	3	0	0	0	4	0	2	100	72.25
SBL	7	3	7	8	3	7	13	7	9	89	
ELC	2	1	0	1	1	2	7	2	4	65	
SAE	3	2	2	1	2	2	8	3	7	90	
PLB	0	0	0	0	2	1	4	0	12	37	
(b) Wavelet coefficients (WC) (Wilks' $\lambda = 0.059$, $p < 0.001$ **)											
MDQ	45	3	1	2	1	4	0	1	0	21	
QQN	3	67	2	1	0	2	0	0	0	11	
CLM	0	2	41	1	0	1	0	2	0	13	
MTH	0	0	2	34	1	1	0	1	0	13	
LCH	0	0	0	1	12	0	0	0	0	8	17.86
SBL	6	2	0	4	4	43	2	3	0	33	
ELC	1	0	1	0	1	1	15	1	0	25	
SAE	2	0	1	0	3	1	0	23	0	23	
PLB	0	0	0	0	0	0	0	0	19	0	

Table 3. *Cont.*

Localities	MDQ	QQN	CLM	MTH	LCH	SBL	ELC	SAE	PLB	Error (%)	Total Error (%)
(c) Fourier descriptors (FD) (Wilks' $\lambda = 0.016$, $p < 0.001$ **)											
MDQ	47	0	1	4	1	2	0	2	0	18	
QQN	0	63	0	3	4	4	0	1	0	16	
CLM	0	2	38	1	2	2	0	2	0	19	
MTH	2	3	0	26	4	3	0	1	0	33	
LCH	0	0	0	0	9	1	2	1	0	31	25.27
SBL	1	1	2	14	3	33	4	6	0	48	
ELC	0	0	1	0	1	2	16	0	0	20	
SAE	0	3	2	2	0	2	0	21	0	30	
PLB	0	0	0	0	0	0	0	0	19	0	
(d) MI + FD + WC (Wilks' $\lambda = 0.002$, $p < 0.001$ **)											
MDQ	55	0	0	0	0	1	0	1	0	4	
QQN	0	72	0	0	0	3	0	0	0	4	
CLM	0	0	45	0	0	1	0	1	0	4	
MTH	1	0	0	36	0	2	0	0	0	8	
LCH	0	0	0	0	11	2	0	0	0	15	7.42
SBL	2	0	0	2	1	57	0	2	0	11	
ELC	0	0	0	0	0	4	16	0	0	20	
SAE	0	0	1	1	0	2	0	26	0	13	
PLB	0	0	0	0	0	0	0	0	19	0	

The highest percentage of total error presented in the cross-classification was obtained by using only the MI as a variable (72, 25% of poorly classified individuals) (Table 3a). The poorly classified individuals were from the localities of LCH, SAE, SBL and MTH. This is also shown in Figure 5a, where no defined clusters are observed between the localities, although there is a North–South gradient. In relation to Fourier descriptors (FDs) the percentage of error in the classified individuals was reduced to 25.27%, (Table 3c). The locality with the highest percentage of classification error was SBL with 48%. Figure 5b–d show that the locality of PLB was separated from the rest by canonical axis 1, while the remaining localities did not form a clear grouping on canonical axis 2. The QQN and MDQ localities were at opposite points of the point cloud (Figure 5c).

The results of the application of wavelet coefficients (WCs), show that the percentage of error in the classified individuals was less than 17.86 (Table 3b). The locality with the highest percentage of classification error was SBL (33%). Figure 5c shows that the otoliths of PLB fish were separated from the rest of the samples by canonical axis 1. In addition, a North–South gradient of the localities of origin of the samples was also observed in the canonical axis 2 (Figure 5c).

The lowest percentage of classification error (7.42%) was obtained by applying the MI, FD and WC together (Table 3d). Figure 5d shows that the locality of PLB was separated from the rest on canonical axis 1, showing a North–South gradient on canonical axis 2 (Figure 5d). This analysis, including the three types of variables, presented a classification more in line with the geographical reality of the localities evaluated, showing, in addition, the lowest percentage of misclassification (Table 3).

PERMANOVA tests were performed on different groups, considering ADL classification errors (Table 2) and groups in the ADL figures (Figure 5). The tests were done over all standardized variables (SI + DF + WC). The PERMANOVA tests showed significant differences among the proposed aggrupation's North (MDQ, QQN, CLM) against South (MTH, LCH, SBL, ELC, SAE, PLB) (*pseudo-f*: 36.76, $p < 0.0001$); also, with North (MDQ, QQN, CLM) against South (MTH, LCH, SBL, ELC, SAE) and PLB alone (*pseudo-f*: 28.54 $p < 0.0001$) and between North localities against PLB and South localities versus PLB (Table 4).

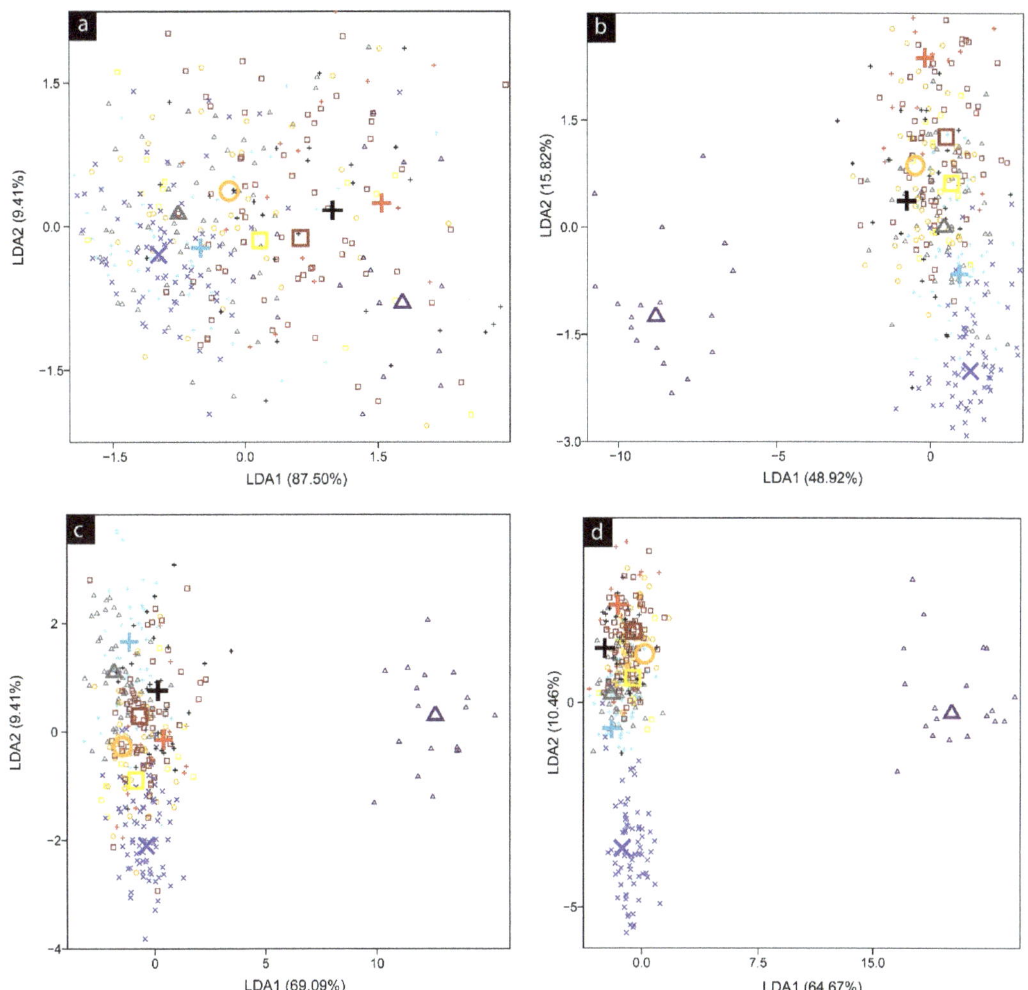

Figure 5. Linear Discriminant Analysis (**a**) Morphometric indices (MI); (**b**) Wavelets coefficients (WC); (**c**) Fourier descriptors (FD); (**d**) MI + WC + FD. MDQ: Mar del Plata; QQN: Quequén; CLM: Claromecó; MTH: Monte Hermoso; LCH: La Chiquita; SBL: San Blas; ELC: El Cóndor; SAE: San Antonio Este; PLB: Puerto Lobos. The centroid of each study area is shown as a larger symbol.

Table 4. Overall PERMANOVA comparison among groups for combined silverside otolith shape indices, Fourier descriptors and Wavelet coefficients, including pairwise comparisons between groups. MDQ: Mar del Plata; QQN: Quequén; CLM: Claromecó; MTH: Monte Hermoso; LCH: La Chiquita; SBL: San Blas; ELC: El Cóndor; SAE: San Antonio Este; PLB: Puerto Lobos. ** Significant *p* value.

Variables	Localities Groups	*F.* Model	*p* Value
Combined Morphometric indices, Fourier descriptors and Wavelet coefficients	MDQ, QQN, CLM vs. MTH, LCH, SBL, ELC, SAE, PLB	36.76	<0.0001 **
	MDQ, QQN, CLM vs. MTH, LCH, SBL, ELC, SAE vs. PLB	28.54	<0.0001 **
	MDQ, QQN, CLM vs. MTH, LCH, SBL, ELC, SAE	29.57	<0.0001 **
	MDQ, QQN, CLM vs. PLB	44.95	<0.0001 **
	MTH, LCH, SBL, ELC, SAE vs. PLB	15.28	<0.0001 **

4. Discussion

Marine and euryhaline coastal species inhabit dynamic environments so they generally show high plasticity [56], such as atherinids, mugilids and sardines [17,37,57–59]. This plasticity can be reflected in otoliths.

The results obtained show that the spatial structure of *O. argentinensis* presents a North–South gradient (38° S–41°51′ S), where three stocks can be accounted for: one in the north with the localities of MDQ, QQN and CLM, another in the south/center with MTH, LCH, SBL, ELC and SAE and the third in the south of the San Matias Gulf in PLB. The latter should be further studied to have more conclusive evidence of its relations. This latitudinal gradient marked differences between the extreme localities of the north (MDQ, QQN), the south/center (SBL, SAE) and the isolated location of PLB. This also coincides with what was recently found by González-Castro et al. [60] when studying molecular diversity. These authors determined the existence of different populations: an isolated population of silversides in the Mar Chiquita lagoon, a population in the coastal area of Mar del Plata, that coincides with the northern group of this study, and another population in San Blas Bay, that would be included in the south/center group of the localities we studied.

On the other hand, González-Castro et al. [60] also found low frequency of haplotypes in the population of Mar del Plata area, which is also consistent with the gradient found in this study using the morphometry of the otoliths. In this sense, Levy et al. [26] found two different stocks of *O. argentinensis* when studying parasite communities: one in Mar del Plata and the other in San Antonio. Their findings agree with the stock spatial structure we propose in this study.

Furthermore, we can establish that the limit between the two stocks is in the zone of MTH, which is consistent with the limits of the El Rincon area and oceanographic region, where the characteristics of the water mass differ from the rest of the coast of the province of Buenos Aires [39,40]. It is worth mentioning that the Puerto Lobos specimens can be established as another stock that was unknown until this study. The Puerto Lobos stock is inside the San Matias gulf, in its southern zone; for example, the gulf has its own isolated stock of Argentinian Hake [61,62]. These findings should be confirmed by more studies with the *O. argentinensis* inside the gulf and further south, because no studies of characteristics were done with the fish, knowing that the distribution of the species continues till Rawson (43°19′ S; 65°02′ W) [27].

Environmental conditions produce a rapid variation in the morphometric and meristic characteristics of other species of silversides, such as *O. regia* [63], which could also generate variations in the morphometry of the otoliths reflected in the morphometric indices of shape of the otoliths.

The plasticity and flexibility of the life history of *O. argentinensis* are manifested in different conditions of salinity and temperature inhabited by the fish, which allows this species to be found in freshwater environments, such as the Uruguay River [64] or the Salada de Pedro Luro [24]. This particularity may occur due to the fact that evolutionarily they come from an ancestral plastic marine population [65].

Heras and Roldán [66], by analyzing the genetics of individuals from different localities of the Argentinian Sea, proposed that *O. argentinensis* may be a model of divergent adaptive selection associated with reproductive events of isolation. This fact could also explain the separation of the shape of the otoliths from the Puerto Lobos specimens studied in this work which is very different from the rest. This could be due to an isolation produced by oceanographic barriers, such as the different bodies of water or the ocean fronts of this locality. This type of oceanographic barrier also affects the spatial distribution of other coastal species, like sciaenids [11,67]. These results coincide with the particular characteristics that the species present in incipient speciation processes in specific areas like those described by Levy et al. [29] and González-Castro et al. [60], applying molecular genetics in silversides of the Mar Chiquita Lagoon. It is important to note that speciation processes in this species can occur on a small geographical scale (<20 km) even without marked geographical barriers [26], so groups that are isolated in particular coastal environments, such as coastal lagoons or areas of different oceanographic features, may be able to differentiate themselves from the rest, as happens with PLB fish.

The values of the morphometric indices of otolith shape obtained by Biolé et al. [37] in silversides of La Lucila del Mar are different in relation to those obtained in this work. These differences may be due to the existence of a latitudinal gradient in the species. The otolith shapes that we found were similar to those referred to by Biolé et al. [37] and Tombari et al. [36], in La Lucila del Mar, Punta Rasa (36°22′ S–56°45′ W) and Miramar (38°16′S–57°50′ W).

The increase in the rate of circularity towards higher latitudes shows the increase in the complexity of otoliths [13,49] towards higher latitudes. This could be caused by environmental changes and the trophic habits of the species.

The shape of the otolith depends on the ecological, evolutionary and phylogenetic characteristics of each species [67–72]. This can be evident in coastal species that inhabit dynamic environments, as is the case with *O. argentinensis*, and can be reflected in the morphometry of the otoliths.

Biolé et al used traditional morphometry and shape analysis by Fourier descriptors [37] in specimens of *O. argentinensis* from La Lucila del Mar in a wider range of sizes than those studied in this work. In this study, as in Biolé et al., Stage III of *O argentinensis* [37], the focus was on sexually mature adult fish. Adults of marine silversides make habitat movements to reproduce which could be reflected in the shape of otoliths as in other fish species [17,58,73].

Methodologically, the application of the analysis of morphometric indices, Fourier descriptors and wavelet coefficients was used in different species of the southwest Atlantic Ocean coast, such as *Coryphaena hippurus* [74], *Menticirrhus americanus* [17,71], *Pagrus pagrus* [75]. The joint application of the three methods referred to by Sadighzadeh et al. [76] was used for the first time regarding this species, with different results for the methods.

Tuset et al. [69] demonstrated that the application of wavelet coefficients on otoliths is a more appropriate methodological option to classify specimens from different localities than morphometric indices. These results are consistent with what was found in this work, since when the wavelet coefficients were incorporated, the percentage of the classification error decreased, while when only the morphometric indices were considered, the classification error was high. This would suggest that the application of joint methodologies (morphometric indices, wavelet coefficients, Fourier descriptors and even otolith chemistry) would be more efficient in otoliths of species that inhabit very environmentally dynamic areas, as has been shown for *Lutjanus* spp and *Chaetodipterus faber* [76,77].

Finally, it is important to highlight that the spatial distribution of the stocks in this study showed three stocks that are under differential fishing pressures. On one hand, the northern stock is found within the fishing area of the most important coastal cities of Argentina, such as Mar del Plata and Necochea, where the fishing for the species is carried out constantly throughout the year, in shores and docks used for this activity. For the south/center stock, fishing takes place mainly in fishing tourism areas, such as San Blas and La Chiquita. In these fishing villages, the fishery is seasonal and the influx of tourism during summer is most important, especially in San Blas [35,78]. On the other hand, the last stock falls under two different jurisdictions, the Province of Buenos Aires, where management measures are carried out by the coastal fishing regulations of Buenos Aires, and minimum catch size and number of pieces for silverside are set, and the southern part, in the localities of El Cóndor, San Antonio Este and also Puerto Lobos, in the Province of Río Negro, which does not have any regulations governing marine recreational fishing [79]. This spatial analysis should be incorporated in the management scenario of silverside fisheries of the marine coast of Argentina [4].

5. Conclusions

The results of this paper allow us to conclude that the spatial structure of *O. argentinensis* in the study area presents three stocks: one includes fish from MDQ localities; QQN and CLM, the second stock includes fish from the localities of MTH, LCH, SBL, ELC and

SAE and the third concerns south of the Gulf of San Matías (PLB). This last stock should be studied further, expanding the sampling to the southern region and increasing the number of individuals in order to strengthen the evidence found in this work. On the other hand, it is important to consider that the limits of these groups of fish are between the localities of MTH and CLM for the north and south/center stock and between SAE and PLB, for the south/central and south stocks.

In relation to management applications of these findings, the three stocks have different problems. The north stock is located under the fishing zone of major coastal cities of Argentina, such as Mar del Plata and Necochea, which implies that this stock could be under larger fishing pressure than the other stocks of this species. On the other hand, the south/center stock is distributed in two jurisdictions, Buenos Aires in the north and Río Negro in the south, and these locales have different management measures; Río Negro does not have any measurement for the management of silverside fishing. In this context, fishery management must be carried out by the two jurisdictions, considering that they share the same resources.

Author Contributions: S.M., conceptualization, methodology, software, data analysis, writing—original draft, writing—review and editing, sampling; P.J.S., conceptualization, methodology, writing—original draft, writing—review and editing, sampling; A.V.V., conceptualization, data analysis, writing—original draft, writing—review and editing. All authors have read and agreed to the published version of the manuscript.

Funding: CONICET (PIP 11220200103264CO, PUE 22920180100047CO), Agencia Nacional de Promoción Científica y Tecnológica (PICT 2019-03888), Universidad de Buenos Aires (UBACYT 20020190100069BA), Universidad Nacional de Rio Negro (PI (40-C-683); PI (40-C-790)).

Institutional Review Board Statement: Not applicable.

Informed Consent Statement: Not applicable.

Data Availability Statement: The data belong to the Ph. Thesis of S. Morawicki and are available in the Centro de Investigaciones y Transferencia de Río Negro repository.

Acknowledgments: Authors thanks to CONICET, Agencia Nacional de Promoción Científica y Tecnológica, Universidad de Buenos Aires, Universidad Nacional de Rio Negro and to the group "Pescadores de Necochea", Daniel Sframeli, Bruno Rodriguez, Eze Duckardt and Marcello Saez, for field support and some of the fish.

Conflicts of Interest: The authors declare no conflict of interest.

References

1. Carr, M.H.; Robinson, S.P.; Wahle, C.; Davis, G.; Kroll, S.; Murray, S.; Schumacker, E.J.; Williams, M. The Central Importance of Ecological Spatial Connectivity to Effective Coastal Marine Protected Areas and to Meeting the Challenges of Climate Change in the Marine Environment. *Aquat. Conserv. Mar. Freshw. Ecosyst.* **2017**, *27*, 6–29. [CrossRef]
2. Dizon, A.E.; Lockyer, C.; Perrin, W.F.; Demaster, D.P.; Sisson, J. Rethinking the Stock Concept: A Phylogeographic Approach. *Conser. Biol.* **1992**, *6*, 24–36. [CrossRef]
3. Begg, G.A.; Waldman, J.R. An Holistic Approach to Fish Stock Identification. *Fish. Res.* **1999**, *43*, 35–44. [CrossRef]
4. Stephenson, R.L. Stock Complexity in Fisheries Management: A Perspective of Emerging Issues Related to Population Sub-Units. *Fish. Res.* **1999**, *43*, 247–249. [CrossRef]
5. Kerr, L.A.; Hintzen, N.T.; Cadrin, S.X.; Clausen, L.W.; Dickey-Collas, M.; Goethel, D.R.; Hatfield, E.M.C.; Kritzer, J.P.; Nash, R.D.M. Lessons Learned from Practical Approaches to Reconcile Mismatches between Biological Population Structure and Stock Units of Marine Fish. *ICES J. Mar. Sci.* **2017**, *74*, 1708–1722. [CrossRef]
6. Côté, C.L.; Gagnaire, P.A.; Bourret, V.; Verreault, G.; Castonguay, M.; Bernatchez, L. Population Genetics of the American Eel (*Anguilla rostrata*): FST = 0 and North Atlantic Oscillation Effects on Demographic Fluctuations of a Panmictic Species. *Mol. Ecol.* **2013**, *22*, 1763–1776. [CrossRef] [PubMed]
7. Abaunza, P.; Murta, A.G.; Campbell, N.; Cimmaruta, R.; Comesaña, A.S.; Dahle, G.; García Santamaría, M.T.; Gordo, L.S.; Iversen, S.A.; MacKenzie, K.; et al. Stock Identity of Horse Mackerel (*Trachurus trachurus*) in the Northeast Atlantic and Mediterranean Sea: Integrating the Results from Different Stock Identification Approaches. *Fish. Res.* **2008**, *89*, 196–209. [CrossRef]

8. Cadrin, S.X.; Bernreuther, M.; Daníelsdóttir, A.K.; Hjörleifsson, E.; Johansen, T.; Kerr, L.; Kristinsson, K.; Mariani, S.; Nedreaas, K.; Pampoulie, C.; et al. Population Structure of Beaked Redfish, *Sebastes mentella*: Evidence of Divergence Associated with Different Habitats. *ICES J. Mar. Sci.* **2010**, *67*, 1617–1630. [CrossRef]
9. Smedbol, R.K.; Wroblewski, J.S. Metapopulation Theory and Northern Cod Population Structure: Interdependency of Subpopulations in Recovery of a Groundfish Population. *Fish. Res.* **2002**, *55*, 161–174. [CrossRef]
10. Park, L.; Moran, P. Developments in Molecular Genetic Techniques in Fisheries. *Rev. Fish Biol. Fish.* **1994**, *4*, 272–299. [CrossRef]
11. Volpedo, A.V.; Fernández Cirelli, A. Otolith Chemical Composition as a Useful Tool for Sciaenid Stock Discrimination in the South-Western Atlantic. *Sci. Mar.* **2006**, *70*, 325–334. [CrossRef]
12. Campana, S.E. Chemistry and Composition of Fish Otoliths: Pathways, Mechanisms and Applications. *Mar. Ecol. Prog. Ser.* **1999**, *188*, 263–297. [CrossRef]
13. Tuset, V.M.; Lozano, I.J.; Gonzĺez, J.A.; Pertusa, J.F.; García-Díaz, M.M. Shape Indices to Identify Regional Differences in Otolith Morphology of Comber, *Serranus cabrilla* (L., 1758). *J. Appl. Ichthyol.* **2003**, *19*, 88–93. [CrossRef]
14. Avigliano, E.; Martinez, C.F.R.; Volpedo, A.V. Combined Use of Otolith Microchemistry and Morphometry as Indicators of the Habitat of the Silverside (*Odontesthes bonariensis*) in a Freshwater-Estuarine Environment. *Fish. Res.* **2014**, *149*, 55–60. [CrossRef]
15. Biolé, F.G.; Thompson, G.A.; Vargas, C.V.; Leisen, M.; Barra, F.; Volpedo, A.V.; Avigliano, E. Fish Stocks of *Urophycis brasiliensis* Revealed by Otolith Fingerprint and Shape in the Southwestern Atlantic Ocean. *Estuar. Coast. Shelf Sci.* **2019**, *229*, 106406. [CrossRef]
16. Tracey, S.R.; Lyle, J.M.; Duhamel, G. Application of Elliptical Fourier Analysis of Otolith Form as a Tool for Stock Identification. *Fish. Res.* **2006**, *77*, 138–147. [CrossRef]
17. de Carvalho, B.M.; Volpedo, A.V.; Fávaro, L.F. Ontogenetic and Sexual Variation in the Sagitta Otolith of *Menticirrhus americanus* (Teleostei; Sciaenidae) (Linnaeus, 1758) in a Subtropical Environment. *Pap. Avulsos Zool.* **2020**, *60*, 1–12. [CrossRef]
18. O'Reilly, K.M.; Horn, M.H. Phenotypic Variation among Populations of *Atherinops affinis* (Atherinopsidae) with Insights from a Geometric Morphometric Analysis. *J. Fish Biol.* **2004**, *64*, 1117–1135. [CrossRef]
19. Tombari, A.D.; Gosztonyi, A.E.; Echeverria, D.D.; Volpedo, A.V. Otolith and Vertebral Morphology of Marine Atherinid Species (Atheriniformes, Atherinopsidae) Coexisting in the Southwestern Atlantic Ocean. *Ciencias Mar.* **2010**, *36*, 213–223. [CrossRef]
20. Lattuca, M.E.; Lozano, I.E.; Brown, D.R.; Renzi, M.; Luizon, C.A. Natural Growth, Otolith Shape and Diet Analyses of *Odontesthes nigricans* Richardson (Atherinopsidae) from Southern Patagonia. *Estuar. Coast. Shelf Sci.* **2015**, *166*, 105–114. [CrossRef]
21. Thompson, G.A.; Volpedo, A.V. Diet Composition and Feeding Strategy of the New World Silverside *Odontesthes argentinensis* in a Temperate Coastal Area (South America). *Mar. Coast. Fish.* **2018**, *10*, 80–88. [CrossRef]
22. Strüssmann, C.A.; Akaba, T.; Ijima, K.; Yamaguchi, K.; Yoshizaki, G.; Takashima, F. Spontaneous Hybridization in the Laboratory and Genetic Markers for the Identification of Hybrids between Two Atherinid Species, *Odontesthes bonariensis* (Valenciennes, 1835) and *Patagonina hatcheri* (Eigenmann, 1909). *Aquac. Res.* **1997**, *28*, 291–300. [CrossRef]
23. Conte-Grand, C.; Sommer, J.; Ortí, G.; Cussac, V. Populations of *Odontesthes* (Teleostei: Atheriniformes) in the Andean Region of Southern South America: Body Shape and Hybrid Individuals. *Neotrop. Ichthyol.* **2015**, *13*, 137–150. [CrossRef]
24. Colautti, D.C.; Miranda, L.; Gonzalez-Castro, M.; Villanova, V.; Strüssmann, C.A.; Mancini, M.; Maiztegui, T.; Berasain, G.; Hattori, R.; Grosman, F.; et al. Evidence of a Landlocked Reproducing Population of the Marine Pejerrey *Odontesthes argentinensis* (Actinopterygii; Atherinopsidae). *J. Fish Biol.* **2020**, *96*, 202–216. [CrossRef]
25. Avigliano, E.; Miller, N.; Volpedo, A.V. Silversides (*Odontesthes bonariensis*) Reside within Freshwater and Estuarine Habitats, Not Marine Environments. *Estuar. Coast. Shelf Sci.* **2018**, *205*, 123–130. [CrossRef]
26. Levy, E.; Canel, D.; Rossin, M.A.; Hernández-Orts, J.S.; González-Castro, M.; Timi, J.T. Parasites as Indicators of Fish Population Structure at Two Different Geographical Scales in Contrasting Coastal Environments of the South-Western Atlantic. *Estuar. Coast. Shelf Sci.* **2019**, *229*, 106400. [CrossRef]
27. Dyer, B.S. Systematic Revision of the South American Silversides (Teleostei, Atheriniformes). *Biocell* **2006**, *30*, 69–88.
28. Di Dario, F.; Macedo dos Santos, V.L.; Maia de Souza Pereira, M. Range Extension of *Odontesthes argentinensis* (Valenciennes, 1835) (Teleostei: Atherinopsidae) in the Southwestern Atlantic, with Additional Records in the Rio de Janeiro State, Brazil. *J. Appl. Ichthyol.* **2014**, *30*, 421–423. [CrossRef]
29. Levy, E.; Canel, D.; Rossin, M.A.; González-Castro, M.; Timi, J.T. Parasite Assemblages as Indicators of an Incipient Speciation Process of *Odontesthes argentinensis* in an Estuarine Environment. *Estuar. Coast. Shelf Sci.* **2021**, *250*, 107168. [CrossRef]
30. De Buen, F. Los Pejerreyes (Familia Atherinidae) En La Fauna Uruguaya, Con Descripción de Nuevas Especies. *Bol. do Inst. Ocean.* **1953**, *4*, 3–80. [CrossRef]
31. Sampaio, L.A. Production of "Pejerrey" *Odontesthes argentinensis* Fingerlings: A Review of Current Techniques. *Biocell* **2006**, *30*, 121–123. [PubMed]
32. Llompart, F.M.; Colautti, D.C.; Maiztegui, T.; Cruz-Jiménez, A.M.; Baigún, C.R.M. Biological Traits and Growth Patterns of Pejerrey *Odontesthes argentinensis*. *J. Fish Biol.* **2013**, *82*, 458–474. [CrossRef]
33. Llompart, F.M.; Colautti, D.C.; Baigún, C.R.M. Conciliating Artisanal and Recreational Fisheries in Anegada Bay, Argentina. *Fish. Res.* **2017**, *190*, 140–149. [CrossRef]
34. Guidi, C.; Baigún, C.R.M.; Ginter, L.G.; Soricetti, M.; Rivas, F.J.G.; Morawicki, S.; Quezada, F.; Bazzani, J.L.; Solimano, P.J. Characteristics, Preferences and Perceptions of Recreational Fishers in Northern Patagonia, Argentina. *Reg. Stud. Mar. Sci.* **2021**, *45*, 101828. [CrossRef]

35. Llompart, F.M.; Colautti, D.C.; Baign, C.R.M. Assessment of a Major Shore-Based Marine Recreational Fishery in the Southwest Atlantic, Argentina. *N. Z. J. Mar. Freshw. Res.* **2012**, *46*, 57–70. [CrossRef]
36. Tombari, A.D.; Volpedo, A.V.; Echeverria, D.D. Desarrollo de La Sagitta En Juveniles y Adultos de *Odontesthes argentinensis* (Valenciennes, 1835) y *O. bonariensis* (Valenciennes, 1835) de La Provincia de Buenos Aires, Argentina (Teleostei: Atheriniformes). *Rev. Chil. Hist. Nat.* **2005**, *78*, 623–633. [CrossRef]
37. Biolé, F.G.; Callicó Fortunato, R.; Thompson, G.A.; Volpedo, A.V. Application of Otolith Morphometry for the Study of Ontogenetic Variations of *Odontesthes argentinensis*. *Environ. Biol. Fishes* **2019**, *102*, 1301–1310. [CrossRef]
38. Kjerfve, B.; Perillo, G.M.E.; Gardner, L.R.; Rine, J.M.; Dias, G.T.M.; Mochel, F.R. Morphodynamics of Muddy Environments along the Atlantic Coasts of North and South America. In *Muddy Coasts of the World: Processes, Deposits and Function*; Healy, T., Wang, Y., Healy, J.A., Eds.; Elsevier: Amsterdam, The Netherlands, 2002.
39. Acha, E.; Orduna, M.; Rodrigues, K.; Militelli, M.; Braverman, M. Caracterización de La Zona de "El Rincón" (Provincia de Buenos Aires) Como Área de Reproducción de Peces Costeros. *Rev. Invest. Desarr. Pesq.* **2012**, *21*, 31–43.
40. Piola, A.R.; Rivas, A.L. Corrientes En La Plataforma Continental. *El Mar Argentino y sus Recur. Pesq.* **1997**, *1*, 119–132.
41. Mancini, M.; Grosman, F.; Dyer, B.; García, G.; Ponti, O.; Sanzano, P.; Salinas, V. *Pejerreyes Del Sur de América: Aportes Al Estado de Conocimiento Con Especial Referencia a Odontesthes Bonariensis*; UniRío Editora: Río Cuarto, Argentina, 2016; ISBN 9789876881968.
42. Libungan, L.A.; Pálsson, S. ShapeR: An R Package to Study Otolith Shape Variation among Fish Populations. *PLoS ONE* **2015**, *10*, e0121102. [CrossRef]
43. John Haines, A.; Crampton, J.S. Improvements to the Method of Fourier Shape Analysis as Applied in Morphometric Studies. *Palaeontology* **2000**, *43*, 765–783. [CrossRef]
44. Nason, G. Wavethresh: Wavelets Statistics and Transforms, R Package. 2016. Available online: https://CRAN.R-project.org/package=wavethresh (accessed on 25 December 2021).
45. Longmore, C.; Fogarty, K.; Neat, F.; Brophy, D.; Trueman, C.; Milton, A.; Mariani, S. A Comparison of Otolith Microchemistry and Otolith Shape Analysis for the Study of Spatial Variation in a Deep-Sea Teleost, *Coryphaenoides rupestris*. *Environ. Biol. Fishes* **2010**, *89*, 591–605. [CrossRef]
46. Agüera, A.; Brophy, D. Use of Saggital Otolith Shape Analysis to Discriminate Northeast Atlantic and Western Mediterranean Stocks of Atlantic Saury, *Scomberesox saurus saurus* (Walbaum). *Fish. Res.* **2011**, *110*, 465–471. [CrossRef]
47. Warnes, G.R.; Bolker, B.; Bonebakker, L.; Gentleman, R.; Liaw, W.H.A.; Lumley, T.; Maechler, M.; Magnusson, A.; Moeller, S.; Schwartz, M.; et al. Gplots: Various R Programming Tools for Plotting Data. R Package Version 2.13.0. 2014. Available online: https://CRAN.R-project.org/package=gplots (accessed on 25 December 2021).
48. Volpedo, A.V.; Vaz-dos-Santos, A.M. *Métodos de Estudios Con Otolitos: Principios y Aplicaciones*; INPA CONICET UBA: Ciudad Autónoma de Buenos Aires, Argentina, 2015.
49. Tuset, V.M.; Lombarte, A.; Assis, C.A. Otolith Atlas for the Western Mediterranean, North and Central Eastern Atlantic. *Sci. Mar.* **2008**, *72*, 7–198. [CrossRef]
50. Lleonart, J.; Salat, J.; Torres, G.J. Removing Allometric Effects of Body Size in Morphological Analysis. *J. Theor. Biol.* **2000**, *205*, 85–93. [CrossRef] [PubMed]
51. Ripley, B.; Venables, B.; Bates, D.M.; Hornik, K.; Gebhardt, A.; Firth, D. Package 'MASS': Support Functions and Datasets for Venables and Ripley's "Modern Applied Statistics with S". 2021. Available online: https://CRAN.R-project.org/package=MASS (accessed on 23 November 2021).
52. Neves, J.; Silva, A.A.; Moreno, A.; Veríssimo, A.; Santos, A.M.; Garrido, S. Population Structure of the European Sardine *Sardina pilchardus* from Atlantic and Mediterranean Waters Based on Otolith Shape Analysis. *Fish. Res.* **2021**, *243*, 106050. [CrossRef]
53. Anderson, M.J. Permutational Multivariate Analysis of Variance (PERMANOVA). In *Wiley StatsRef: Statistics Reference Online*; Balakrishnan, N., Colton, T., Everitt, B., Piegorsch, W., Ruggeri, F., Teugels, J.L., Eds.; John Wiley & Sons, Ltd.: Hoboken, NJ, USA, 2017; pp. 1–15, ISBN 9781118445112.
54. Oksanen, J.; Blanchet, F.G.; Kindt, R.; Legendre, P.; Minchin, P.; O'Hara, R.B. Vegan: Community Ecology Package, Version 2.0–7. R Package. 2021. Available online: https://CRAN.R-project.org/package=vegan (accessed on 25 December 2021).
55. Peters, A.; Hothorn, T.; Ripley, B.D.; Therneau, T.; Atkinson, B. Ipred: Improved Predictors. R Package. 2021. Available online: https://CRAN.R-project.org/package=ipred (accessed on 25 December 2021).
56. Donelson, J.M.; Sunday, J.M.; Figueira, W.F.; Gaitán-Espitia, J.D.; Hobday, A.J.; Johnson, C.R.; Leis, J.M.; Ling, S.D.; Marshall, D.; Pandolfi, J.M.; et al. Understanding Interactions between Plasticity, Adaptation and Range Shifts in Response to Marine Environmental Change. *Philos. Trans. R. Soc. B Biol. Sci.* **2019**, *374*, 20180186. [CrossRef]
57. Callicó Fortunato, R.; Benedito Durà, V.; Volpedo, A.V. Otolith Morphometry and Microchemistry as Habitat Markers for Juvenile *Mugil cephalus* Linnaeus 1758 in Nursery Grounds in the Valencian Community, Spain. *J. Appl. Ichthyol.* **2017**, *33*, 163–167. [CrossRef]
58. Callicó Fortunato, R.; González-Castro, M.; Reguera Galán, A.; García Alonso, I.; Kunert, C.; Benedito Durà, V.; Volpedo, A. Identification of Potential Fish Stocks and Lifetime Movement Patterns of *Mugil liza* Valenciennes 1836 in the Southwestern Atlantic Ocean. *Fish. Res.* **2017**, *193*, 164–172. [CrossRef]
59. Thompson, G.; Callico Fortunato, R.; Chiesa, I.; Volpedo, A. Trophic Ecology of *Mugil liza* at the Southern Limit of Its Distribution (Buenos Aires, Argentina). *Brazilian J. Oceanogr.* **2015**, *63*, 271–278. [CrossRef]

60. González-Castro, M.; Rosso, J.J.; Delpiani, S.M.; Mabragaña, E.; Díaz de Astarloa, J.M. Inferring Boundaries among Fish Species of the New World Silversides (Atherinopsidae; Genus *Odontesthes*): New Evidences of Incipient Speciation between Marine and Brackish Populations of *Odontesthes argentinensis*. *Genetica* **2019**, *147*, 217–229. [CrossRef] [PubMed]
61. Di Giácomo, E.E.; Calvo, J.; Perier, M.R.; Morriconi, E. Spawning Aggregations of *Merluccius hubbsi*, in Patagonian Waters: Evidence for a Single Stock? *Fish. Res.* **1993**, *16*, 9–16. [CrossRef]
62. Sardella, N.H.; Timi, J.T. Parasites of Argentine Hake in the Argentine Sea: Population and Infracommunity Structure as Evidence for Host Stock Discrimination. *J. Fish Biol.* **2004**, *65*, 1472–1488. [CrossRef]
63. Deville, D.; Sanchez, G.; Barahona, S.; Yamashiro, C.; Oré-Chávez, D.; Bazán, R.Q.; Umino, T. Morphological Variation of the Sea Silverside *Odontesthes regia* in Regions with Dissimilar Upwelling Intensity along the Humboldt Current System. *Ocean Sci. J.* **2020**, *55*, 33–48. [CrossRef]
64. D'Anatro, A.; González-Bergonzoni, I.; Vidal, N.; Tesitore, G.; de Mello, F.T. Confirmation of the Occurrence of *Odontesthes argentinensis* (Valenciennes, 1835) (Atheriniformes, Atherinopsidae) in the Rio Uruguay. *Panam. J. Aquat. Sci.* **2020**, *15*, 100–105.
65. Hughes, L.C.; Cardoso, Y.P.; Sommer, J.A.; Cifuentes, R.; Cuello, M.; Somoza, G.M.; González-Castro, M.; Malabarba, L.R.; Cussac, V.; Habit, E.M.; et al. Biogeography, Habitat Transitions and Hybridization in a Radiation of South American Silverside Fishes Revealed by Mitochondrial and Genomic RAD Data. *Mol. Ecol.* **2020**, *29*, 738–751. [CrossRef]
66. Heras, S.; Roldán, M.I. Phylogenetic Inference in *Odontesthes* and *Atherina* (Teleostei: Atheriniformes) with Insights into Ecological Adaptation. *C. R.—Biol.* **2011**, *334*, 273–281. [CrossRef]
67. Volpedo, A.V.; Miretzky, P.; Fernández-Cirelli, A. Stocks Pesqueros de *Cynoscion guatucupa* y *Micropogonias furnieri* (Pisces, Sciaenidae), En La Costa Atlántica de Sudamérica: Comparación Entre Métodos de Identificación. *Mem. De La Fund. La Salle De Cienc. Nat.* **2007**, *2007*, 115–130.
68. Tuset, V.M.; Farré, M.; Otero-Ferrer, J.L.; Vilar, A.; Morales-Nin, B.; Lombarte, A. Testing Otolith Morphology for Measuring Marine Fish Biodiversity. *Mar. Freshw. Res.* **2016**, *67*, 1037–1048. [CrossRef]
69. Tuset, V.M.; Otero-Ferrer, J.L.; Siliprandi, C.; Manjabacas, A.; Marti-Puig, P.; Lombarte, A. Paradox of Otolith Shape Indices: Routine but Overestimated Use. *Can. J. Fish. Aquat. Sci.* **2021**, *78*, 681–692. [CrossRef]
70. Volpedo, A.; Echeverría, D.D. Ecomorphological Patterns of the Sagitta in Fish on the Continental Shelf off Argentine. *Fish. Res.* **2003**, *60*, 551–560. [CrossRef]
71. De Carvalho, B.M.; Spach, H.L.; Vaz-Dos-Santos, A.M.; Volpedo, A.V. Otolith Shape Index: Is It a Tool for Trophic Ecology Studies? *J. Mar. Biol. Assoc. U. K.* **2019**, *99*, 1675–1682. [CrossRef]
72. Assis, I.O.; da Silva, V.E.L.; Souto-Vieira, D.; Lozano, A.P.; Volpedo, A.V.; Fabré, N.N. Ecomorphological Patterns in Otoliths of Tropical Fishes: Assessing Trophic Groups and Depth Strata Preference by Shape. *Environ. Biol. Fishes* **2020**, *103*, 349–361. [CrossRef]
73. de Carvalho, B.M.; Vaz-dos-Santos, A.M.; Spach, H.L.; Volpedo, A.V. Ontogenetic Development of the Sagittal Otolith of the Anchovy, *Anchoa tricolor*, in a Subtropical Estuary. *Sci. Mar.* **2015**, *79*, 409–418. [CrossRef]
74. Almeida, P.R.C.; Monteiro-Neto, C.; Tubino, R.A.; Costa, M.R. Variations in the Shape of the Sagitta Otolith of *Coryphaena hippurus* (Actinopterygii: Coryphaenidae) in a Resurgence Area on the Southwest Coast of the Atlantic Ocean. *Iheringia—Ser. Zool.* **2020**, *110*, 1–11. [CrossRef]
75. Kikuchi, E.; García, S.; da Costa, P.A.S.; Cardoso, L.G.; Haimovici, M. Discrimination of Red Porgy *Pagrus pagrus* (Sparidae) Potential Stocks in the South-Western Atlantic by Otolith Shape Analysis. *J. Fish Biol.* **2021**, *98*, 548–556. [CrossRef] [PubMed]
76. Sadighzadeh, Z.; Tuset, V.M.; Valinassab, T.; Dadpour, M.R.; Lombarte, A. Comparison of Different Otolith Shape Descriptors and Morphometrics for the Identification of Closely Related Species of *Lutjanus* spp. from the Persian Gulf. *Mar. Biol. Res.* **2012**, *8*, 802–814. [CrossRef]
77. Soeth, M.; Spach, H.L.; Daros, F.A.; Adelir-Alves, J.; de Almeida, A.C.O.; Correia, A.T. Stock Structure of Atlantic Spadefish *Chaetodipterus faber* from Southwest Atlantic Ocean Inferred from Otolith Elemental and Shape Signatures. *Fish. Res.* **2019**, *211*, 81–90. [CrossRef]
78. Colautti, D.; Suquele, P.; Calvo, S. *La Pesca Artesanal En La Zona Sur de La Bahía Anegada, Provincia de Buenos Aires*; Ministerio de Asuntos Agrarios Dirección Provincial de Pesca: La Plata, Argentina, 2009.
79. Venerus, L.A.; Cedrola, P.V. Review of Marine Recreational Fisheries Regulations in Argentina. *Mar. Policy* **2017**, *81*, 202–210. [CrossRef]

Article

Regional Population Structure of the European Eel at the Southern Limit of Its Distribution Revealed by Otolith Shape Signature

Ana Moura [1,*], Ester Dias [1], Rodrigo López [1] and Carlos Antunes [1,2]

[1] Interdisciplinary Centre of Marine and Environmental Research (CIIMAR), University of Porto, Terminal de Cruzeiros do Porto de Leixões, Avenida General Norton de Matos S/N, 4450-208 Matosinhos, Portugal; esterdias@ciimar.up.pt (E.D.); rodrigo_lyc@hotmail.com (R.L.); cantunes@ciimar.up.pt (C.A.)

[2] Aquamuseu do Rio Minho, Parque do Castelinho, 4920-290 Vila Nova de Cerveira, Portugal

* Correspondence: anacatarinamoura@hotmail.com

Abstract: Given the European eel population's marked decrease since the 1980s, it has become urgent to collect information describing its regional population structure to improve management plans. The Minho River (NW-Portugal, SW-Europe) is an important basin for the eel at the southern limit of its distribution, but the species is poorly described. Thus, we aimed to study the structure of the European eel population in the Minho River using otolith shape analysis, which has proven to be effective in discriminating fish groups experiencing different environmental conditions through ontogeny. Our results showed complete discrimination between the two main types of habitats studied (tributaries and estuaries). Otoliths of eels from the estuary were rectangular and elliptic, whereas in the tributaries they presented a more round and circular form. Eels collected in both habitats were mostly yellow-stage eels with a similar age range, but the eels from the tributaries showed smaller length-at-age and lower body condition than those collected in the estuary. Additionally, the sex ratio was skewed towards males in the tributaries and females in the estuary. This study reveals that there are at least two distinct groups of eels in this basin, likely with different development characteristics.

Keywords: *Anguilla anguilla*; fish biology; ecosystem variability; minho river; stock assessment

Citation: Moura, A.; Dias, E.; López, R.; Antunes, C. Regional Population Structure of the European Eel at the Southern Limit of Its Distribution Revealed by Otolith Shape Signature. *Fishes* **2022**, *7*, 135. https://doi.org/10.3390/fishes7030135

Academic Editor: Josipa Ferri

Received: 10 May 2022
Accepted: 2 June 2022
Published: 7 June 2022

Publisher's Note: MDPI stays neutral with regard to jurisdictional claims in published maps and institutional affiliations.

1. Introduction

The European eel *Anguilla anguilla* (Linnaeus, 1758) population is currently outside of safe biological limits [1]; the number of eels reaching the European coastal areas has decreased by 90% since the 1980s [2–4]. Eel fishing still occurs throughout most of Europe and across most life stages [5], even though this species is considered critically endangered by the IUCN's Red List and listed in Appendix II of the CITES and Bonn Convention [6–8]. The lack of basic knowledge about the status of regional stocks and its vast distribution range has precluded the development of adequate management plans [5,9–12]. Eels grow and adapt to habitats, from freshwater to saltwater, presenting a high degree of flexibility in habitat use patterns [13–15]. Identifying regional population structure is of outmost importance to unravel how each river basin is contributing to the spawning population.

The European eel (hereafter, eel) is a semelparous and semi-catadromous fish species with a single randomly mating population [13,15,16]. It has a remarkable life cycle which includes migration of ca. 6000 km between the spawning grounds in the Sargasso Sea on the NW-Atlantic Ocean and the nursery areas distributed along the coasts of Europe and Northern Africa [17–19]. Leptocephali (larvae) drift across the Atlantic Ocean and metamorphose into glass eels before entering the continental shelf [20]. They then use divergent colonization migratory tactics depending on the time of arrival and the settlement habitat [21,22]. The growth phase in coastal waters typically lasts from 2 to 20 years (i.e., the yellow eel stage), increasing with decreasing temperatures [23]. After this period, they metamorphose into silver eels and achieve sexual maturation as they swim back to their spawning grounds [24,25].

At the scale of the species distribution, eel growth rates decrease from the south to the north of Europe [23]. Additionally, as the result of the latitudinal gradients of temperature, photoperiod, hydrology, and productivity, eels achieve larger sizes with increasing latitude and distance from the spawning areas [26]. Within river basins, eel development can vary according to environmental drivers such as temperature, salinity, depth, and food availability [15,27,28]. Downstream brackish and marine environments promote higher growth rates in eels than upstream freshwater or riverine habitats [29]. Demographical factors such as sex and density are also known to influence individual development [30–32]. There is a predominance of males in Southern Europe and a predominance of females in Northern Europe [23]. Males generally occupy the most downstream areas in a river, grow fast, and mature earlier at a smaller size, whereas females develop slowly in upstream areas and mature later at larger sizes [23,33,34]. Eel growth is usually low at higher densities due to increased intraspecific competition for resources or habitat loss [35,36]. Males tend to predominate in such environments [28,30,37], which seems to contradict the general idea of faster growth in males [18,38]. Nonetheless, several aspects of a riverine environment are known to affect trophic interactions, fish distribution, and development, such as productivity [39], ecosystem size [40,41], and disturbance [41,42]. The interactions between these factors differ between basins, which will then reflect on the dynamic and river-specific eel biology/ecology, not following the specific expected patterns.

The observed variations in development across the eel's distribution range could be reflected, for example, in the otolith formation [43–45]. Otoliths are metabolically inert structures less vulnerable to post-depositional chemical and structural modification, as they grow through uptake from the water masses through which a fish passes during its lifetime [44,46,47], being good records of its life. Otoliths are widely used for age estimation, but their microstructure can also be used to investigate habitat characteristics and regional differences or similarities between fish populations [48,49]. They have proven to successfully identify the population structure of several fish species [50–52]. Endogenous and exogenous factors determine both otoliths' overall shape and growth patterns [53,54], functioning as good phenotypic markers, as their formation is influenced by feeding regime and differences in body condition and growth [55–57]. Otolith shape has been studied for the European eel in the Mediterranean and successfully discriminated eels that grow in different habitat types, proving to be a valuable tool to study this species ontogeny [58,59].

In this study, we aimed to investigate for the first time the population structure of the European eel in the Minho River using the otolith shape signature. The eels' biological characteristics were recorded to explore and discuss possible differences between eel groups. This river basin is characterized by different habitats, such as upstream tributary and downstream estuary ecosystems, whose dynamics and specific characteristics will determine the ecological assemblage and its influence on eel life history. We hypothesize that eels present distinct development strategies in these different environments, which will reflect in the otolith shape characteristics. The results of this study are a step forward in understanding the population structure of the European eels in their Southern Atlantic distribution area.

2. Materials and Methods

2.1. Population Characterization

2.1.1. Study Area

Sampling took place between November 2017 and October 2020 in fixed locations considered the most relevant areas for the occurrence of this species in the Portuguese border of the International Minho river basin (unpublished results). This river is located in the NW-Iberian Peninsula (SW-Europe). The river is 343 km long, and the last 76 km serve as the north-western Portuguese–Spanish border [60]. The limit of tidal influence is about 40 km inland [61], and the uppermost 30 km are tidal freshwater wetlands (TFW) [62,63].

The estuary has 23 km^2 and is partially mixed, but during periods of high floods it tends to evolve towards a salt wedge estuary [64].

The Minho River is the only area in Portugal where glass eel fishery still legally occurs, and until the ban imposed in 2011, the yellow and silver eel fisheries were also economically, socially, and culturally important. Minho has a separated eel management plan from the rest of the Portuguese rivers; therefore, the data provided here can help inform the success of its implementation. It also has a great ecological value, with several areas classified as a Natura 2000 site. In the European context, this ecosystem is of reference, as it integrates an extensive monitoring program that provides ICES with estimates for the status of the European eel in Portugal, and is of relevance as a particular example of cross-border eel management. This study was conducted under the Sudoang project, which aimed at developing common management and assessment tools, obtaining coordinated and long-term monitoring strategies, and reinforcing cooperation between stakeholders in the Southwest Europe (Sudoe) area. The Minho River was part of an eel sampling network that included 10 pilot basins in the Mediterranean and Atlantic which were representative of the different ecosystems of the Sudoe area.

In the Minho basin, the habitat available to migratory fish decreased from 17,000 km^2 (at the beginning of the 20th century) to 1000 km^2 [65]. The river continuum is interrupted at 76 km from the mouth of the river due to the presence of the Frieira dam, which impairs upstream fish migration. There are 30 main tributaries on the Portuguese side, with water line extensions varying from around 2000 to 46,000 m. Anthropogenic pressures, such as human or industrial pollution, water capture, water flow control, construction of small physical obstacles, or agricultural land, influence the water masses in these areas.

2.1.2. Sampling

Temperature, salinity (Aquaread aquaprobe AP2000), and river depth were determined during each sampling event. All individuals were measured ($\pm$0.1 cm), weighed ($\pm$0.1 g), and classified macroscopically as yellow or silver eels using three criteria: the color of the back and belly, presence of a well-marked lateral line, and eye diameter [66]. For eels above 30 cm, additional biometric measures were recorded for the Durif Silvering Index estimation, including the pectoral fin length and the vertical/horizontal eye diameter [67,68]. Fulton's condition factor (FU) was calculated as a fitness indicator based on the weight and length of each eel following Fulton [69]. As normality (Shapiro–Wilk test) and homogeneity of variances (Levene's test) assumptions were met, One-Way Analysis of Variance (One-Way ANOVA) was used to test for regional differences in FU (0.05 level of significance).

A total of 1428 eels were sampled in eight tributaries (Figure 1), which are freshwater shallow areas (mean depth varying between 0.1 and 0.7 m) with temperatures ranging from 10 °C to 20 °C throughout the year. Eels were caught during spring and autumn by electrofishing using two different gears, Hans Grassl model EL62 IIG and Electrocatch International model WFC911, in deeper or shallow tributaries, respectively. This is the primary sampling method used to monitor freshwater fish abundance in rivers and streams, as it samples populations quickly, is non-lethal, and can capture the entire size range of fish present at a given location [70].

A total of 2676 eels were sampled in the estuary (Figure 1) in four fixed stations with similar depths (two to four meters): one located near the river mouth (mesohaline to euhaline; salinity varies with tides), with water temperatures reaching a maximum of 20 °C, and three located in the tidal freshwater (fresh to oligohaline), with water temperatures reaching a maximum of 26 °C [63,71]. Eels were collected using fyke nets of 10 mm mesh size, 7 m total length, 0.7 m mouth diameter, and 3.5 m central wing. This type of gear is selective for large eel individuals (>25 cm).

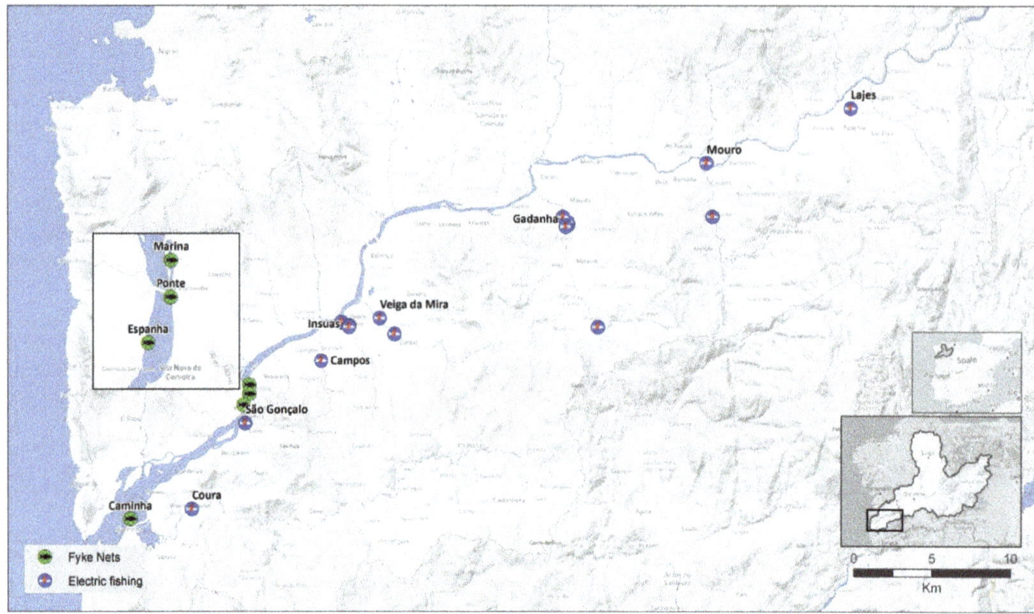

Figure 1. Map showing the sampling locations of the *Anguilla anguilla* individuals collected from November 2017 to October 2020 in the Minho River basin. Green dots denote the estuary sampling locations, and blue dots show the tributaries' sampling locations.

2.2. Otolith Analysis

2.2.1. Age Estimation

The sagittal otoliths from eels euthanized with an anesthetic overdosage were extracted. Macroscopic inspection of gonads was also undertaken for sex determination: eels with thin, regularly lobed organs were classified as males, and eels with broad, folded, curtain-like gonads were classified as females [28].

Eel age was assigned by counting the annual growth increments on the sagittal otoliths [72,73]. For this, all otoliths were immersed in a clearing agent (ethanol and glycerol, 1:1) to enhance their transparency during reading and examined using a stereomicroscope (Nikon SMZ800) against a dark background with polarized reflected light (Figure 2a). An age of 0 years was attributed to the glass eels to consider only the continental age. Otoliths of eels older than 5 years old (140 individuals) were treated to enhance rings' visualization. The process consisted of embedding the otolith in epoxy resin and mounting it on a glass slide for sagittal grinding. Each otolith was grounded manually using a decreasing range in the coarseness of silicon carbide wet–dry papers (600–4000 grit) (Figure 2b). The otolith plane was etched with a drop of 5% EDTA for 3 min and rinsed with running water, stained with a drop of 1% Toluidine blue until it dried, and rinsed again (Figure 2c). Otoliths were observed with a stereomicroscope and polarized reflected light, and age was estimated. Three independent readings were undertaken, and only otoliths with consistent concordance in age were used (a total of 420 otoliths, 210 from the tributaries and 210 from the estuary). As linearity assumptions were not met, the relationship between age and length was investigated using the Spearman rank–order correlation test.

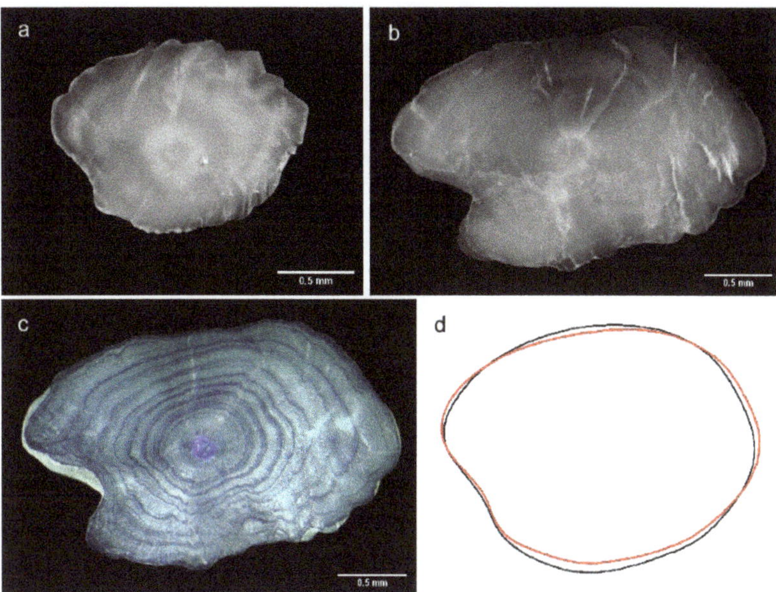

Figure 2. Right sagittal otoliths: in the first image (**a**), the otolith rings are visible with the clearing agent, followed by a non-readable ground otolith (**b**) and the resulting rings' visualization after coloration (**c**). The last image (**d**) is the averaged outline contour of the Elliptic Fourier analysis, in red for the estuary and black for the tributaries.

2.2.2. Shape Descriptors

Orthogonal two-dimensional digital images of the right sagittal otoliths were captured using a stereomicroscope coupled with a digital camera (Digital Sight DS-5M) at 2× magnification. Otoliths were all photographed in the same position with reflected light and dark backgrounds.

Binary otolith images were measured using the program ImageJ v. 1.50 (Bethesda, MD, USA) to assess the morphometric size parameters, otolith length (OL, mm), otolith width (OW, mm), otolith area (OA, mm^2), and otolith perimeter (OP, mm) [74]. With these variables, it was possible to determine the Shape Indices (SI), Form factor (FF, $(4\pi OA)/OP^2$), Roundness (RO, $(4OA)/(\pi OL^2)$), Circularity (CI, OP^2/OA), Ellipticity (EL, $(OL - OW)/(OL + OW)$), and Rectangularity (RE, $OA/(OL \times OW)$), which describe the otolith plane [75].

The Elliptic Fourier analysis was also used. This analysis fits a closed curve to an ordered set of data points, decomposes the contour into a sum of harmonically related ellipses, and indicates the contribution of each harmonic to the total otolith shape [76–78]. The program Shape v.1.3 (Tsukuba, Japan) was used to capture the otolith contour and to determine the number of Elliptic Fourier Descriptors (EFD) required to describe the otolith outline adequately. A level of 95% of the accumulated variance was used to select the minimum number of harmonics [79] (Figure 2d). The first 6 harmonics reached >95% of the cumulative power, excluding coefficients c6 and d6. Moreover, after normalization to the first harmonic (EFD invariant to otolith size), the first three coefficients (a1, b1, and c1) were constant and excluded from subsequent analyses [80]. Thus, 19 Fourier coefficients (d1 to b6) could adequately explain otolith shape.

2.2.3. Statistical Analysis

Shape Indices (SI) were checked for normality (Shapiro–Wilk test) and homogeneity of variances (Levene's test). These assumptions were met after $\log_{10}$ (e.g., CI) or square root (e.g., FF) transformations. Differences in fish length distributions can disrupt location-specific characteristics, as the otolith shape relates to the fish development [76]. The relationship between SI and otolith size was investigated to ensure that differences in fish size among samples did not confound habitat-specific differences in otolith SI. Because all SI correlated with OL, the variables were corrected using the formula Vadj = V − (β × covariate), where Vadj is the adjusted sample value, V is the original sample value, and β is the ANCOVA slope value (sampling location as factor and OL as a covariate) [81]. Statistical analyses were performed using R studio v.3.5.1 (Boston, MA, USA) with a 0.05 level of significance.The statistical analyses described in the following paragraphs were performed for two groups, for the undifferentiated eels only, and for all the eels combined, in order to remove potential sex-related effects, which may exist due to sexual growth dimorphism [43]. The results were similar for both cases, and thus differences were presented for all eels. Comparisons were made between different habitats (estuary vs. tributary) and between locations within each habitat (sampling locations within the estuary and the tributaries).

One-Way ANOVA was used to test for regional differences in individual shape variables, followed by a Tukey post-hoc test for significant differences. Multivariate Analyses of Variance (MANOVA) tested regional differences using all SI and EFD variables. For MANOVA, the approximate F-ratio statistic was reported for the most robust test of multivariate statistics (Pillai's trace), followed by multivariate pairwise comparisons using the Hotelling's T-squared test. Statistical analyses were performed using R studio v.3.5.1 (Boston, MA, USA) with a 0.05 level of significance.

A Linear Discriminant Analysis (LDA) was used to examine the reclassification accuracy of eels to their original location, verified through the percentage of correct reclassification accuracies of the discriminant functions using a jackknifed classification matrix. A Canonical Variates Analysis (CVA) (with Mahalanobis distances) was used to visualize differences and to identify the variables that contributed most to the discrimination. These results were presented in a two-dimensional biplot, with ellipses representing 95% confidence. Both analyses were conducted using the software PAST v.4.05 (Oslo, Norway).

3. Results

3.1. Population Characterization

The information regarding the 4104 eels caught between November 2017 and October 2020 is summarized in Table 1 and Figure 3. The size of the eels sampled in the tributaries (total = 1428 eels) varied between 6 cm and 39 cm, with an average (±SD) of 18 ± 7.32 cm, and 1.5% were macroscopically classified as silver eels. In the estuary, the size of the sampled eels (total = 2676 eels) varied between 6 cm and 88 cm (36 ± 10.64 cm), and the proportion of eels macroscopically classified as silver eels was 3.2%. Overall, size distribution was right-skewed, and there was a higher proportion of yellow-stage eels than silver eels.

The majority of eels from the tributaries Coura (a), S.Gonçalo (b), Lajes (g), and Mouro (h) presented sizes lower than 20 cm, while those at Campos (c) and Insuas (d) were smaller than 15 cm (Figure 3). The majority of eels collected at Gadanha (f) presented sizes varying between 10 cm and 25 cm, while the sizes of the eels collected at V. Mira (e) were more evenly distributed (Figure 3). Based on the Durif Silvering Index, 97% of eels were yellow-stage eels, and silver individuals were all males (SMII) with lengths between 29 and 39 cm.

The majority of eels collected in Caminha (i), Espanha (j), Ponte (k), and Marina (l) presented sizes varying between 25 and 35 cm, and the largest eels were collected at Ponte (k), the only location with the presence of eels larger than 80 cm (Figure 3). In the estuary, 84% were yellow-stage eels, 14.5% of which were females, and the rest were classified as silver eels (4.5% females and 11.5% males). The number of females and silver eels in the estuary slightly increased with increasing distance from the river mouth (Figure 3).

Table 1. Eels caught in the monitoring campaigns in the Minho River between November 2017 and October 2020. Data presented include: total sample size (N total), the number of silver eels classified macroscopically (N silver), minimum, maximum, and mean (±standard deviation, SD), total length (cm) of all eels, and the mean (cm ± SD) total length (cm) of the individuals characterized using the Durif Silvering Index as yellow resident (SI and SFII), female pre-migrant (SFIII), silver female (SFIV and SFV), and as silver male (SMII).

Sampling		N		Length (cm)			Durif Index-Length (cm, Mean ± SD)					
Habitat	**Location**	**Total**	**Silver**	**Min**	**Max**	**Mean ± SD**	**SI**	**SMII**	**SFII**	**SFIII**	**SFIV**	**SFIV**
Tributary	Coura	127	3	6	34	17 ± 7.37	16 ± 6.86	33 ± 1.04	-	-	-	-
	S.Gonçalo	50	0	6	31	19 ± 6.72	19 ± 6.72	-	-	-	-	-
	Campos	184	0	6	34	14 ± 6.24	14 ± 6.24	-	-	-	-	-
	Insuas	249	3	7	36	14 ± 8.07	14 ± 7.34	34 ± 1.46	-	-	-	-
	V.Mira	99	0	6	33	19 ± 7.70	19 ± 7.70	-	-	-	-	-
	Gadanha	298	12	9	39	23 ± 6.39	22 ± 6.09	33 ± 2.13	-	-	-	-
	Mouro	284	3	10	35	19 ± 5.37	19 ± 5.06	31 ± 2.50	-	-	-	-
	Lajes	137	1	11	35	19 ± 5.84	18 ± 5.40	32 ± 2.80	-	-	-	-
Estuary	Caminha	84	2	15	64	33 ± 8.80	29 ± 4.27	36 ± 3.85	-	63 ± 3.49	-	-
	Espanha	551	18	8	73	35 ± 11.21	30 ± 6.01	34 ± 3.09	49 ± 5.68	53 ± 11.76	56 ± 8.46	60 ± 7.00
	Ponte	967	50	6	88	36 ± 11.92	30 ± 5.05	36 ± 3.68	49 ± 4.38	54 ± 13.72	57 ± 8.63	61 ± 6.28
	Marina	1074	18	19	71	34 ± 8.95	30 ± 5.21	35 ± 3.74	46 ± 1.56	48 ± 3.83	55 ± 4.08	57 ± 6.00

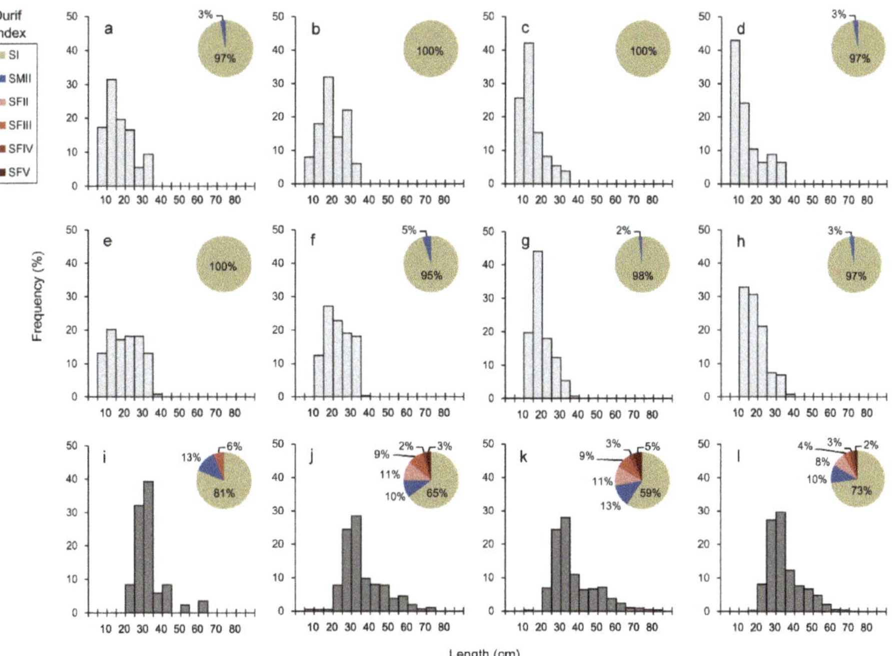

Figure 3. Catch length composition of the eels collected in the tributaries ((**a**) Coura, (**b**) S.Gonçalo, (**c**) Campos, (**d**) Insuas, (**e**) V.Mira, (**f**) Gadanha, (**g**) Mouro, and (**h**) Lajes) and in the estuaries ((**i**) Caminha, (**j**) Espanha, (**k**) Ponte, and (**l**) Marina). Pie charts present the proportion of individuals sampled according to the life stage classification based on the Durif Silvering Index (Durif et al. 2009): yellow resident (SI and SFII), female pre-migrant (SFIII), silver female (SFIV and SFV), and silver male (SMII).

3.2. Otolith Analysis

The information regarding the 420 eels used for the otolith analysis is summarized in Table 2 and Figure 4. The age of the eels varied between 0 and 10 years and sizes between 6 cm and 60 cm. Eels caught in the tributaries (Figure 4a) and estuary (Figure 4b) had similar length-at-age for the age classes 0 and 1. From the age of 2 years onwards, the eels collected in the estuary were 25% to 40% longer (mean total length) than those collected in the tributaries. In the tributaries, most of the individuals belonged to the 2, 3, and 4 age classes, whereas in the estuary most of the eels belonged to the 4, 5, and 6 age classes. The eels collected in the estuary presented higher length dispersion over an age class (Spearman correlation; rs = 0.6797, $p < 0.05$) than those collected in the tributaries (Spearman correlation; rs = 0.9642, $p < 0.05$; Figure 4). The condition of the eels collected in the tributaries only differed between Mouro and Gadanha (One-Way ANOVA, $F_{(6, 203)} = 4.55$, $p < 0.05$; Tukey Test, $p < 0.05$; Table 2). The condition of the eels collected in the estuary varied between stations (One-Way ANOVA, $F_{(3, 206)} = 4.79$, $p < 0.05$), with eels sampled in Caminha presenting the lowest mean FU value (Tukey Test, $p < 0.05$). Eels collected in the estuary presented higher FU values than those collected in the tributaries (One-Way ANOVA, $F_{(1, 418)} = 109.6$, $p < 0.05$).

Table 2. Number of eels collected in the Minho River according to habitat and sampling location (N), number of macroscopically sexed eels (males *), minimum and maximum age, weight (W), length (L), and mean (±standard error, SE), Fulton's condition factor (FU). Data also include age, W, L, and silvering stage of undifferentiated, male, and female eels according to the habitat.

Habitat	Location	N	Sexed	Age	W (g)	L (cm)	FU (Mean ± SE)
Tributary	Coura	30	2 *	1–10	1.53–69.4	10.6–34.9	0.146 ± 0.004
	S.Gonçalo	15	0	0–8	0.23–70.8	6.3–32.7	0.151 ± 0.004
	Insuas	30	2 *	0–10	0.4–75.6	7–36	0.136 ± 0.005
	V.Mira	30	0	0–8	1.25–55.3	9.5–31	0.144 ± 0.003
	Gadanha	15	12 *	2–10	3.02–75.6	14.9–34.6	0.154 ± 0.006
	Mouro	60	3 *	1–10	1.6–73.6	11.6–35.2	0.125 ± 0.003
	Lajes	15	1 *	1–9	3.55–90.5	14.4–35.9	0.150 ± 0.007
Estuary	Caminha	30	3/1 *	3–7	16.3–78.2	23.0–36.2	0.156 ± 0.004
	Espanha	45	12	0–9	0.67–276	8.4–53	0.160 ± 0.004
	Ponte	45	25	0–9	0.57–367	6.7–54.6	0.173 ± 0.003
	Marina	90	40/3 *	2–9	15.4–483	20.5–60.5	0.168 ± 0.004
Habitat	**Sex**			**Age**	**W(g)**	**L(cm)**	**Silvering**
Tributary	Undifferentiated			0–10	0.23–69.4	6.3–34.9	All Yellow
	Male			6–10	32.1–90.5	29–36	All Silver
Estuary	Undifferentiated			0–7	0.57–98	6.7–36.5	All Yellow
	Female			4–9	57.6–483	31.1–60	2 Silver
	Male			4–7	46.9–87	30.8–37.6	All Silver

Shape Indicies (SI) and Elliptic Fourier Descriptors (EFD) values and variable individual results are in Tables S1 and S2, respectively. In the tributaries, otolith shape differed between a few locations in the tributaries: Coura from Insuas and Lajes, Mouro from Insuas and Gadanha, and S.Gonçalo from Lajes (MANOVA, Pillai´s Trace, $F_{1.6049} = 0.999$, $p < 0.05$). The LDA, through the Jackknife reclassification matrix, showed that Lajes was the site with the best reclassification value (40%), but the overall reclassification success was low (26%; Table 3. The lack of specific groups can be visualized in the CVA plot (Figure 5a). Vectors' overlay shows the RO as the most prominent variable for group characterization. The otolith shape of eels collected in the brackish sampling location (Caminha) was different from those collected in the TFW (MANOVA, Pillai´s Trace, $F_{1.799} = 0.589$, $p < 0.05$). The LDA, through the Jackknife reclassification matrix, showed good reclassification success for Caminha (57%), but an overall poor reclassification success for the eels collected in the estuary (34%; Table 3). The discrimination of Caminha can be visualized in the CVA plot, where vector overlays show the prevalence of the otolith outline EFD variables d1, b2, and b3 (Figure 5b). Otolith shape proved to be successful in differentiating eels from tributary vs. estuary habitats (MANOVA, Pillai´s Trace, $F_{83.883} = 0.842$, $p < 0.05$). The LDA, through the Jackknife reclassification matrix, showed an overall reclassification success of 98%, with complete reclassification success for the tributary habitat (100%; Table 3). The discrimination of both habitats can be visualized in the CVA plot (Figure 5c). Additionally, the CVA plot of the undifferentiated eels shows similar results (Figure 5d). Vector overlays show that the SI variables RO and CI describe the otoliths of the individuals collected in the tributaries, whereas EL and RE describe the otoliths of the individuals collected in the estuary. Otolith outline EFD variables d1, b2, c2, b4, d3, d4, and b3 also contribute to the discrimination between groups.

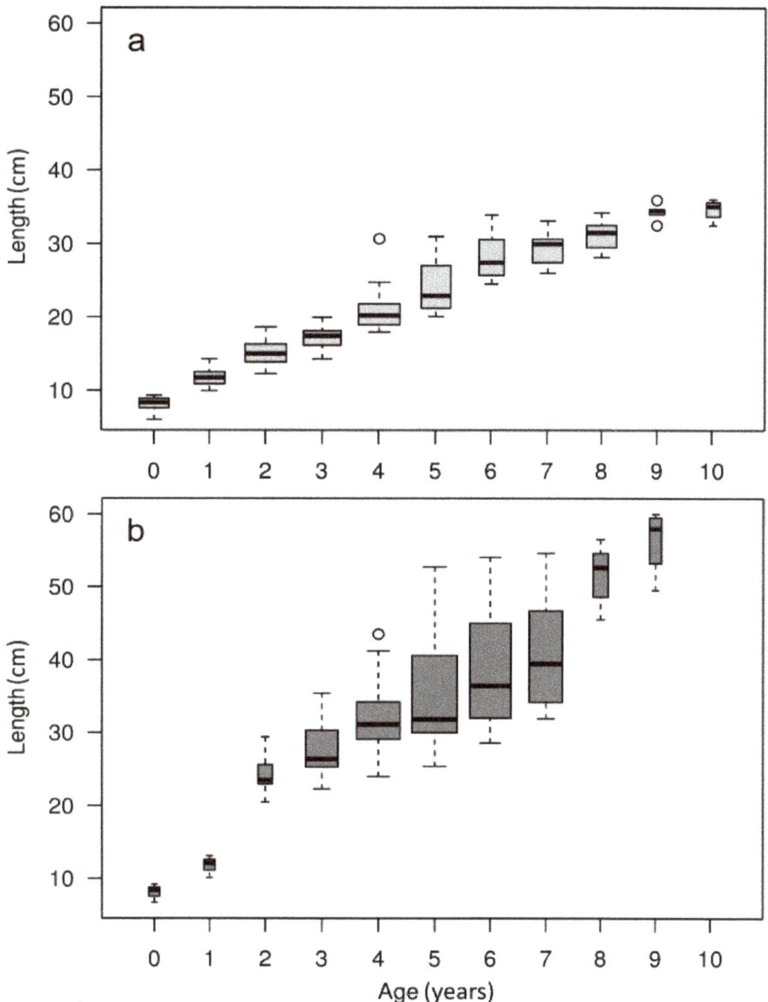

Figure 4. Boxplot of age-at-length data for eels collected in the (**a**) tributaries and (**b**) estuary. Thicker boxes indicate a higher number of eels for the corresponding age class.

Table 3. Jackknife reclassification matrix of the otolith shape signatures of the eels collected in the Minho River according to (a) the tributary sampling locations, (b) the estuary sampling locations, and (c) habitat.

(a) Tributary

Original Locations	Coura	S.Gonçalo	Insuas	V.Mira	Gadanha	Mouro	Lajes	% Correct
Coura	8	4	5	6	4	1	2	**27**
S.Gonçalo	2	3	1	3	2	3	1	**20**
Insuas	1	5	9	1	5	4	5	**30**
V.Mira	5	7	1	5	7	1	4	**17**
Gadanha	5	3	2	5	9	3	3	**30**
Mouro	7	5	6	11	11	15	5	**17**
Lajes	2	0	1	3	1	2	6	**40**
Total	30	27	25	34	39	29	26	**26**

(b) Estuary

Original Locations	Caminha	Ponte	Espanha	Marina	% Correct
Caminha	17	5	6	2	**57**
Ponte	6	12	12	15	**27**
Espanha	9	11	15	10	**33**
Marina	13	28	21	28	**31**
Total	45	56	54	55	**34**

(c) Habitat

Original Locations	Tributary	Estuary	% Correct
Tributary	210	0	**100**
Estuary	8	202	**96**
Total	218	202	**98**

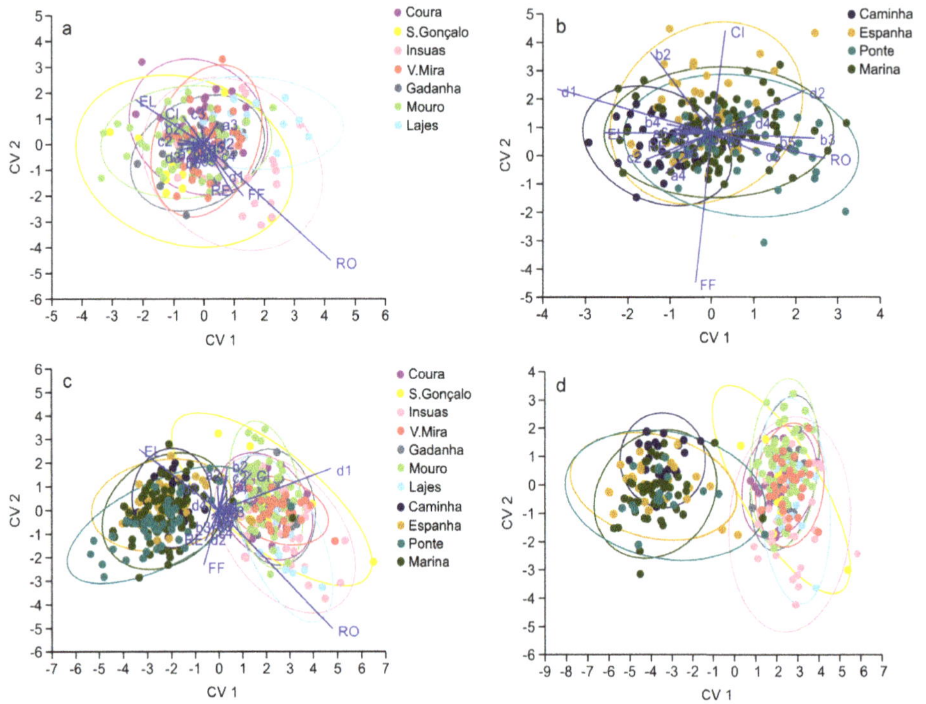

Figure 5. Canonical Variates Analysis (CVA) plot of the otolith shape signature for eels collected in the (**a**) tributaries, (**b**) estuary, (**c**) tributaries and estuary combined, and (**d**) undifferentiated individuals only from the tributaries and estuary.

4. Discussion

This study was the first attempt to characterize the European eel population structure in the Minho River basin. Otolith shape results revealed the existence of two distinct groups of eels associated with different ecosystems: tributaries vs. estuaries. The otoliths of eels from the tributaries were round and circular, whereas the otoliths of eels from the estuary were elliptical and rectangular. Moreover, the otoliths' outline of eels from the brackish location slightly differed from those of the tidal freshwater wetlands within the estuary. Eels collected in the tributary and estuary habitats were mostly yellow-stage eels with a similar age range, but eels from the tributaries showed lower length-at-age and lower body condition than those collected in the estuary. The sex ratio was skewed towards males in the tributaries and females in the estuary. Additionally, in the estuary, the brackish location presented a higher percentage of males and eels with a lower-body condition than those from TFW locations.

Previous studies have shown that variability in fish development can be mirrored in the shape of otoliths. Several abiotic characteristics of the environment, such as temperature fluctuations, salinity, depth, and food availability, are responsible for that variability [44,82–84]. Moreover, otolith shape indices have proven to be useful in discriminating eels growing in different habitats at the regional level [58,59]. For instance, Capoccioni et al. [58] found similar results to those obtained during this study, including elongated otoliths with a trimmed outline, evidence of growth effects in estuarine habitats, and maintenance of the initial glass eel circular shape throughout the eel life when growing in tributaries. The different strategies in development for the eel in a variety of river basins have been shown to reflect the complexity of the riverine habitats, which allows the understanding of habitat suitability, recruitment success, and eel productivity [85,86].

There are no studies on the effects of environmental fluctuations and differences in habitat suitability on eel development in the Minho River. Despite that, these relations were found for other basins containing similar habitat types. The smaller length-at-age and lower body condition of the eels collected in the tributaries, compared to those collected in the estuary, may indicate that tributaries are less suitable habitats for eels to grow. Previous studies have shown that growth tends to be higher in brackish environments and in freshwater marshes closer to the sea than in freshwater upstream habitats [27,87,88] due to increased productivity and food quality [29]. Although there are no estimates for the productivity in the tributaries of the Minho River, previous studies indicate that the food webs in the tributaries are mainly supported by aquatic- and terrestrial-derived detritus [89]. The detrital pathway is usually considered less efficient when compared to the phytoplankton pathway [90], which in this case, is more associated with the estuary [63]. Additionally, extreme hydrological regimes in the tributaries, characterized by droughts during the summer and torrential flooding during the winter, can impact fish responses, promoting stress conditions [91]. Moreover, eels in freshwater environments tend to present lower levels of fat accumulation and higher prevalences of the parasite *Anguillicola crassus*, which, in tandem with decreases in habitat quality and complexity due to human interventions, may impose further restrictions on eel growth and development with consequences for their performance, health, and survival [92–94]. In the Minho River, the prevalence of the parasite *Anguillicola crassus* has been increasing through the years [95], which could impair eel development.

The proportion of males was higher in the tributaries than in the estuary; this may indicate further habitat restrictions. One possible explanation could be related to the existence of the Frieira dam 76 km from the river mouth, which is the first main obstacle to the upstream migration of eels in the Minho River. As a result, high numbers of yellow eels concentrate in areas below the dam, a behavior already observed in other ecosystems [92,96,97]. This promotes competition for food and shelter, parasite dissemination, and long-term effects such as spawning biomass reduction due to sex ratio changes [31,97,98], specifically skewness towards males [28,30,37]. It was estimated that the habitat reduction for migratory species due to dams in this ecosystem is around 90%, and because eels cannot migrate to and from upstream areas, they are caught in the dam and distributed mainly in the downstream tributaries. The habitat reduction, both in extension and quality, and the environmental variability in tributaries may be responsible for the poorer development of eels compared to the conditions in the estuary since those in the tributaries presented lower body conditions, smaller length at a certain age, and a skewness towards males. Further studies are necessary to relate the type of habitat and environmental variability (e.g., productivity, pollution, and extreme events) with eel condition (e.g., stress, fat content, and prevalence of parasites).

Eels collected in brackish estuarine habitat (i.e., Caminha) presented a lower body condition when compared to the eels collected in the upstream tidal freshwater wetlands. This result is unexpected, as productivity tends to be higher in brackish waters. However, this could mean that eels invest more in somatic growth than in energy storage. Previous records of lighter individuals in coastal areas compared to those in inland waters were reported [28]. Although the movement patterns of eels in the Minho River are unknown, we hypothesize that this behavior could be linked to the conditions eels experienced during early life [99]. The settlement of eels in brackish habitats can be condition-dependent, with low-body-condition glass eels preferring high-salinity habitats [22]. Early migrants usually present better body condition and migrate further upstream than the late migrants, which are usually in the worst conditions and settle in the lower estuary [100–102]. Similar compensatory-growth behaviors were observed in other fish and invertebrate species, where they were able to increase the growth rate but maintained a low body condition when transiting to habitats with high food availability [103,104]. Another explanation, which is not mutually exclusive, could be related to movements throughout the yellow eel stage, from freshwater to the brackish estuary, as observed for this species in other ecosystems [15]. Still, additional information combining individual growth and habitat shift events is necessary to understand the influence of different environments on eels' growth strategies and condition throughout ontogeny.

The areas where eels were larger and had the highest body condition values were located in the tidal freshwater estuary. These are transition areas between the brackish estuary and the non-tidal freshwater streams, and tend to be less susceptible to variations in salinity or temperature and depth than the brackish estuary or the tributaries, respectively [105]. Positive effects on eel development were associated with high temperatures and depths [27,106]. In the Minho basin, the TFW area reaches maximum water temperatures of 26 °C during the summer, salinity is below 0.5 throughout most of the year, and depth is relatively constant, varying between 2 and 4 m across this area. The salinity in the brackish estuary varies daily with tides and along the year with the river flow, while in the tributaries, depths can be as low as 0.1–0.7 m during spring and autumn, and water temperature in both areas usually does not exceeds 20 °C. Thus, the relatively stable abiotic conditions and higher temperature and depth values in the TFW may promote a steady development environment, allowing eels to grow to larger sizes. However, several unknown variables could be acting together, limiting further conclusions about the influence of the environment on development [27]. Eels have sex-specific life-history strategies, with females showing long maturation periods to produce eggs, thus having higher energetic demands than males, which leads to a sexual dimorphism based on differences in length at maturity [28,68]. Males migrate at a lower length, around 35–45 cm, minimizing the duration of their yellow stage, while females migrate at sizes of 40–130 cm, optimizing their size to reach a higher fecundity [23,28,68]. This trade-off strategy in sex differentiation depends indirectly on the effects of environmental conditions on growth [107,108]. Thus, the higher proportion of females in the estuarine TFW may indicate that this habitat is more favorable for eel development.

5. Conclusions

Our study reveals the existence of at least two distinct groups of eels in the Minho River, associated with different types of habitats: tributaries vs. estuaries. Otolith shape analysis completely discriminated these two groups; the otoliths were more round and circular in the tributaries and more elliptical and rectangular in the estuary. Upstream tributary environments likely offer poorer development conditions for eels than estuarine areas, as these eels were smaller and in worse condition than those in estuaries. Moreover, eels collected from tributaries skewed towards males. The estuary was dominated by larger eels than the tributaries, skewed towards females, and presented eels with higher body condition, especially in the tidal freshwater wetlands. This information is of outmost importance to begin understanding the development strategies of the eel in the Minho River and the contribution of this river to the spawning population. Continued European eel research in this species' wide distribution area is urgent in order to improve stock management.

Supplementary Materials: The following supporting information can be downloaded at: https://www.mdpi.com/article/10.3390/fishes7030135/s1, Table S1: Shape Indices ANOVA results; Table S2: Elliptical Fourier Descriptors ANOVA results.

Author Contributions: Conceptualization, A.M. and C.A.; methodology, A.M. and C.A.; software, A.M.; validation, A.M. and C.A; formal analysis, A.M. and ED.; investigation, A.M., R.L., and C.A.; resources, C.A.; data curation, A.M.; writing—original draft preparation, A.M.; writing—review and editing, A.M., E.D., and C.A.; visualization, A.M.; supervision, C.A.; project administration, C.A.; funding acquisition, C.A. All authors have read and agreed to the published version of the manuscript.

Funding: This study was supported by European funds, through the SUDOANG project (SOE2/P5/E0617) within the Interreg Sudoe Program and Interreg MIGRAMIÑO-MINHO (0016_MIGRA_MINHO_1_E) project, and by national funds through the FCT-Foundation for Science and Technology within the scope of UIDB/04423/2020 and UIDP/04423/2020. Additionally, Aquamuseu do Rio Minho provided most-needed resource support.

Institutional Review Board Statement: The study was conducted according to the guidelines of the Declaration of Helsinki and approved by the CIIMAR's Ethical Committee along with the CIIMAR´S Animal Welfare Body (ORBEA) in compliance with the European Directive 2010/63/EU, on the protection of animals used for scientific purposes, and its transposition to the Portuguese law. Animals were sacrificed according to the recommendations provided by the Bioterium of Aquatic Organisms (BOGA) at CIIMAR, which is certified by "Direção Geral de Alimentação e Veterinária (DGAV)" issued under Article 21°, of Decree-Law N.° 113/2013 of 7th August. The last author (C.A) has category B certification by Direção Geral de Alimentação e Veterinária (DGAV)", which is in compliance with the Federation of European Laboratory Animal Science (FELASA) corresponding to functions (a), (c), (e), and (d) defined in national and EU Directive 2010/63/EU. Electric fishing procedures were approved by the Portuguese Nature Conservation Institute and Forestry (licenses N.° 373/2018, N.° 263/2019 and N.° 161/2020, CAPT CREDENCIAL PESCA N.° 36/2018, N.° 7/2019, and N.° 21/2020, respectively). Fyke nets sampling procedures were approved by the local Captaincy of the Port of Caminha. Otoliths used in this study came from eels sacrificed for a variety of undergoing studies, ensuring that the lethal procedure would result in gathering all the available biological information possible. The number of otoliths chosen was believed to be necessary to fairly represent the riverine systems in this study.

Data Availability Statement: Data from this study are available from the corresponding author upon request (AM.: anacatarinamoura@hotmail.com).

Acknowledgments: The authors would like to thank Eduardo Martins, Patrício Bouça, Mafalda Fernandes, and Ana Lages from Aquamuseu do Rio Minho for their assistance in the field.

Conflicts of Interest: The authors declare no conflict of interest.

References

1. ICES. *Report of the Joint EIFAAC/ICES/GFCM Working Group on Eel (WGEEL)*; ICES Document CM 2015/ACOM: 18; ICES: Antalya, Turkey, 2015.
2. Moriarty, C.; Dekker, W. Management of European eel fisheries. *Irish Fish. Bull.* **1997**, *15*, 108.
3. Dekker, W. Status of the European eel stock and fisheries. In *Eel Biology*; Springer: Tokyo, Japan, 2003.
4. Bornarel, V.; Lambert, P.; Briand, C.; Antunes, C.; Belpaire, C.; Ciccotti, E.; Diaz, E.; Diserud, O.; Doherty, D.; Domingos, I.; et al. Modelling the recruitment of European eel (*Anguilla anguilla*) throughout its European range. *ICES J. Mar. Sci.* **2018**, *75*, 541–552. [CrossRef]
5. Dekker, W. The history of commercial fisheries for European eel commenced only a century ago. *Fish. Manag. Ecol.* **2019**, *26*, 6–19. [CrossRef]
6. Jacoby, D.; Gollock, M. *The IUCN Red List of Threatened Species*; IUCN: Gland, Switzerland, 2014.
7. Jacoby, D.M.; Casselman, J.M.; Crook, V.; DeLucia, M.B.; Ahn, H.; Kaifu, K.; Kurwie, T.; Sasal, P.; Silfvergrip, A.M.; Smith, K.G.; et al. Synergistic patterns of threat and the challenges facing global anguillid eel conservation. *Glob. Ecol. Conserv.* **2015**, *4*, 321–333. [CrossRef]
8. Miller, M.J.; Feunteun, E.; Tsukamoto, K. Did a "perfect storm" of oceanic changes and continental anthropogenic impacts cause northern hemisphere anguillid recruitment reductions? *ICES J. Mar. Sci.* **2016**, *73*, 43–56. [CrossRef]
9. Dekker, W.; Beaulaton, L. Climbing back up what slippery slope? Dynamics of the European eel stock and its management in historical perspective. *ICES J. Mar. Sci.* **2016**, *73*, 5–13. [CrossRef]
10. ICES. *Joint EIFAAC/ICES/GFCM Working Group on Eels (WGEEL)*; ICES: Antalya, Turkey, 2019.
11. The European Commission. Commission Decision 2010/93/EU of 18 December 2009 adopting a multiannual Community program for the collection, management and use of data in the fisheries sector for the period 2011–2013 (notified under document C (2009) 10121). *Off. J. Eur. Union L* **2009**, *41*, 8–71.
12. ICES. *Report of the Workshop on Designing an Eel Data Call (WKEELDATA)*; ICES CM 2017/SGIEOM: 30; ICES: Rennes, France, 2017.
13. Tsukamoto, K.; Nakai, I. Do all freshwater eels migrate? *Nature* **1998**, *396*, 635–636. [CrossRef]
14. Limburg, K.E.; Wickstrom, H.; Svedang, H.; Elfman, M.; Kristiansson, P. Do stocked freshwater eels migrate? Evidence from the Baltic suggests "yes". In *American Fisheries Society Symposium*; American Fisheries Society: New York, NY, USA, 2003; pp. 275–284.
15. Daverat, F.; Limburg, K.E.; Thibault, I.; Shiao, J.C.; Dodson, J.J.; Caron, F.; Tzeng, W.N.; Iizuka, Y.; Wickström, H. Phenotypic plasticity of habitat use by three temperate eel species, *Anguilla anguilla*, A japonica and A. rostrata. *Mar. Ecol. Prog. Ser.* **2006**, *308*, 231–241. [CrossRef]
16. Als, T.D.; Hansen, M.M.; Maes, G.E.; Castonguay, M.; Riemann, L.; Aarestrup, K.I.M.; Munk, P.; Sparholt, H.; Hanel, R.; Bernatchez, L. All roads lead to home: Panmixia of European eel in the Sargasso Sea. *Mol. Ecol.* **2011**, *20*, 1333–1346. [CrossRef]
17. Schmidt, J. On the distribution of the freshwater eels (Anguilla) throughout the world. I. Atlantic Ocean and adjacent regions. *Medd. Fra Komm. Havunders. Seri Fisk.* **1909**, *3*, 1–45.
18. Tesch, F.W. *Biology and Management of Anguillid Eels*; CRC Press: Boco Raton, FL, USA, 1977.
19. Dekker, W. On the distribution of the European eel (*Anguilla anguilla*) and its fisheries. *Can. J. Fish. Aquat. Sci.* **2003**, *60*, 787–799. [CrossRef]
20. Antunes, C.; Tesch, F.W. A critical consideration of the metamorphosis zone when identifying daily rings in otoliths of European eel, *Anguilla anguilla* (L.). *Ecol. Freshw. Fish* **1997**, *6*, 102–107. [CrossRef]
21. Edeline, E.; Dufour, S.; Elie, P. Role of glass eel salinity preference in the control of habitat selection and growth plasticity in *Anguilla anguilla*. *Mar. Ecol. Prog. Ser.* **2005**, *304*, 191–199. [CrossRef]
22. Edeline, E.; Lambert, P.; Rigaud, C.; Elie, P. Effects of body condition and water temperature on *Anguilla anguilla* glass eel migratory behavior. *J. Exp. Mar. Biol. Ecol.* **2006**, *331*, 217–225. [CrossRef]
23. Vøllestad, L.A. Geographic variation in age and length at metamorphosis of maturing European eel: Environmental effects and phenotypic plasticity. *J. Anim. Ecol.* **1992**, *61*, 41–48. [CrossRef]
24. Bertin, L. *Eels: A Biological Study*; Cleaver-Hume Press: London, UK, 1956.
25. Van Den Thillart, G.E.E.J.; Van Ginneken, V.; Körner, F.; Heijmans, R.; Van der Linden, R.; Gluvers, A. Endurance swimming of European eel. *J. Fish Biol.* **2004**, *65*, 312–318. [CrossRef]
26. Helfman, G.S.; Facey, D.E.; Hales, L.S., Jr.; Bozeman, E.L., Jr. Reproductive ecology of the American eel. In *American Fisheries Society Symposium*; American Fisheries Society: New York, NY, USA, 1987; Volume 1, pp. 42–56.
27. Daverat, F.; Beaulaton, L.; Poole, R.; Lambert, P.; Wickström, H.; Andersson, J.; Aprahamian, M.; Hizem, B.; Elie, P.; Yalçın-Özdilek, S.; et al. One century of eel growth: Changes and implications. *Ecol. Freshw. Fish* **2012**, *21*, 325–336. [CrossRef]
28. Tesch, F.W. *The Eel*, 5th ed.; Blackwell Publishing: Oxford, UK, 2003.
29. Jessop, B.M.; Shiao, J.C.; Iizuka, Y.; Tzeng, W.N. Variation in the annual growth, by sex and migration history, of silver American eels Anguilla rostrata. *Mar. Ecol. Prog. Ser.* **2004**, *272*, 231–244. [CrossRef]
30. Krueger, W.H.; Oliveira, K. Evidence for environmental sex determination in the American eel, Anguilla rostrata. *Environ. Biol. Fishes* **1999**, *55*, 381–389. [CrossRef]
31. Bevacqua, D.; Melia, P.; De Leo, G.A.; Gatto, M. Intra-specific scaling of natural mortality in fish: The paradigmatic case of the European eel. *Oecologia* **2011**, *165*, 333–339. [CrossRef]

32. Boulenger, C.; Acou, A.; Gimenez, O.; Charrier, F.; Tremblay, J.; Feunteun, E. Factors determining survival of European eels in two unexploited sub-populations. *Freshw. Biol.* **2016**, *61*, 947–962. [CrossRef]
33. Leo, G.D.; Gatto, M. A size and age-structured model of the European eel (*Anguilla anguilla* L.). *Can. J. Fish. Aquat. Sci.* **1995**, *52*, 1351–1367. [CrossRef]
34. Poole, W.R.; Reynolds, J.D. Variability in growth rate in European eel *Anguilla anguilla* (L.) in a western Irish catchment. In *Biology and Environment: Proceedings of the Royal Irish Academy*; Royal Irish Academy: Dublin, Ireland, 1998; pp. 141–145.
35. Parsons, J.; Vickers, K.U.; Warden, Y. Relationship between elver recruitment and changes in the sex ratio of silver eels *Anguilla anguilla* L. migrating from Lough Neagh, Northern Ireland. *J. Fish Biol.* **1977**, *10*, 211–229. [CrossRef]
36. Walsh, C.T.; Pease, B.C.; Booth, D.J. Variation in the sex ratio, size and age of long finned eels within and among coastal catchments of south-eastern Australia. *J. Fish Biol.* **2004**, *64*, 1297–1312. [CrossRef]
37. Kettle, A.J.; Asbjørn Vøllestad, L.; Wibig, J. Where once the eel and the elephant were together: Decline of the European eel because of changing hydrology in southwest Europe and northwest Africa? *Fish Fish.* **2011**, *12*, 380–411. [CrossRef]
38. Vøllestad, L.A.; Jonsson, B. A 13-year study of the population dynamics and growth of the European eel *Anguilla anguilla* in a Norwegian river: Evidence for density-dependent mortality, and development of a model for predicting yield. *J. Anim. Ecol.* **1988**, *57*, 983–997. [CrossRef]
39. Sakai, M.; Iwabuchi, K. Unique habitat and macroinvertebrate assemblage structures in spring-fed streams: A comparison among lowland tributaries and mainstreams in northern Japan. *bioRxiv* **2020**, *22*, 193–202. [CrossRef]
40. Post, D.M.; Pace, M.L.; Hairston, N.G. Ecosystem size determines food-chain length in lakes. *Nature* **2000**, *405*, 1047–1049. [CrossRef] [PubMed]
41. Maceda-Veiga, A.; Mac Nally, R.; de Sostoa, A. Environmental correlates of food-chain length, mean trophic level and trophic level variance in invaded riverine fish assemblages. *Sci. Total Environ.* **2018**, *644*, 420–429. [CrossRef]
42. Pimm, S.L.; Lawton, J.H. Number of trophic levels in ecological communities. *Nature* **1977**, *268*, 329–331. [CrossRef]
43. Panfili, J.; Ximénès, M.C.; Crivelli, A.J. Sources of variation in growth of the European eel (*Anguilla anguilla*) estimated from otoliths. *Can. J. Fish. Aquat. Sci.* **1994**, *51*, 506–515. [CrossRef]
44. Campana, S.E.; Thorrold, S.R. Otoliths, increments, and elements: Keys to a comprehensive understanding of fish populations? *Can. J. Fish. Aquat. Sci.* **2001**, *58*, 30–38. [CrossRef]
45. Lin, Y.J.; Ložys, L.; Shiao, J.C.; Iizuka, Y.; Tzeng, W.N. Growth differences between naturally recruited and stocked European eel *Anguilla anguilla* from different habitats in Lithuania. *J. Fish Biol.* **2007**, *71*, 1773–1787. [CrossRef]
46. Thorrold, S.R.; Campana, S.E.; Jones, C.M.; Swart, P.K. Factors determining δ13C and δ18O fractionation in aragonitic otoliths of marine fish. *Geochim. Cosmochim. Acta* **1997**, *61*, 2909–2919. [CrossRef]
47. Elsdon, T.S.; Gillanders, B.M. Reconstructing migratory patterns of fish based on environmental influences on otolith chemistry. *Rev. Fish Biol. Fish.* **2003**, *13*, 217–235. [CrossRef]
48. Begg, G.A.; Campana, S.E.; Fowler, A.J.; Suthers, I.M. Otolith research and application: Current directions in innovation and implementation. *Mar. Freshw. Res.* **2005**, *56*, 477–483. [CrossRef]
49. Campana, S.E. Otolith science entering the 21st century. *Mar. Freshw. Res.* **2005**, *56*, 485–495. [CrossRef]
50. Moreira, C.; Froufe, E.; Vaz-Pires, P.; Correia, A.T. Otolith shape analysis as a tool to infer the population structure of the blue jack mackerel, Trachurus picturatus, in the NE Atlantic. *Fish. Res.* **2019**, *209*, 40–48. [CrossRef]
51. Moura, A.; Muniz, A.A.; Mullis, E.; Wilson, J.M.; Vieira, R.P.; Almeida, A.A.; Pinto, E.; Brummer, G.J.A.; Gaever, P.V.; Gonçalves, J.M.S.; et al. Population structure and dynamics of the Atlantic mackerel (Scomber scombrus) in the North Atlantic inferred from otolith chemical and shape signatures. *Fish. Res.* **2020**, *230*, 105621. [CrossRef]
52. Muniz, A.A.; Moura, A.; Triay-Portella, R.; Moreira, C.; Santos, P.T.; Correia, A.T. Population structure of the chub mackerel (Scomber colias) in the North-east Atlantic inferred from otolith shape and body morphometrics. *Mar. Freshw. Res.* **2020**, *72*, 341–352. [CrossRef]
53. Jones, C.M. Development and application of the otolith increment technique. In *Otolith Microstructure Examination and Analysis*; Canadian Special Publication of Fisheries and Aquatic Sciences 117; Publishing Supply and Services Canada: Ottawa, ON, Canada, 1992; Volume 117, pp. 1–11.
54. Lombarte, A.; Torres, G.J.; Morales-Nin, B. Specific Merluccius otolith growth patterns related to phylogenetics and environmental factors. Marine Biological Association of the United Kingdom. *J. Mar. Biol. Assoc.* **2003**, *83*, 277. [CrossRef]
55. Bacha, M.; Jemaa, S.; Hamitouche, A.; Rabhi, K.; Amara, R. Population structure of the European anchovy, Engraulis encrasicolus, in the SW Mediterranean Sea, and the Atlantic Ocean: Evidence from otolith shape analysis. *ICES J. Mar. Sci.* **2014**, *71*, 2429–2435. [CrossRef]
56. Vieira, A.R.; Neves, A.; Sequeira, V.; Paiva, R.B.; Gordo, L.S. Otolith shape analysis as a tool for stock discrimination of forkbeard (Phycis phycis) in the Northeast Atlantic. *Hydrobiologia* **2014**, *728*, 103–110. [CrossRef]
57. Jemaa, S.; Bacha, M.; Khalaf, G.; Dessailly, D.; Rabhi, K.; Amara, R. What can otolith shape analysis tell us about population structure of the European sardine, Sardina pilchardus, from Atlantic and Mediterranean waters? *J. Sea Res.* **2015**, *96*, 11–17. [CrossRef]
58. Capoccioni, F.; Costa, C.; Aguzzi, J.; Menesatti, P.; Lombarte, A.; Ciccotti, E. Ontogenetic and environmental effects on otolith shape variability in three Mediterranean European eel (*Anguilla anguilla*, L.) local stocks. *J. Exp. Mar. Biol. Ecol.* **2011**, *397*, 1–7. [CrossRef]

59. Milošević, D.; Bigović, M.; Mrdak, D.; Milašević, I.; Piria, M. Otolith morphology and microchemistry fingerprints of European eel, *Anguilla anguilla* Linnaeus, 1758) stocks from the Adriatic Basin in Croatia and Montenegro. *Sci. Total Environ.* **2021**, *786*, 147478. [CrossRef]
60. Antunes, C.; Araújo, M.J.; Braga, C.; Roleira, A.; Carvalho, R.; Mota, M. Valorização dos recursos naturais da bacia hidrográfica do rio Minho. In *Final Report from the Project Natura Miño-Minho*; Centro interdisciplinar de Investigação Marinha e Ambiental; Universidade do Porto: Porto, Portugal, 2011.
61. Vilas, F.; Somoza, L. El estuario del rio Miño: Observaciones previas de su dinâmica. *Thalassas* **1984**, *2*, 87–92.
62. Sousa, R.; Dias, S.C.; Guilhermino, L.; Antunes, C. Minho River tidal freshwater wetlands: Threats to faunal biodiversity. *Aquat. Biol.* **2008**, *3*, 237–250. [CrossRef]
63. Dias, E.; Morais, P.; Cotter, A.M.; Antunes, C.; Hoffman, J.C. Estuarine consumers utilize marine, estuarine and terrestrial organic matter and provide connectivity among these food webs. *Mar. Ecol. Prog. Ser.* **2016**, *554*, 21–34. [CrossRef]
64. Sousa, R.; Guilhermino, L.; Antunes, C. Molluscan fauna in the freshwater tidal area of the River Minho estuary, NW of Iberian Peninsula. *Ann. Limnol.-Int. J. Limnol.* **2005**, *41*, 141–147. [CrossRef]
65. Mota, M.; Rochard, E.; Antunes, C. Status of the diadromous fish of the Iberian Peninsula: Past, present and trends. *Limnetica* **2016**, *35*, 1–18.
66. Feunteun, E.; Acou, A.; Laffaille, P.; Legault, A. European eel (*Anguilla anguilla*): Prediction of spawner escapement from continental population parameters. *Can. J. Fish. Aquat. Sci.* **2000**, *57*, 1627–1635. [CrossRef]
67. Durif, C.; Dufour, S.; Elie, P. The silvering process of *Anguilla anguilla*: A new classification from the yellow resident to the silver migrating stage. *J. Fish Biol.* **2005**, *66*, 1025–1043. [CrossRef]
68. Durif, C.; Guibert, A.; Elie, P. Morphological discrimination of the silvering stages of the European eel. In *American Fisheries Society Symposium*; American Fisheries Society: New York, NY, USA, 2009; Volume 58, pp. 103–111.
69. Fulton, T.W. The rate of growth of fishes. In *Twenty-Second Annual Report*; Fisheries Board of Scotland: Edinburgh, UK, 2009; pp. 141–241.
70. Howard, S.W.; Crow, S.K.; Jellyman, P.G. Site-Specific Selectivity of Electric-Fishing Gear. New Zealand Fisheries Assessment Report. 2019. Available online: https://www.google.com.hk/url?sa=t&rct=j&q=&esrc=s&source=web&cd=&ved=2ahUKEwi5zODSr5r4AhXzg2MGHff_B2cQFnoECAgQAQ&url=https%3A%2F%2Fwww.mpi.govt.nz%2Fdmsdocument%2F33166-far-201906-site-specific-selectivity-of-electric-fishing-gear&usg=AOvVaw3dPD5unARJBiQ69KpXsxdG (accessed on 22 May 2022).
71. Souza, A.T.; Dias, E.; Nogueira, A.; Campos, J.; Marques, J.C.; Martins, I. Population ecology and habitat preferences of juvenile flounder Platichthys flesus (Actinopterygii: Pleuronectidae) in a temperate estuary. *J. Sea Res.* **2013**, *79*, 60–69. [CrossRef]
72. ICES. *Workshop on Age Reading of European and American Eel (WKAREA)*; ICES CM 2009\ACOM; ICES: Bordeaux, France, 2009; p. 48.
73. ICES. *Third Workshop on Age Reading of European and American Eel (WKAREA3)*; ICES: Bordeaux, France, 2020.
74. Rasband, W. *ImageJ v. 1.50i*; National Institute of Health: Bethesda, MD, USA, 2009. Available online: http://imagej.nih.gov/ij/ (accessed on 22 May 2022).
75. Tuset, V.M.; Lozano, I.J.; González, J.A.; Pertusa, J.F.; García-Díaz, M.M. Shape indices to identify regional differences in otolith morphology of comber, Serranus cabrilla (L., 1758). *J. Appl. Ichthyol.* **2003**, *19*, 88–93. [CrossRef]
76. Campana, S.E.; Casselman, J.M. Stock discrimination using otolith shape analysis. *Can. J. Fish. Aquat. Sci.* **1993**, *50*, 1062–1083. [CrossRef]
77. Kuhl, F.P.; Giardina, C.R. Elliptic Fourier features of a closed contour. *Comput. Graph. Image Process.* **1982**, *18*, 236–258. [CrossRef]
78. Ponton, D. Is geometric morphometrics efficient for comparing otolith shape of different fish species? *J. Morphol.* **2006**, *267*, 750–757. [CrossRef] [PubMed]
79. Ferguson, G.J.; Ward, T.M.; Gillanders, B.M. Otolith shape and elemental composition: Complementary tools for stock discrimination of mulloway (Argyrosomus japonicus) in southern Australia. *Fish. Res.* **2011**, *110*, 75–83. [CrossRef]
80. Pothin, K.; Gonzalez-Salas, C.; Chabanet, P.; Lecomte-Finiger, R. Distinction between Mulloidichthys flavolineatus juveniles from Reunion Island and Mauritius Island (south-west Indian Ocean) based on otolith morphometrics. *J. Fish Biol.* **2006**, *69*, 38–53. [CrossRef]
81. Campana, S.E.; Chouinard, G.A.; Hanson, J.M.; Frechet, A.; Brattey, J. Otolith elemental fingerprints as biological tracers of fish stocks. *Fish. Res.* **2000**, *46*, 343–357. [CrossRef]
82. Wilson, R.R., Jr. Depth-related changes in sagitta morphology in six macrourid fishes of the Pacific and Atlantic Oceans. *Copeia* **1985**, *4*, 1011–1017. [CrossRef]
83. Campana, S.E.; Neilson, J.D. Microstructure of fish otoliths. *Can. J. Fish. Aquat. Sci.* **1985**, *42*, 1014–1032. [CrossRef]
84. Morales-Nin, B. Review of the growth regulation processes of otolith daily increment formation. *Fish. Res.* **2000**, *46*, 53–67. [CrossRef]
85. Schiavina, M.; Bevacqua, D.; Melia, P.; Crivelli, A.J.; Gatto, M.; De Leo, G.A. A user-friendly tool to assess management plans for European eel fishery and conservation. *Environ. Model. Softw.* **2015**, *64*, 9–17. [CrossRef]
86. Bevacqua, D.; Melià, P.; Schiavina, M.; Crivelli, A.J.; De Leo, G.A.; Gatto, M. A demographic model for the conservation and management of the European eel: An application to a Mediterranean coastal lagoon. *ICES J. Mar. Sci.* **2019**, *76*, 2164–2178. [CrossRef]
87. Edeline, E.; Elie, P. Is salinity choice related to growth in juvenile eel *Anguilla anguilla*. *Cybium* **2004**, *28*, 77–82.

88. Daverat, F.; Tomas, J. Tactics and demographic attributes in the European eel *Anguilla anguilla* in the Gironde watershed, SW France. *Mar. Ecol. Prog. Ser.* **2006**, *307*, 247–257. [CrossRef]
89. Dias, E.; Miranda, M.L.; Sousa, R.; Antunes, C. Riparian vegetation subsidizes sea lamprey ammocoetes in a nursery area. *Aquat. Sci.* **2019**, *81*, 1–13. [CrossRef]
90. Rooney, N.; McCann, K.S. Integrating food web diversity, structure and stability. *Trends Ecol. Evol.* **2011**, *27*, 40–46. [CrossRef] [PubMed]
91. Rytwinski, T.; Harper, M.; Taylor, J.J.; Bennett, J.R.; Donaldson, L.A.; Smokorowski, K.E.; Clarke, K.; Bradford, M.J.; Ghamry, H.; Olden, J.D.; et al. What are the effects of flow-regime changes on fish productivity in temperate regions? A systematic map. *Environ. Evid.* **2020**, *9*, 1–26.
92. Domingos, I.; Costa, J.L.; Costa, M.J. Factors determining length distribution and abundance of the European eel, *Anguilla anguilla*, in the River Mondego (Portugal). *Freshw. Biol.* **2006**, *51*, 2265–2281. [CrossRef]
93. Gravato, C.; Guimarães, L.; Santos, J.; Faria, M.; Alves, A.; Guilhermino, L. Comparative study about the effects of pollution on glass and yellow eels (*Anguilla anguilla*) from the estuaries of Minho, Lima and Douro Rivers (NW Portugal). *Ecotoxicol. Environ. Saf.* **2010**, *73*, 524–533. [CrossRef] [PubMed]
94. Marohn, L.; Jakob, E.; Hanel, R. Implications of facultative catadromy in *Anguilla anguilla.* Does individual migratory behaviour influence eel spawner quality? *J. Sea Res.* **2013**, *77*, 100–106. [CrossRef]
95. Pereira, L.; Braga, A.C.; Moura, A.; Antunes, C. Prevalence of the Anguillicola crassus parasite in the International Minho River. *Environ. Smoke* **2021**, *17*, 64–74. [CrossRef]
96. Feunteun, E.; Acou, A.; Guillouët, J.; Laffaille, P.; Legault, A. Spatial distribution of an eel population (*Anguilla anguilla* L.) in a small coastal catchment of Northern Brittany (France). Consequences of hydraulic works. *Bulletin Français de la Pêche et de la Pisciculture* **1998**, *349*, 129–139. [CrossRef]
97. Félix, P.M.; Costa, J.L.; Monteiro, R.; Castro, N.; Quintella, B.R.; Almeida, P.R.; Domingos, I. Can a restocking event with European (glass) eels cause early changes in local biological communities and its ecological status? *Glob. Ecol. Conserv.* **2020**, *21*, e00884. [CrossRef]
98. Davey, A.J.; Jellyman, D.J. Sex determination in freshwater eels and management options for manipulation of sex. *Rev. Fish Biol. Fish.* **2005**, *15*, 37–52. [CrossRef]
99. Taborsky, B. The influence of juvenile and adult environments on life-history trajectories. *Proc. R. Soc. B Biol. Sci.* **2005**, *273*, 741–750. [CrossRef] [PubMed]
100. Elie, P. Contribution a L'etude des Montees de Civelles *Anguilla anguilla* Linne (Poisson, Teleosteen, Anguilliforme), Dans L'estuaire de la Loire: Peche, Ecologie, Ecophysiologie et Elevage. Ph.D. Thesis, University of Rennes, Brittany, France, 1979.
101. Charlon, N.; Blanc, J.M. Etude des civelles d´*Anguilla anguilla* L. dans la region du bassin de l´adour. I. Caracteristiques biometriques de longueur et poids en function de la pigmentation. *Arch. Hydrobiol.* **1982**, *93*, 238–255.
102. Edeline, E.; Dufour, S.; Elie, P. Proximate and ultimate control of eel continental dispersal. In *Spawning Migration of the European Eel*; Springer: Dordrecht, The Netherlands, 2009; pp. 433–461.
103. Johansen, S.J.S.; Ekli, M.; Stangnes, B.; Jobling, M. Weight gain and lipid deposition in Atlantic salmon, Salmo salar, during compensatory growth: Evidence for lipostatic regulation? *Aquac. Res.* **2002**, *32*, 963–974. [CrossRef]
104. Zeller, M.; Koella, J.C. Effects of food variability on growth and reproduction of Aedes aegypti. *Ecol. Evol.* **2016**, *6*, 552–559. [CrossRef]
105. Whigham, D.F.; Baldwin, A.H.; Barendregt, A. Tidal Freshwater Wetlands. In *Coastal Wetlands*; Elsevier: Amsterdam, The Netherlands, 2019; pp. 619–640.
106. Laffaille, P.; Feunteun, E.; Baisez, A.; Robinet, T.; Acou, A.; Legault, A.; Lek, S. Spatial organization of European eel (*Anguilla anguilla* L.) in a small catchment. *Ecol. Freshw. Fish* **2003**, *12*, 254–264. [CrossRef]
107. Melià, P.A.C.O.; Bevacqua, D.; Crivelli, A.J.; Panfili, J.; De Leo, G.A.; Gatto, M. Sex differentiation of the European eel in brackish and freshwater environments: A comparative analysis. *J. Fish Biol.* **2006**, *69*, 1228–1235. [CrossRef]
108. Geffroy, B.; Bardonnet, A. Sex differentiation and sex determination in eels: Consequences for management. *Fish Fish.* **2016**, *17*, 375–398. [CrossRef]

MDPI

Article

Length–Weight Relationships, Growth Models of Two Croakers (*Pennahia macrocephalus* and *Atrobucca nibe*) off Taiwan and Growth Performance Indices of Related Species

Shu-Chiang Huang [1,†], Shui-Kai Chang [1,*], Chi-Chang Lai [2,†], Tzu-Lun Yuan [3], Jinn-Shing Weng [2] and Jia-Sin He [2]

[1] Graduate Institute of Marine Affairs, National Sun Yat-sen University, Kaohsiung 804, Taiwan
[2] Coastal and Offshore Resources Research Center, Fisheries Research Institute, Council of Agriculture, Kaohsiung 806, Taiwan
[3] Department of Applied Mathematics, Tunghai University, Taichung 974, Taiwan
* Correspondence: skchang@faculty.nsysu.edu.tw; Tel.: +88-675-250-050
† These authors contributed equally to this work.

Abstract: Information on age and growth is essential to modern stock assessment and the development of management plans for fish resources. To provide quality otolith-based estimates of growth parameters, this study performed five types of analyses on the two important croakers that were under high fishing pressure in southwestern Taiwan: *Pennahia macrocephalus* (big-head pennah croaker) and *Atrobucca nibe* (blackmouth croaker): (1) Estimation of length–weight relationships (LWR) with discussion on the differences with previous studies; (2) validation of the periodicity of ring formation using edge analysis; (3) examination of three age determination methods (integral, quartile and back-calculation methods) and selection of the most appropriate one using a k-fold cross-validation simulation; (4) determination of the representative growth models from four candidate models using a multimodel inference approach; and, (5) compilation of growth parameters for all *Pennahia* and *Atrobucca* species published globally for reviewing the clusters of estimates using auximetric plots of logged growth parameters. The study observed that features of samples affected the LWR estimates. Edge analysis supported the growth rings were formed annually, and the cross-validation study supported the quartile method (age was determined as the number of opaque bands on otolith plus the quartile of the width of the marginal translucent band) provided more appropriate estimates of age. The multimodel inference approach suggested the von Bertalanffy growth model as the optimal model for *P. macrocephalus* and logistic growth model for *A. nibe*, with asymptotic lengths and relative growth rates of 18.0 cm TL and 0.789 year^{-1} and 55.21 cm, 0.374 year^{-1}, respectively. Auximetric plots of global estimates showed a downward trend with clusters by species. Growth rates of the two species were higher than in previous studies using the same aging structure (otolith) and from similar locations conducted a decade ago, suggesting a possible effect of increased fishing pressure and the need to establish a management framework. This study adds updated information to the global literature and provides an overview of growth parameters for the two important croakers.

Keywords: otolith; age determination; growth curves; edge analysis; growth performance index; quartile of marginal growth band method; k-fold cross-validation; back-calculation

Citation: Huang, S.-C.; Chang, S.-K.; Lai, C.-C.; Yuan, T.-L.; Weng, J.-S.; He, J.-S. Length–Weight Relationships, Growth Models of Two Croakers (*Pennahia macrocephalus* and *Atrobucca nibe*) off Taiwan and Growth Performance Indices of Related Species. *Fishes* **2022**, *7*, 281. https://doi.org/10.3390/fishes7050281

Academic Editor: Josipa Ferri

Received: 3 September 2022
Accepted: 7 October 2022
Published: 11 October 2022

1. Introduction

Information on age and growth is essential to modern stock assessment and the development of management plans for fish resources; it has many functions such as converting input catch estimates from biomass to numbers and converting output numbers into biomass [1,2] and can affect the estimation of natural mortality and the mean age at maturity [3–5]. Establishing appropriate aging procedures and selecting representative growth models are thus important steps in developing stock assessments [1,6]. Even if the growth curve has been estimated previously, reviewing the change in key parameters by a timely repeat of estimation is also beneficial to monitoring the stock status under fishing pressure and climate change [7–11].

Several approaches to estimate the growth curve have been used, including length-frequency analysis and growth bands analysis of fin rays, vertebrae, otoliths and scales. Otoliths are calcium carbonate structures formed inside the inner ears and grow with the deposition of continuous calcium carbonate layers, which can respond to both daily and seasonal changes. For teleost fishes, otoliths analysis is the most commonly used and reliable approach to age determination [12–14]. Age estimation using otoliths however requires high precision in laboratory techniques and estimating considerations, including validation of the periodicity of growth-rings formation, selection of age determination method (expression of age in integral count of growth rings or estimated value), and determination of representative growth curves.

Accuracy of age estimates is crucial in deriving age-based population parameters for making the right management decisions [15–17]. Edge analysis and marginal increment analysis (MIA) have been the most commonly used method to verify the increment periodicity of the otolith [18–21]. Edge analysis examines the timing of the translucent zone formation, and the ages are considered validated if only one opaque and one translucent band is formed annually [22–25]. The method is relatively easy to conduct compared to MIA, which plots the proportion of completion of the outermost increment against the month of capture to examine if it shows a yearly sinusoidal cycle [13,26,27] and can be used alone for the validation of annual periodicity of growth increment [18,24].

When determining the age under the microscope, several methods can be used for teleost fishes, including the commonly used integral counting method (integral method) where the age is the count of opaque bands, ignoring the growth within a year reflected in the width of the outmost translucent bands [28–31], and the back-calculate method where the age was determined as the proportion of the hypothesized hatching date to the sampling date plus the number of opaque bands [32,33]. Daily increments count has also been used [34]. Another method that can be explored is the quartile of marginal growth band method (quartile method) [35] to enhance the integral counting method by adding a quartile of the width of the marginal translucent band to the age. Ages estimates resulting from applying these methods are different and the accuracy needs to be investigated in advance.

The von Bertalanffy growth model (VBGM) has been widely used for fitting growth curves to the estimated age and the length data, but the standard VBGM is not always appropriate for all species since the ontogenetic changes in growth rates vary by species, and so alternative models should be explored [6,36]. The multimodel inference approach [37,38] is a useful way to select the most representative growth model from a suite of candidates for the species [27,39,40].

Plotting the estimated growth parameters together with those of other studies of the same species or related species can provide an overview of the similarities and differences between the current study and other global studies, and a chance to explore the differences [41–43]. Pauly and Munro [44] proposed a growth performance index based on logarithmized mean asymptotic length and relative growth rate obtained from the growth model, which allowed the identification of the bias in growth parameter estimates within a species while also revealing the difference between species. The auximetric plot created using these two parameters can help to visualize the clusters formed by the same species [43,45]. It has been applied widely to compare estimated growth parameters from different studies [27,46–48].

Sciaenidae includes 70 genera with around 282 species distributed in the Atlantic, Indian, and Pacific Oceans [49] and are important to coastal fisheries of many countries. There were 23 species in 12 genera of Sciaenidae recorded in Taiwan waters [50]. The high economic value of the Sciaenidae and its characteristics of mass migration and concentration during the breeding season make it easy to be caught in large quantities. Sciaenidae species are an important target to the trawl fishery of Taiwan. Among them, *Pennahia macrocephalus* ([51]; big-head pennah croaker) and *Atrobucca nibe* ([52]; blackmouth croaker) are two of the most abundant commercial species in southwestern Taiwan (Figure 1). They are widely distributed across Indo-Pacific waters with depth between 20 to 200 m, feeds on small invertebrates and crustaceans. Total landing of *Pennahia* sp. once reached a peak of 8.6 thousand tons in 1996 and declined to 852 tons in 2015, while that of *A. nibe* was 2.5 thousand tons in 1993 and declined to less than 500 tons after 2011 [50]. Lack of logbook from coastal fisheries due to exemption of submission responsibility [53] has hindered the assessment of the stock using catch data; studies of biological parameters are considered as alternatives for providing supplemental information on the stock.

Figure 1. Two important croaker species occurred in Taiwan. *Pennahia macrocephalus* (big-head pennah croaker, age = 3.5) and *Atrobucca nibe* (blackmouth croaker, age = 2).

As indicated above, we suppose that different age determination methods may affect the results of growth estimates, the most optimal growth model of these two species may not be the VBGM, and same fish species may have a relatively similar growth performance [44]. Therefore, five types of analyses were performed on the two species in this study: (1) estimation of length–weight relationships (LWR) which is also a key element in the examination of fish biology and contains valuable information to assess the general health condition of fish species [54–56]; (2) validation of the periodicity of ring formation using edge analysis; (3) examination of three age determination methods and selection of the most appropriate one using a k-fold cross-validation simulation study [57,58]; (4) determination of the best fitting growth models from four candidate models; and, (5) compilation of growth parameters for all *Pennahia* and *Atrobucca* species published globally for reviewing the clusters (groups) of estimates using auximetric plots. Most growth studies on the two species were conducted a decade ago and the majority of them were published in grey literature or in Chinese. This study adds updated information to the global literature and provides an overview of growth parameters for the two important croakers.

2. Materials and Methods

2.1. Data Sources and Estimation of Length-Weight Relationships (LWR)

Biological data: total length (TL) in cm, body weight (BW) in g, sex, and otolith samples of *P. macrocephalus* in southwestern Taiwan waters were collected monthly from July to October 2018 and February to August 2021 in the fish markets of Zihquan and Tungkang (Figure 2). Biological data of *A. nibe* were collected monthly from July 2017 to August 2018 at the fish markets of, or from fishing boats based in, Zihquan, Tungkang, Fangliao and Liuqiu (Figure 2). Excluding samples abandoned due to damage caused during otolith processing or disagreements on the number of ring marks between two age readings, a total of 359 *P. macrocephalus* and 386 *A. nibe* samples were used in this study.

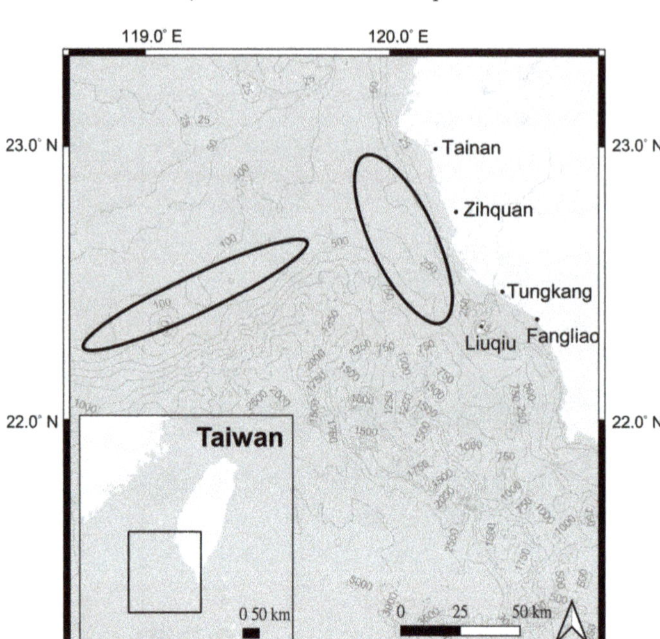

Figure 2. Black circles indicate the major fishing areas of *P. macrocephalus* and *A. nibe* from the trawlers based in Zihquan, Tungkang, Fangliao and other ports in southwestern Taiwan. The market samples used in this study came from these fishing areas.

The length–weight relationship (LWR) for each species was estimated using a power function of the form $BW = a \times TL^b$ where a is the coefficient of the power function, and b is the allometric coefficient. Parameters a and b were estimated with a linear regression of $\ln(BW)$ against $\ln(TL)$. Size distribution and LWR by sex for each species were also examined. The difference of the parameters from the linear regression fits between sexes for both species were tested using ANOVA [59].

2.2. Otolith Processing, Periodicity of Increment and Age Determination

Otoliths (sagittae, lapilli, and asterisci, Supplementary Figure S1) were extracted from the fish samples. Sagittae, the largest of the three otoliths, are commonly used for estimating fish age [12] and were used for this study. They were firstly washed with water and air-dried for at least 24 h. Next, the otoliths were prepared for transverse sections by embedding the whole otolith in epoxy resin and epoxy hardener mixture (with the ratio 10:0.95), placing them at 60 °C for 2 h, and sectioning transversely with a low-speed diamond wheel saw (approximately 500 μm thick). Finally, the otolith sections were polished with silicone abrasive grinding paper (1200 to 4000 grit) until the rings became clear. The otoliths were examined using a microscope with transmitted light, by which the periods of fast growth accreted otolith material are translucent whereas it is opaque during slow growth periods, which is different from the observation viewed under a dissecting microscope with reflected light [60]. A translucent zone and subsequent opaque zone are referred to as an 'increment'. Ring marks on the otoliths were counted and measured by at least two different readers or three times by the same reader. The results were accepted when the readings matched; otherwise, the results were abandoned and were not included in this study.

Edge analysis was used to examine the periodicity of increment formation (the timing of the translucent zone formation) and was calculated through the monthly proportion of the translucent zone on the outer edge of the sectioned otolith [23,24]. Subsamples for edge analysis were selected by month from the most abundant age groups, with efforts to restrict the number of age groups [13] (n = 296 for *P. macrocephalus* with age 2–4, TL 13.7–19.6 cm and 186 for *A. nibe* with age 4–6, TL 23.2–53.8 cm).

Three methods for determining the annual increments were conducted and evaluated for selecting the optimal one. (1) Integral counting method (integral method), where the opaque bands were counted from the nucleus to the edge of the otolith. This method was commonly used for many fish species, but since the marginal translucent band was ignored, this method may underestimate the actual age of the fish (Figure 3a). (2) Quartile of marginal growth band method (quartile method). The age was determined in the same way as the integral counting method when the edge of the otolith is an opaque band. If the edge of the otolith is a translucent band, the width of the marginal translucent band was compared to the width of the previous complete annulus to the approximate value of 0.25, 0.5 or 0.75; and the age was determined as the quartile plus the number of opaque bands (Figure 3b) [35]. (3) Back-calculation method. The age was determined as the proportion of the hypothesized hatching date to the sampling date plus the number of opaque bands [32].

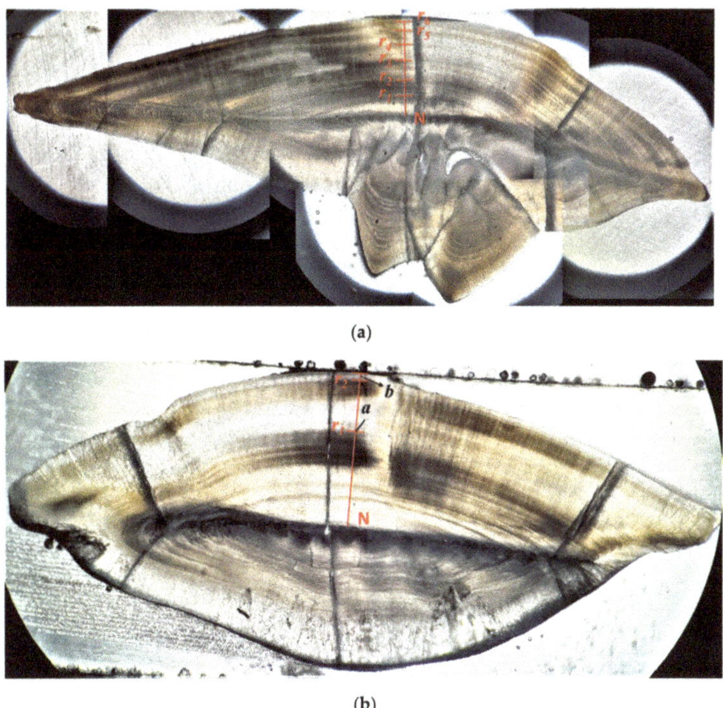

Figure 3. Sample images of sagittal otolith (transverse sections) showing the determination of ages by different methods. (**a**) Integral counting method (sample of *A. nibe*; total length 41.5 cm): r_n indicate the ring marks on the outer margin of the opaque bands used for determining age class n. (N: nucleus). For this case, the age is six. (**b**) Quartile of marginal growth band method (quantile method, sample of *P. macrocephalus*; total length 16.8 cm): r_1 and r_2 indicate the ring marks on the outer margin of the opaque bands. a is the width of the most recent formed annulus, and b is the width of the hyaline band extended from the margin of the outmost opaque band to the edge. Ages were determined as the number of opaque bands plus approximate quartile value of b against a. For this case, the age is 2.25.

A k-fold cross-validation simulation study [57,58] was performed to evaluate the goodness of fit of the growth model to the age-length data estimated by each of the three methods. The most commonly used VBGM was used as the reference model. The age-length data, with length data and age estimates from the three methods, were firstly split into four groups based on the results of the integral counting method (e.g., ages 0–1, age 2, age 3, and ages 4–5). Data within each age group were then randomly divided into five sub-groups, or folds, of approximately equal size. Each fold contains data from all age groups. The first fold (across the four age groups) was treated as a validation set, and the VBGM was fit on the remaining four folds (training set). The mean square error (MSE) was calculated from the observed length of the validation set, and the predicted length was estimated from the VBGM developed using the training set. Afterwards, the other folds were treated as validation sets, and the VBGM-developing and MSE-estimation were repeated. This loop was repeated 100 times to finally obtain the mean and standard deviation (sd) of the MSE for each type of age determination method.

2.3. *Growth Parameters Estimation*

An information-theoretic, multi-model inference approach was used to determine the optimal growth model for both species [37,38]. The candidate growth models examined include non-sigmoidal VBGM and three sigmoidal models: Gompertz, Logistic, and Richards models [61–64]. The first three are the most common three-parameter models [65], and the last is a four-parameter model.

$$\text{Von Bertalanffy model}: \; L_t = L_\infty\left(1 - e^{-K(t-t_0)}\right) \tag{1}$$

$$\text{Gompertz model}: \; L_t = L_\infty e^{-e^{-K(t-t_0)}} \tag{2}$$

$$\text{Logistic model}: \; L_t = L_\infty\left(1 + e^{-K(t-t_0)}\right)^{-1} \tag{3}$$

$$\text{Richards model}: \; L_t = L_\infty\left(1 + \frac{1}{p}e^{-K(t-t_0)}\right)^{-p} \tag{4}$$

where L_t is the length (TL in cm) at age t, L_∞ is the mean asymptotic length, p is a dimensionless parameter of Richards model, and t_0 is the theoretical age when length equals zero for the VBGM or the age at the inflection point of the growth curve for the other models. Parameter k (year^{-1}) is a relative growth rate, describing how quickly the asymptotic length is approached or the rate of exponential decrease of the relative growth rate with age [37,66].

The length-at-age data (combined and sex-specific) were fitted to the four candidate models using non-linear least squares in R version 4.1.2 [67]. The small-sample bias-corrected form (*AICc*) of the AIC (Akaike Information Criterion) was used for model selection [65,68,69]. The model with the smallest *AIC* value ($AIC_{c,\,min}$) was selected as the 'best' among the models tested. *AIC* differences $\Delta = AIC_{c,\,min} - AIC_{c,\,i}$ were computed overall candidate models i. The Akaike weight, w_i, of each model was then calculated using these differences to quantify the plausibility, which is considered as the weight of evidence in favor of model i being the best of the available set of models [37,69].

$$w_i = \frac{e^{-0.5\Delta_i}}{\sum_{i=1}^{4} e^{-0.5\Delta_i}} \tag{5}$$

2.4. *Global Estimate Compilation*

A literature (e.g., books, peer-reviewed articles and project reports) search was conducted on the estimation of growth parameters for *Pennahia* and *Atrobucca* species; however, only four *Pennahia* and two *Atrobucca* species (i.e., *P. macrocephalus, P. argentata, P. pawak, P. anea, A. nibe* and *A. alcocki*) were available. Global estimates of growth parameters of the six species were compiled from the literature in Table 1 and Supplementary Table S1.

Estimates of k were in year^{-1}; estimates of L_∞ were in cm in either standard length (SL), fork length (FL), total length (TL), body length (BL) or not available (NA). Most of the L_∞ were in TL; thus, the estimates were converted to TL using the formulae listed below [70–72] or assumed to be TL for the cases of NA.

$$\begin{gathered} \text{TL} = 0.651 + 1.162\ \text{SL}\ (R^2 = 0.997, \text{for } P.\ macrocephalus) \\ \text{TL} = -0.37531 + 1.225\ \text{SL}\ (R^2 = 0.965, \text{for } P.\ argentata) \\ \text{TL} = \text{FL}\ (\text{For } P.\ anea) \end{gathered} \tag{6}$$

Table 1. Global estimates of growth parameters of *Pennahia and Atrobucca* spp. from the literature. Lengths are in the unit of cm and k in year^{-1}. Refer to Table S1 for detailed information.

Code	Species	Region	Material (Method)	L_∞ (TL)	k (cm y^{-1})	φ	Ref
m_1	*P. macrocephalus*	Beibu Gulf, SCS	Otolith (I)	27.371	0.408	2.485	[29]
m_2	*P. macrocephalus*	Beibu Gulf, SCS	Len. freq.	28.899	0.520	2.638	[70]
m_3	*P. macrocephalus*	Beibu Gulf, SCS	Scale	31.502	0.590	2.768	[73]
m_4	*P. macrocephalus*	Southwest, TW	Otolith (D)	27.028	0.596	2.639	[34]
m_5	*P. macrocephalus*	Yunlin, TW	Otolith (B)	21.358	0.371	2.228	[32]
mc_1	*P. macrocephalus*	Southwest, TW	Otolith (I)	19.592	0.328	2.099	This study
mc_2	*P. macrocephalus*	Southwest, TW	Otolith (Q)	18.004	0.789	2.408	This study
mc_3	*P. macrocephalus*	Southwest, TW	Otolith (B)	18.258	0.617	2.313	This study
a_1	*P. argentata*	TW strait	Scale	30.887	0.978	2.970	[74]
a_2	*P. argentata*	TW strait	Scale	35.713	0.375	2.680	[74]
a_3	*P. argentata*	Southern, ECS	Scale	28.878	1.207	3.003	[74]
a_4	*P. argentata*	Southern, ECS	Scale	34.733	0.315	2.580	[74]
a_5	*P. argentata*	Beibu Gulf, SCS	Scale	30.500	0.350	2.513	[75]
a_6	*P. argentata*	Beibu Gulf, SCS	Scale	28.230	0.500	2.600	[76]
a_7	*P. argentata*	Northern, SCS	Len. freq.	38.200	0.420	2.787	[71]
a_8	*P. argentata*	Beibu Gulf, SCS	Len. freq.	31.500	0.350	2.541	[71]
a_9	*P. argentata*	TW strait	Otolith	36.301	0.377	2.696	[30]
a_10	*P. argentata*	Southern Sea, SK	Otolith	35.529	0.380	2.681	[31]
a_11	*P. argentata*♀	Ariake Sound, JP	Otolith	31.200	0.384	2.573	[77]
a_12	*P. argentata*♂	Ariake Sound, JP	Otolith	29.400	0.360	2.493	[77]
a_13	*P. argentata* ♀	Tachibana Bay, JP	Otolith	35.400	0.272	2.533	[28]
a_14	*P. argentata* ♂	Tachibana Bay, JP	Otolith	29.000	0.342	2.459	[28]
a_15	*P. argentata*♀	Omura Bay, JP	Otolith	37.000	0.256	2.545	[28]
a_16	*P. argentata*♂	Omura Bay, JP	Otolith	37.700	0.181	2.410	[28]
a_17	*P. argentata*♀	The Sea of Goto, JP	Otolith	44.500	0.241	2.679	[28]
a_18	*P. argentata*♂	The Sea of Goto, JP	Otolith	45.200	0.195	2.600	[28]
a_19	*P. argentata*	Seto Inland Sea	Scale	39.56	0.307	5.68	[78]
p_1	*P. pawak*	Beibu Gulf, SCS	Scale (Logistic)	22.030	0.580	2.449	[79]
p_2	*P. pawak*	Beibu Gulf, SCS	Len. freq.	24.150	0.390	2.357	[80]
p_3	*P. pawak*	Beibu Gulf, SCS	Len. freq.	22.050	0.320	2.192	[80]
an_1	*P. anea*	Paradeep, IN	Len. freq.	30.300	0.860	2.897	[81]
an_2	*P. anea*	San Miguel Bay, PHI	Len. freq.	20.000	0.600	2.380	[82]
an_3	*P. anea*	Manila Bay, PHI	Len. freq.	26.500	1.400	2.993	[83]
an_4	*P. anea*	Penang/Perak, MA	NA	34.200	0.400	2.670	[84]
an_5	*P. anea*	Bombay, IN	Len. freq.	24.500	0.640	2.585	[85]
an_6	*P. anea*	Bombay, IN	Len. freq.	26.000	1.200	2.909	[86]
an_7	*P. anea*	Rameswaram, IN	Len. freq.	23.300	1.260	2.835	[87]
an_8	*P. anea*	Mandapam, IN	Len. freq.	26.000	0.980	2.821	[88]
an_9	*P. anea*	Bombay, IN	Len. freq.	27.300	1.940	3.160	[89]
an_x	*P. anea*	Andhra Pradesh, IN	Len. freq.	33.000	0.700	2.882	[90]
an_y	*P. anea*	Indonesia	Len. freq.	23.890	0.840	2.681	[91]
n_1	*A. nibe*	Formosa Strait	NA	58.500	0.116	2.599	[92]
n_2	*A. nibe*	North, TW	Len. freq.	58.700	0.145	2.699	[92]
n_3	*A. nibe*	North, TW	Len. freq.	37.100	0.354	2.688	[92]
n_4	*A. nibe*	North, TW	Len. freq.	37.100	0.523	2.857	[92]
n_5	*A. nibe*	North, TW	Len. freq.	57.300	0.177	2.764	[92]
n_6	*A. nibe*	South, SK	Len. freq.	50.900	0.227	2.769	[92]
n_7	*A. nibe*	South, SK	Len. freq.	46.900	0.297	2.815	[92]
n_8	*A. nibe*	TW	Scale	56.290	0.120	2.580	[93]
n_9	*A. nibe*	ECS	Scale	43.200	0.252	2.672	[94]
n_10	*A. nibe*	Oman Sea	Otolith	50.000	0.200	2.699	[95]
n_11	*A. nibe*	Guei-Shan Island, TW	Otolith	48.060	0.275	2.803	[96]

Table 1. *Cont.*

Code	Species	Region	Material (Method)	L_∞ (TL)	k (cm y^{-1})	φ	Ref
n_12	*A. nibe*	Guei-Shan Island, TW	Otolith	64.910	0.147	2.792	[96]
n_13	*A. nibe*♀	NA	Len. freq.	45.000	0.225	2.659	[92]
n_14	*A. nibe*♂	NA	Len. freq.	40.100	0.238	2.583	[92]
n_15	*A. nibe* ♀	Southwest, TW	Otolith	53.384	0.183	2.717	[97]
n_16	*A. nibe* ♂	Southwest, TW	Otolith	52.577	0.187	2.713	[97]
n_17	*A. nibe* ♀	Northeast, TW	Otolith	49.527	0.227	2.746	[98]
n_18	*A. nibe*♂	Northeast, TW	Otolith	48.417	0.201	2.673	[98]
nc_1	*A. nibe*	Southwest, TW	Otolith (VBGM)	68.149	0.130	−0.919	This study
nc_2	*A. nibe*	Southwest, TW	Otolith (Logistic)	53.111	0.394	2.958	This study
al	*A. alcocki*	Pakistani waters	Len. freq.	47.250	0.180	2.604	[99]

Note: $L\infty$ (TL), the $L\infty$ value in total length (TL, might be converted from the original value). For Sex: M, male; F, female; C, sex-combined. For Region: SCS, the South China Sea; TW, Taiwan; ECS, the East China Sea; SK, South Korea; PHI, Philippines; MA, Malaysia; IN, India; JP, Japan. For method: I, integral counting method; Q, quartile method; B, back-calculated method; D, daily growth ring method. NA, information not available.

Growth parameters were estimated using different methods: the aging of otoliths (sagittae) or scales, length-frequency data (LFD), or unknown method. Growth parameters for the two species in this study were compiled into the global estimates table (62 estimates in total) for classification analysis. All the parameters were from VBGM, except for the records coded as p_1 and nc_2, which were Logistic.

The two variables, $\log(L_\infty)$ and $-\log(k)$, of the growth performance index formula, $\varphi = 2\log(L_\infty) + \log(k)$ [44], were used to identify the clusters of growth estimates from the global summary using the analysis of the auximetric plot [43,45,46]. Ellipses of 95% confidence intervals were also produced in the plot except for *P. pawak* with three records and *A. alcocki* with one record only.

3. Results

3.1. Samples and Length-Weight Relationships (LWR)

For *P. macrocephalus*, 359 individuals with total lengths and bodyweight ranging from 12.20 to 20.13 cm and from 21.70 to 104.82 g, respectively, were examined (Supplementary Table S2 with raw data and summary statistics). Males ranged from 12.20 to 20.13 cm TL and 21.70 to 104.70 g (n = 207) and females from 13.45 to 19.62 cm TL and 27.17 to 104.82 g (n = 152). The ANOVA test results indicated that there is no significant difference of the LWR parameters between sexes (p = 0.095, df = 355). Therefore, the sex combined LWR equation is presented as $BW = 1.122 \times 10^{-3} \cdot TL^{3.061}$ (R^2 = 0.892) (Figure 4) (or $BW = 2.770 \times 10^{-2}\ SL^{2.937}$). To examine if the coefficients have been biased by possible overweighting by more samples from the breeding season (June to September), an additional LWR estimation was performed using limited sub-samples: all data in a month were used when the monthly sample size was <30, or 30 fish were randomly selected when the monthly sample size was >30. The resulted average was b = 3.122 (n = 10), suggesting that reducing sample size from the breeding season does not reduce the estimation of b and that the current estimates using all sample data are representative.

For *A. nibe*, 386 individuals with total lengths and bodyweight ranging from 17.02 to 56.6 cm and from 46.30 to 1941.60 g, respectively, were examined (Supplementary Table S3 with raw data and summary statistics). Males ranged from 17.01 to 52.70 cm TL and from 46.30 to 1235.07 g (n = 152) and females from 18.65 to 56.60 cm TL and from 78.37 to 1941.60 g (n = 234). The ANOVA test results indicated that there is no significant difference of the LWR parameters between sexes (p = 0.156, df = 382). The length–weight relationship equation is thus presented as $BW = 1.448 \times 10^{-2} \cdot TL^{2.886}$ (R^2 = 0.985) for sex combined.

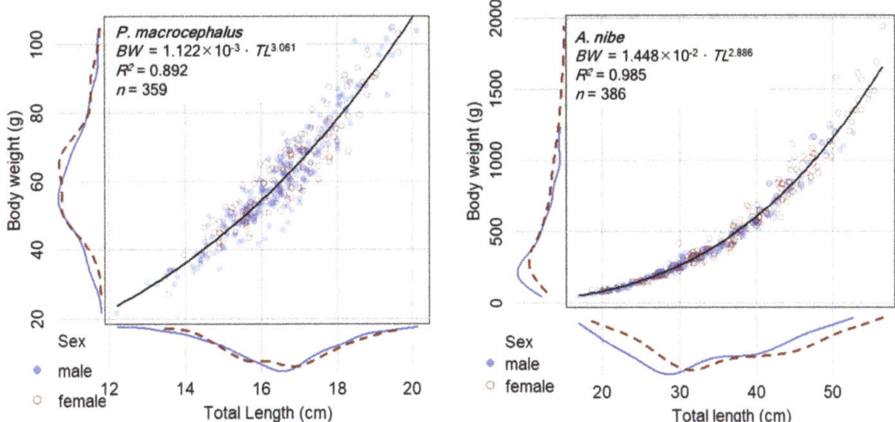

Figure 4. Total length (TL) and body weight (BW) data and the estimated length-weight equation of *P. macrocephalus* and *A. nibe*. The blue lines and circles are male, and the red lines and circles are female.

3.2. Periodicity of Increment Formation—Edge Analysis

Monthly percent occurrence of the translucent band at the edge for the two croakers is depicted in Figure 5. For *P. macrocephalus*, we were not able to obtain samples for November to January since it did not coincide with the active fishing season. However, the trend from the lowest proportion of translucent bands in February to the high level from June to October could still be observed. Although there was data deficiency during the winter period, the apparent trend suggested that the translucent bands started to form once a year near the end of winter and the early spring.

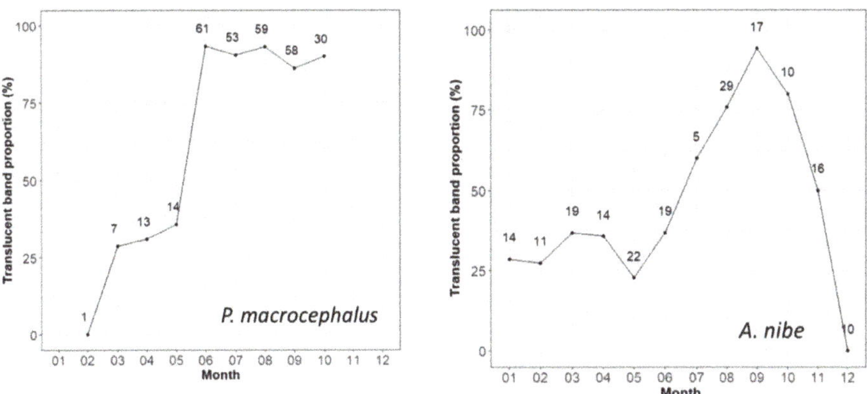

Figure 5. Monthly percent occurrence of the translucent band at the edge of samples from *P. macrocephalus* and *A. nibe*. Numbers indicate the sample sizes.

On the other hand, *A. nibe* can be caught throughout the year. The result of the edge analysis showed that the proportion of translucent bands started to increase from January to September (100%), with a drop in May (likely caused by sampling issues) followed by a sharp decline to 0% in December. The clear trend from edge analysis suggested that the translucent bands started to form once a year near the end of winter and during the spring.

3.3. Age Determination Methods and Growth Parameters Estimation

For *P. macrocephalus*, mean ± sd of MSE from the k-folds cross-validation runs for the three age determination methods was 1.722 ± 0.010 for the integral method, 1.572 ± 0.048 for the quartile method, and 1.730 ± 0.011 for the back-calculation method. The quartile method has the lowest mean MSE compared to the other two methods and was considered the optimal method. A similar conclusion was obtained for *A. nibe* that the quartile method has the lowest mean MSE, with mean ± sd of MSE of 31.233 ± 5.005 for the integral method, 30.538 ± 4.011 for the quartile method, and 33.783 ± 1.760 for the back-calculation method. The quartile method was thus selected as the age determination method for both species.

For *P. macrocephalus*, the Richards model could not converge in the growth fitting analysis. The growth model that fits the length and age data best was VBGM with *AICc* differences $\Delta < 2$ (indicating substantial support as the best model) [38] and Akaike weight w (the expected weight of evidence in favor of the model being the best among the four models) of 47% with sex combined (Figure 6, Table 2). The growth parameters of VBGM (L_∞, k, t_0) were: 18.0 cm TL, 0.789 year^{-1}, and −0.872 year for sex-combined; 18.673 cm, 0.495 year^{-1}, −2.095 year for female; and, 17.864 cm, 0.892 year^{-1}, −0.571 year for male.

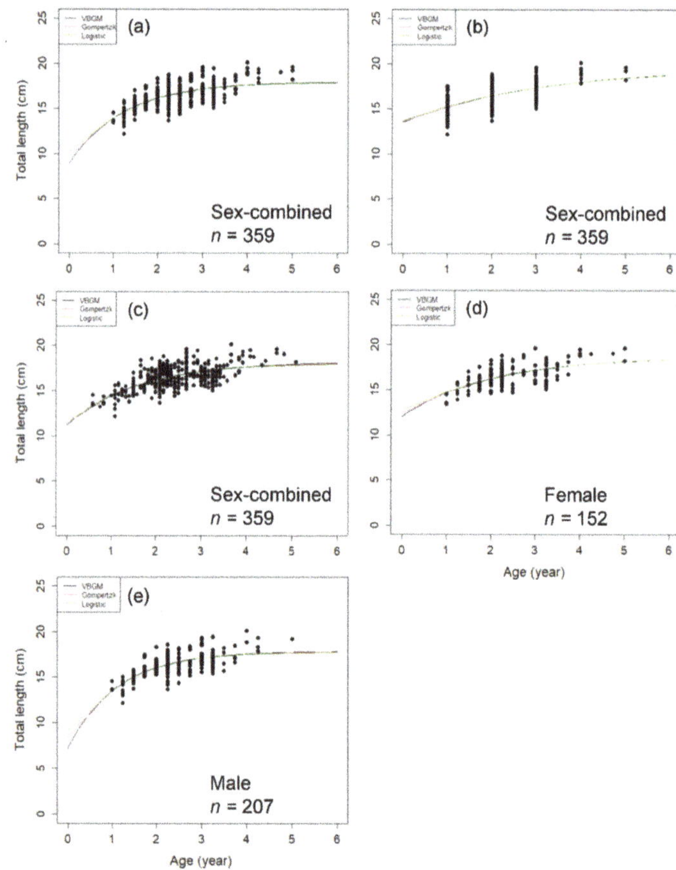

Figure 6. Age and length of sex-combined and sex-separated data and fitted growth curves of the three growth models for *P. macrocephalus* off southwestern Taiwan. (**a**,**d**,**e**) are the quartile of the marginal growth band method; (**b**) is the integral counting method; (**c**) is the back-calculation method.

Table 2. Parameter estimates (±standard error) from four candidate growth models, by sex, for *P. macrocephalus* and *A. nibe* off Southwestern Taiwan. The best-fit models are shown in bold.

Species	Model	Sex	L_∞ (cm TL)	k (Year^{-1})	t_0 (Year)	p	$AICc$	$\Delta AICc$	w
P. macrocephalus	VBGM	C	18.004 (0.315)	0.789 (0.157)	−0.872 (0.368)	-	1029.04	0	0.468
P. macrocephalus	VBGM	F	18.673 (0.982)	0.495 (0.239)	−2.095 (1.280)	-	445.08	0	0.386
P. macrocephalus	VBGM	M	17.864 (0.352)	0.892 (0.202)	−0.571 (0.367)	-	585.84	0	0.390
P. macrocephalus	Gompertz	C	17.921 (0.285)	0.879 (0.162)	−0.564 (0.294)	-	1029.81	0.774	0.318
P. macrocephalus	Gompertz	F	18.647 (0.955)	0.532 (0.243)	−1.693 (1.015)	-	445.39	0.311	0.330
P. macrocephalus	Gompertz	M	17.777 (0.316)	1.002 (0.210)	−0.286 (0.290)	-	586.17	0.328	0.331
P. macrocephalus	Logistic	C	17.850 (0.262)	0.970 (0.168)	−0.311 (0.241)	-	1030.59	1.556	0.215
P. macrocephalus	Logistic	F	18.643 (0.947)	0.565 (0.246)	−1.354 (0.808)	-	445.69	0.612	0.284
P. macrocephalus	Logistic	M	17.704 (0.288)	1.112 (0.219)	−0.055 (0.236)	-	586.52	0.678	0.278
A. nibe	VBGM	C	68.149 (8.182)	0.130 (0.033)	−0.919 (0.413)	-	2373.40	6.380	0.029
A. nibe	VBGM	F	60.548 (5.835)	0.186 (0.047)	−0.266 (0.439)	-	1443.83	4.090	0.075
A. nibe	VBGM	M	90.094 (37.291)	0.070 (0.046)	−1.926 (0.870)	-	911.61	1.736	0.180
A. nibe	Gompertz	C	57.162 (3.512)	0.263 (0.036)	1.919 (0.194)	-	2369.53	2.514	0.202
A. nibe	Gompertz	F	54.732 (3.226)	0.318 (0.052)	1.853 (0.160)	-	1441.39	1.648	0.255
A. nibe	Gompertz	M	62.594 (9.428)	0.197 (0.050)	2.313 (0.720)	-	910.56	0.679	0.306
A. nibe	Logistic	C	53.111 (2.294)	0.394 (0.040)	2.958 (0.219)	-	2367.02	0	0.709
A. nibe	Logistic	F	52.142 (2.306)	0.448 (0.052)	2.804 (0.160)	-	1439.74	0	0. 582
A. nibe	Logistic	M	55.313 (5.285)	0.322 (0.054)	3.372 (0.615)	-	909.88	0	0.429
A. nibe	Richards	C	55.345 (2.904)	0.282 (0.035)	1.843 (0.162)	-3.661×10^6 (3.477×10^6)	2371.95	4.927	0.060
A. nibe	Richards	F	54.349 (3.260)	0.322 (0.053)	1.843 (0.183)	-1.067×10^5 (3.060×10^6)	1443.53	3.788	0.088
A. nibe	Richards	M	57.794 (6.277)	0.224 (0.049)	1.944 (0.409)	-6.994×10^6 (6.283×10^6)	913.12	3.248	0.085

Note: For sex: M, male; F, female; C, sex-combined.

For *A. nibe*, the best growth model was logistic with *AICc* differences $\Delta < 2$ and Akaike weight w of 75% with sex combined (Figure 7, Table 2). The growth parameters of logistic (L_∞, k, t_0) were: 53.11 cm, 0.39 year^{-1}, and 2.96 year for sex-combined; 52.14 cm, 0.45 year^{-1}, 2.80 year for female; and, 55.31 cm, 0.32 year^{-1}, 3.37 year for male.

The selection of the best model for each species was based on the quality of statistical fit assuming that the data could represent the potential species-specific growth pattern. Table 2 provides estimates of all the tested models for optional choices when considering the representativeness of the data set and the theoretical growth pattern from biological details of the fish [38,100].

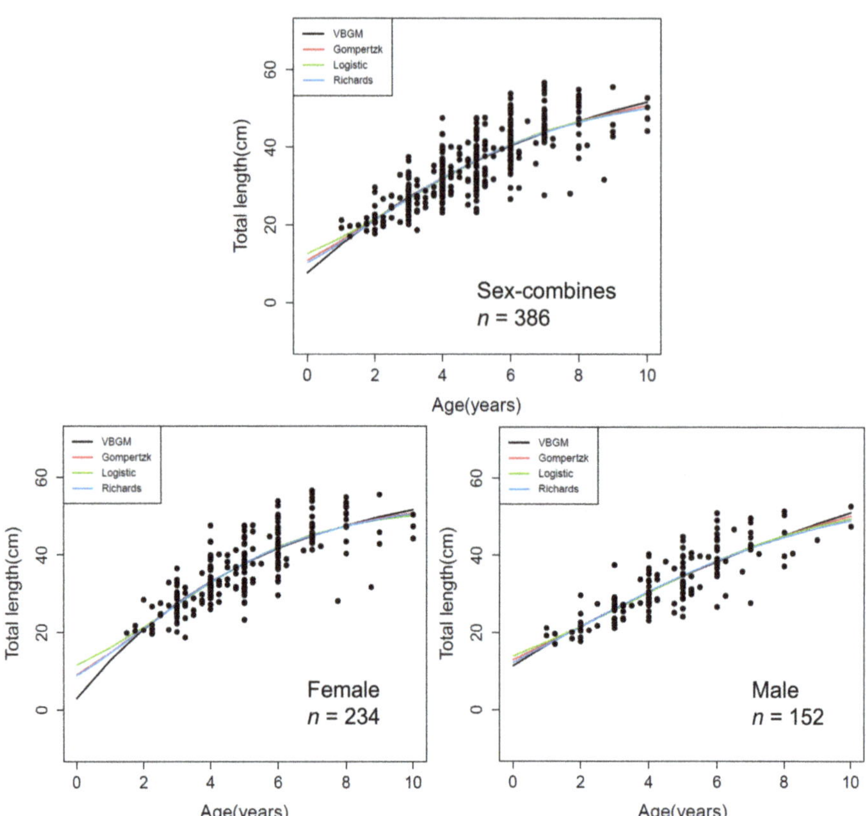

Figure 7. Age and length of sex-combined and sex-separated data and fitted growth curves of the three growth models for *A. nibe* off southwestern Taiwan.

3.4. Global Estimates Compilation

Few studies on *P. macrocephalus* and *A. nibe* were found in the literature. Especially for *A. nibe*, many were reported in grey literature and were cited in academic publications without detailed information (e.g., many were cited in Pauly [92], and information on the data sources are not available online). Due to the lack of studies in the two targeted species, studies from the same genus documented previously were also included in the analyses. A total of 62 growth records were obtained covering six species from ten regions with three estimation methods, including otolith, scale and length frequency (Table 1).

For *Pennahia* species, the age ranged from 0 to 10 years, and the growth parameters of VBGM ranged from 15.15–45.20 cm TL for L_∞ and 0.181–1.94 year^{-1} for k. The growth performance index was 2.10 to 3.16 (mean = 2.61), and for *P. macrocephalus* was between 2.10 to 2.77 (mean = 2.43). For *Atrobucca* species, the age ranged from 1 to 13 years, and the growth parameters ranged from 37.10–68.15 cm TL for L_∞ and 0.116–0.523 year^{-1} for k. The growth performance index was 2.58–3.05 (mean = 2.75), and for *A. nibe* was in the same range (mean = 2.76).

The two variables, log(L_∞) and log(k), of the growth performance index formula and the results of the growth performance index, were shown in the auximetric plots (Figure 8). The same species tend to be in the same cluster with similar φ.

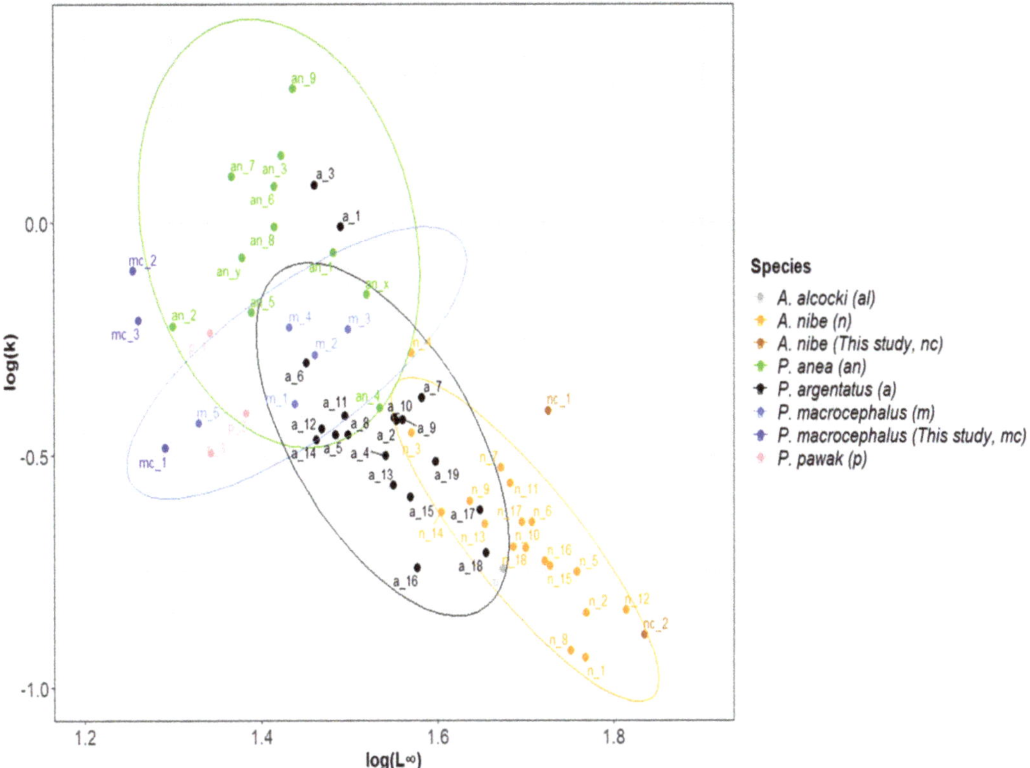

Figure 8. Auximetric plot of the two variables, log(L_∞) and log(k), of the growth performance index formula obtained from the previous studies globally of Sciaenidae family. Ellipses with a 95% confidence interval. Code referred to Table 1.

4. Discussion

4.1. Length-Weight Relationships

In order to compare results from this study with past studies, the LWR in TL was transformed to a LWR in SL for *P. macrocephalus* using the length-length relationship [70]. Divergencies were noted among the LWR estimates in Table 3. Although the reasons for the divergencies have not been fully explored, in addition to possible sampling errors in data collection methods [101], the following differences in data handling may, in part, account for the observed differences. (1) Different logarithmic functions are used for the LWR parameter estimation. Both base-10 logarithmic (Log_{10}) [54,70] and natural logarithmic (Ln) functions [59,102–104] have been used to linearize the LWR equation for regression estimation, and most of the studies did not specify the logarithmic function they used. The two functions will result in different parameter values while the *b* parameter remains the same. (2) Uneven distribution of the samples by fish size, such as oversampling of medium-sized fish, can affect the regression fitness. For example, large fish were less fitted in the LWR of Tasi (Figure 19 in reference [34]) (Table 3), which might cause an overestimation of the allometric coefficient (*b* value). (3) Uneven distribution of sampling size among months, e.g., over-weighted samples were from the months when the fish grow faster or slower, could cause estimation differences since LWR may change seasonally with different levels of gonad maturity and other parameters [105]. Nhuận and Khuong [73] (Table 3) showed that *b* values estimated in the breeding season (third season) tended to be higher than in other seasons. For this study, most of the samples were from the breeding season, and thus, the *b* value was likely overestimated to some extent.

Table 3. The coefficients *a* and *b* of the length–weight relationship equation (LWR) of *P. macrocephalus* (standard length) and *A. nibe* (total length) of the current and previous studies.

Species	Region	*a*	*b*	Reference
P. macrocephalus	Beibu Gulf, SCS	2.37×10^{-5}	3.248	[29]
P. macrocephalus	Beibu Gulf, SCS	2.16×10^{-2}	3.032	[70]
P. macrocephalus	Beibu Gulf, SCS	0.01×10^{-6}~0.01×10^{-7}	2.95~3.57	[73]
P. macrocephalus	Yunlin, TW	4.30×10^{-2}	2.747	[32]
P. macrocephalus	Southwest, TW ♀	4.70×10^{-2}	2.730	[106]
P. macrocephalus	Southwest, TW ♂	4.08×10^{-2}	2.786	[106]
P. macrocephalus	Southwest, TW	9.90×10^{-3}	3.337	[34]
P. macrocephalus	Southwest, TW	2.77×10^{-2}	2.937	This study
A. nibe	Yellow Sea and ECS	2.03×10^{-3}	3.463	[107]; cited in [108]
A. nibe	ECS	2.84×10^{-3}	3.220	[94]; cited in [108]
A. nibe	North, TW	6.39×10^{-4}	3.207	[109]; cited in [108]
A. nibe	Northeast, TW	1.41×10^{-3}	2.934	[98]; cited in [108]
A. nibe	Northeast, TW	6.92×10^{-3}	3.100	[96]
A. nibe	Southwest, TW	1.21×10^{-3}	2.968	[97]
A. nibe	Southwest, TW	2.00×10^{-3}	2.907	[110]
A. nibe	Oman Sea ♀	1.21×10^{-2}	2.939	[95]
A. nibe	Oman Sea ♂	7.5×10^{-3}	3.074	[95]
A. nibe	Southwest, TW	1.45×10^{-2}	2.886	This study

Note: For Region: SCS, the South China Sea; TW, Taiwan; ECS, the East China Sea. ♀: female, ♂male.

For *A. nibe*, differences in LWR were also observed (Table 3), which might result from several possible reasons, including seasonal or environmental changes, as proposed in Schneider [111] and De Giosa [112] and the abovementioned data handling issues. Sample distributions by fish size and month were evenly distributed in this study, so the LWR estimation was considered representative.

4.2. Periodicity of Increment Formation—Edge Analysis

Both edge analysis and marginal increment analysis (MIA) are commonly used methods to verify the increment periodicity of the otolith [20,21], with edge analysis being a qualitative approach and MIA a quantitative approach [40]. While MIA is generally considered more robust than edge analysis, edge analysis (or marginal analysis) only requires recording the otolith margin type, thus is relatively affordable both in terms of equipment and time [26] and was chosen in this study to examine the periodicity of increment formation. For *P. macrocephalus*, the occurrence of the translucent bands was lowest in February, increased in March and April, and then jumped to a high level from June to October (Figure 5), suggesting that the translucent bands were formed once a year near the end of winter and the early spring. This pattern was supported by an additional MIA (Supplementary Figure S2), which showed a continuously increasing trend of marginal increment from February to October.

The spawning season of this fish is from April to October (also the main fishing season), and the gonadosomatic index (GSI) value declines significantly from September to October [113]. The fish migrate to Taiwan mainly for spawning, and many large spawning fish species inhabit deeper depths after spawning (e.g., white-mouth croaker [114]) and thus are more difficult to catch by fisheries. Though sampling could not be performed year-round, this study assumed that most fish were hatched before or in October and that the lowest proportion of the translucent bands might be in the month without fish samples (November to January). The MIA of Attaqi [32] also showed a single sinusoidal cycle in which the highest appeared from August to September and decreased sharply in October. Results from this study and Attaqi [32] both support ring marks being formed once a year.

On the other hand, consecutive monthly samples were available for *A. nibe*, and the edge analysis showed a continuous increase in the occurrence of the translucent band from the lowest month in December to the highest month in September (Figure 5). The single sinusoidal cycle pattern also supported the annual formation of the ring mark.

The edge analysis and MIA are valuable tools for the validation of age estimates, but both have their limitations (refer to Campana [13] for a detailed discussion on the topic). Therefore, although the current analyses supported annual formation of the ring mark for the two species, further analyses are still needed for definite validations; parameters that would possibly affect ring formation (e.g., fluctuations of environmental factors and timing of high feeding sources) are also worth further investigation.

4.3. Age Determination Methods

A growth ring is a part of an otolith that consists of one translucent zone and one opaque zone [115]. When the annual formation of the growth ring is validated, the age is usually determined by the number of ring counts [20,116,117]. However, this method ignores valuable information contained within a year's growth. Although age estimates for older fish may be best approximated by integer ages because the edge proportion is usually hard to identify for older fish (as seen in Figure 7), information on decimal ages can be gathered in younger fish. The quartile method could retain the lost information and be almost as fast as the integral method when conducting. Cross-validation simulation demonstrated that the quartile method could provide more appropriate estimates of age than the integral method.

The back-calculation method has been applied in many studies [32,33]. It needs an assumption of the birthdate of the fish. Vanderkooy [33] described the age from this method as the biological age, derived from the information of capture month (sampling month), number of annuli, and accepted birthdate estimate. This method could provide better distribution of age estimates when developing a growth model than the integral method if the study has representative samples of the sizes closest to the breeding and juvenile stages to determine the correct birthdate. In this regard, this method needs beforehand studies on fish reproduction or daily-rings counts to provide birthdate information. For *P. macrocephalus*, Attaqi [32] set 1st August as the hypothesized hatching date based on Wang [118], but the study from He [113] suggested that the spawning season could be as long as seven months since April. In this case, the difficulty in determining the birthdate would cause uncertainty in the estimate of age. Comparatively, the quartile method is more straightforward in interpreting the age of fish otoliths with both efficiency and accuracy.

Only Tsai [34] estimated the growth parameters of *P. macrocephalus* based on "daily rings", while others used annuli. The study suggested one ring formed every two days ("two-day ring") in the otolith based on the comparisons of the back-calculated birthdate and the GSI trend, and so the maximum age of this species was estimated to be 2.2 years. Besides lacking validations of the periodicity of increment formation and the first increment [27], the study also needs to consider the possibility that nondaily deposition can occur in suboptimal or extreme environmental conditions [119]. With these issues being addressed, daily rings for growth studies could provide more accurate estimates than the three methods described here, although the high cost of time and human resources need to be considered.

4.4. Growth Parameter Estimates—Global and Local

The global estimates of growth parameters for the Sciaenidae genera, together with those of this study, were collated and plotted for an overview (Figure 8) for the first time. The overall trend is downwards, from *P. anea* in the top-right corner to *P. macrocephalus*, *P. pawak*, *P. argentata*, and *A. nibe* in the bottom-left corner with comparatively higher L_∞ and smaller k.

Estimates of each species were clustered as a group, demonstrating that each species has a relatively similar growth performance [44], which is also noted for tuna species [43,45]. The elongated clusters were also found in catfish (Ariidae) and sardines (Clupeidae), despite the different sampling regions or aging methods [41,42]. Estimates of the same species but outside the group might have specific reasons or need further investigation [27,47,48]. For example, estimates of a_1 and a_3 of *P. argentata* are far away from the cluster of the species, and further examining the estimates could find that the two estimates were specifically on smaller fish (SL $\leq$ 26.59 cm for a_1 and SL $\leq$ 24.47 cm for a_3) that have smaller L_∞ but much higher k. As demonstrated in other studies, growth parameter estimates would be significantly affected by the lack of very young or old individuals [120], and the lack of older fish might underestimate asymptotic length [121].

The type of aging structure has been observed to significantly affect the estimation results [27,48]. However, although otolith, scale or length-frequency data were used for growth studies on the species in Table 1, differences in the estimates using different structures were not obvious for unknown reasons.

There were not many growth studies for *P. macrocephalus*. As explained earlier, Tsai's [34] study (m_4) was likely biased. Excluding this estimate, the estimates from the South China Sea (m_1 to m_3) departed from those from Taiwan waters (m_5 and mc_1–mc_3), suggesting possible regional effects (different environmental productivity or fishing pressure) on the estimation (Figure 8). Diverse results using different age determination methods in this study were observed in the figure. The recommended estimate with higher precision (using quartile method, mc_2) departs from Attaqi's [32] estimate that studied the fish in a similar location with smaller L_∞ and higher k. In addition to many possible technical reasons in the estimation [27], the phenomenon might affect high fishing pressure. The data used in Attaqi [32] was from 2010–2011, about a decade before the data used for this study (2018–2021). Total landings of the category of "white mouth croaker" (*P. macrocephalus* composed the majority of the catch) in the market had declined drastically from 800 tons in 2011 to <200 tons after 2018 [122]. High fishing pressure can lead to less competition for food and space in the fish population and thus produce higher growth coefficients [7–9,123]. On the other hand, high fishing pressure can cause the loss of larger individuals and a decrease of relative abundance of target species which have the capacity to grow to large size, and thus is attributed to a smaller estimate of asymptotic length [7,124]. The same inference was drawn in Yi [80] for *P. pawak* in which high fishing pressure contributed to the decline of L_∞ from 24.15 cm in 2008–2009 to 22.05 cm in 2018–2019, a decade later. These findings suggested the necessity to establish a management framework on the species to reduce fishing pressures.

For *A. nibe*, the growth estimates of VBGM (nc_2) from this study and from the previous studies (n_1 to n_18) formed a cluster, and all were within the 95% confidence interval. The best estimate in this study was the logistic model (nc_1); however, this estimate deviated from the 95% confidence interval ellipse, which was likely due to the difference in growth models. Except for one of the estimates from a study in the Oman Sea (n_10), the rest were all from the northeast Pacific Ocean (Taiwan, Japan, Korea). However, since this species is not a highly migrating species, those fish in different waters might belong to different stocks, so the growth parameters were not comparable.

Otolith, scale and length-frequency data have been used to estimate the parameters. Estimates using length-frequency data tend to have lower k than estimates using hard structures for dolphinfish and flyingfishes [27,48]; however, this pattern has not been observed for *A. nibe*. Compared to the estimates from Taiwan using the same aging material of otolith (n_11, n_12, n_15–n_18), the k estimates of this study were much higher. The obvious difference between these studies was that the data of previous studies were conducted about a decade ago (1993 and 2008). Based on market data, the annual landing of the species has dropped 75% from the early 2010s to the late 2020s and hence fishing pressure might have contributed to the increase in k estimates. The other difference was that this study included more samples of older fish (ages 9 and 10), which could make the estimates of L_∞ higher and k lower [121]. However, care needs to be taken that the quality of this study might be affected by the lack of small fish samples (age 1), as suggested in the following.

Although the VBGM was selected as the best model for *P. macrocephalus* because of the lowest $\Delta AICc$, the differences among models were not substantial (Table 2). The growth curves at Figure 6 also showed little differences in fit among the models where data are present; the differences occur mainly beyond the data (e.g., age 0). In addition, all models underestimate lengths of the oldest ages (4+). For *A. nibe*, the logistic model fits the age groups 2–8 better but diverted in age 0–1 and age 9+. Meanwhile, L_∞ estimates for both species fall beyond the range of observed data. These indicated the limitation of the lack of old and very young (0–1) age group samples in this study. To refine the growth estimates (especially the L_∞ estimates), further sample collections targeted at larger fish and very young (age 0–1) fish are needed in the future.

5. Conclusions

P. macrocephalus and *A. nibe* are among the most abundant commercial Sciaenidae species in southwestern Taiwan. Landings of the two species have dropped rapidly since 1998, presumably due to overfishing. Most studies on the estimation of growth parameters for the two species were conducted a decade ago, and the majority of them were published in the grey literature or in Chinese. To fill the information gap, this study provides reliable growth estimates and length–weight relationships (in Section 3.1 and Table 2) as updated information to the global literature and provides an overview of growth parameters for the two important croakers (in Table 1), which could be used for the development of a sound stock assessment on the resources. Comparison of the resulted growth parameters with those estimated a decade ago suggested that these two species have undergone high fishing pressures and induced consequent higher growth coefficients and smaller asymptotic length, and thus suggested the need for establishing a management framework to reduce fishing pressures.

Methodology analyses of this study suggested that the quartile of marginal growth band method could provide more appropriate estimates of age from otoliths than the other two methods. The results suggested the VBGM as the optimal growth model for *P. macrocephalus* and the logistic model for *A. nibe*, based on the samples collected. However, both results need further refinements by collecting more samples of large fish and very young fish. Finally, the auximetric plot of this study supported the argument that the same fish species has a similar growth performance and could be used as a quick tool for identifying outliers of growth estimates for further investigations.

Supplementary Materials: The following supporting information can be downloaded at: https://www.mdpi.com/article/10.3390/fishes7050281/s1, Table S1: Detail information on the estimates of growth parameters for *Pennahia* and *Atrobucca* species from studies published globally; Table S2: Length and weight raw data for *Pennahia macrocephalus*; Table S3: Length and weight raw data for *Atrobucca nibe;* Figure S1: Images of otoliths (sagitta, lapillus, and asteriscus) of *P. macrocephalus* and *A. nibe*; Figure S2: The marginal increment ratio (MIR) of *P. macrocephalus.*

Author Contributions: Conceptualization, S.-C.H. and S.-K.C.; Data curation, S.-C.H. and J.-S.H.; Formal analysis, S.-C.H. and T.-L.Y.; Funding acquisition, S.-K.C. and J.-S.W.; Investigation, S.-C.H. and C.-C.L.; Methodology, S.-C.H., S.-K.C. and T.-L.Y.; Project administration, S.-K.C. and C.-C.L.; Resources, S.-K.C.; Software, S.-C.H. and T.-L.Y.; Supervision, S.-K.C. and J.-S.W.; Validation, J.-S.H.; Visualization, S.-C.H.; Writing—original draft, S.-C.H.; Writing—review & editing, S.-K.C. and C.-C.L. All authors have read and agreed to the published version of the manuscript.

Funding: This research was funded by the Ministry of Science and Technology (MOST), Taiwan, grant number MOST 106-2611-M-110-006, MOST 108-2611-M-110-002, MOST 109-2611-M-110-004.

Institutional Review Board Statement: Not applicable.

Data Availability Statement: Data of this study is included in the report, the Supplementary Materials or may be made available by the authors.

Acknowledgments: The authors appreciate the technical support from Ying-Chu Chen in the otolith ring reading and the financial support of the Ministry of Science and Technology for the open access publication fee.

Conflicts of Interest: The authors declare no conflict of interest.

References

1. Maunder, M.N.; Piner, K.R. Contemporary Fisheries Stock Assessment: Many Issues Still Remain. *ICES J. Mar. Sci.* **2015**, *72*, 7–18. [CrossRef]
2. Francis, R.I.C.C.; Aires-da-Silva, A.M.; Maunder, M.N.; Schaefer, K.M.; Fuller, D.W. Estimating Fish Growth for Stock Assessments Using Both Age–Length and Tagging-Increment Data. *Fish. Res.* **2016**, *180*, 113–118. [CrossRef]
3. Then, A.Y.; Hoenig, J.M.; Hall, N.G.; Hewitt, D.A. Handling editor: Ernesto Jardim Evaluating the Predictive Performance of Empirical Estimators of Natural Mortality Rate Using Information on over 200 Fish Species. *ICES J. Mar. Sci.* **2015**, *72*, 82–92. [CrossRef]
4. Farley, J.; Eveson, P.; Krusic-Golub, K.; Sanchez, C.; Roupsard, F.; Nicol, S.; Leroy, B.; Smith, N.; Chang, S.-K. *Project 35: Age, Growth and Maturity of Bigeye Tuna in the Western and Central Pacific Ocean*; CSIRO: Rarotonga, Cook Islands, 2017; p. 51.
5. McKechnie, S.; Pilling, G.; Hampton, J. Stock Assessment of Bigeye Tuna in the Western and Central Pacific Ocean. In Proceedings of the Thirteenth Regular Session of the Scientific Committee, Rarotonga, Cook Islands, 9–17 August 2017; p. 149.
6. Maunder, M.N.; Crone, P.R.; Punt, A.E.; Valero, J.L.; Semmens, B.X. Growth: Theory, Estimation, and Application in Fishery Stock Assessment Models. *Fish. Res.* **2016**, *180*, 1–3. [CrossRef]
7. Jennings, S.; Greenstreet, S.P.R.; Reynolds, J.D. Structural Change in an Exploited Fish Community: A Consequence of Differential Fishing Effects on Species with Contrasting Life Histories. *J. Anim. Ecol.* **1999**, *68*, 617–627. [CrossRef]
8. Polacheck, T.; Eveson, J.P.; Laslett, G.M. Increase in Growth Rates of Southern Bluefin Tuna (*Thunnus maccoyii*) over Four Decades: 1960 to 2000. *Can. J. Fish. Aquat. Sci.* **2004**, *61*, 307–322. [CrossRef]
9. Reznick, D.N.; Ghalambor, C.K. Can Commercial Fishing Cause Evolution? Answers from Guppies (*Poecilia reticulata*). *Can. J. Fish. Aquat. Sci.* **2005**, *62*, 791–801. [CrossRef]
10. Kimura, D.K. Extending the von Bertalanffy Growth Model Using Explanatory Variables. *Can. J. Fish. Aquat. Sci.* **2008**, *65*, 13. [CrossRef]
11. Lavin, C.P.; Gordó-Vilaseca, C.; Stephenson, F.; Shi, Z.; Costello, M.J. Warmer Temperature Decreases the Maximum Length of Six Species of Marine Fishes, Crustacean, and Squid in New Zealand. *Environ. Biol. Fishes* **2022**. [CrossRef]
12. Campana, S.E.; Neilson, J.D. Microstructure of Fish Otoliths. *Can. J. Fish. Aquat. Sci.* **1985**, *42*, 1014–1032. [CrossRef]
13. Campana, S.E. Accuracy, Precision and Quality Control in Age Determination, Including a Review of the Use and Abuse of Age Validation Methods. *J. Fish Biol.* **2001**, *59*, 197–242. [CrossRef]
14. Morat, F.; Wicquart, J.; Schiettekatte, N.M.D. Individual Back-Calculated Size-at-Age Based on Otoliths from Pacific Coral Reef Fish Species. *Sci. Data* **2020**, *7*, 370. [CrossRef] [PubMed]
15. Lai, H.L.; Gunderson, D.R. Effects of Ageing Errors on Estimates of Growth, Mortality, and Yield per Recruit for Walleye Pollock (*Theragra chlacogramma*). *Fish. Res.* **1987**, *5*, 287–302.
16. Beamish, R.J.; McFarlane, G.A. A. A Discussion of the Importance of Ageing Errors, and an Application to Walleye Pollock: The World's Largest Fishery. In *Recent Developments in Fish Otolith Research*; Secor, D.H., Dean, J.M., Campana, S.E., Eds.; University of South Carolina Press: Columbia, SC, USA, 1995; pp. 545–565.
17. Porta, M.J.; Snow, R.A. Validation of Annulus Formation in White Perch Otoliths, Including Characteristics of an Invasive Population. *J. Freshw. Ecol.* **2017**, *32*, 489–498. [CrossRef]
18. Froeschke, B.; Allen, L.G.; Pondella, D.J. Life History and Courtship Behavior of Black Perch, *Embiotoca jacksoni* (Teleostomi: Embiotocidae), from Southern California. *Pac. Sci.* **2007**, *61*, 521–531. [CrossRef]
19. Okamura, H.; Punt, A.E.; Semba, Y.; Ichinokawa, M. Marginal Increment Analysis: A New Statistical Approach of Testing for Temporal Periodicity in Fish Age Verification. *J. Fish Biol.* **2013**, *82*, 1239–1249. [CrossRef]
20. Smith, J. Age Validation of Lemon Sole (*Microstomus kitt*), Using Marginal Increment Analysis. *Fish. Res.* **2014**, *157*, 41–46. [CrossRef]
21. Hidalgo-de-la-Toba, J.A.; Morales-Bojórquez, E.; González-Peláez, S.S.; Bautista-Romero, J.J.; Lluch-Cota, D.B. Modeling the Temporal Periodicity of Growth Increments Based on Harmonic Functions. *PLoS ONE* **2018**, *13*, 0196189. [CrossRef]
22. Prince, E.D.; Lee, D.W.; Berkeley, S.A. Use of Marginal Increment Analysis to Validate the Anal Spine Method for Ageing Atlantic Swordfish and Other Alternatives for Age Determination. *Col. Vol. Sci. Pap. ICCAT* **1988**, *27*, 194–201.
23. Pearson, D.E. Timing of Hyaline-Zone Formation as Related to Sex, Location, and Year of Capture in Otoliths of the Widow Rockfish, *Sebastes entomelas*. *Fish. Bull.* **1996**, *94*, 190–197.
24. Laidig, T.E.; Pearson, D.E.; Sinclair, L.L. Age and Growth of Blue Rockfish (*Sebastes mystinus*) from Central and Northern California. *Fish Bull* **2003**, *101*, 800–808.

25. Young, J.; Drake, A.; Groisson, A.-L. *Age and Growth of Broadbill Swordfish (Xiphias glades) from Eastern Australian Waters—Preliminary Results;* CSIRO, Division of Marine Research: Hobart, Australia, 2003.
26. Chang, S.-K.; Chou, Y.-T.; Hoyle, S.D. Length-Weight Relationships and Otolith-Based Growth Curves for Brushtooth Lizardfish off Taiwan with Observations of Region and Aging-Material Effects on Global Growth Estimates. *Front. Mar. Sci.* **2022**, *9*, 921594. [CrossRef]
27. Chang, S.-K.; Yuan, T.-L.; Hoyle, S.D.; Farley, J.H.; Shiao, J.-C. Growth Parameters and Spawning Season Estimation of Four Important Flyingfishes in the Kuroshio Current off Taiwan and Implications from Comparisons with Global Studies. *Front. Mar. Sci.* **2022**, *8*, 747382. [CrossRef]
28. Yamaguchi, A.; Kume, G.; Takita, T. Geographic Variation in the Growth of White Croaker, *Pennahia argentata*, off the Coast of Northwest Kyushu, Japan. *Environ. Biol. Fishes* **2004**, *71*, 179–188. [CrossRef]
29. Yan, Y.R.; Hou, G.; Lu, H.S.; Li, Z.L. Growth Characteristics and Population Composition of Big-Head Pennah Croaker, *Pennahia macrocephalus* in the Beibu Gulf. *J. Fish. Sci. China* **2010**, *40*, 61–68.
30. Ju, P.L.; Yang, L.; Lu, Z.B.; Yang, S.Y.; Du, J.G.; Zhong, H.Q.; Chen, J.; Xiao, J.M.; Chen, M.R.; Zhang, C.Y. Age, Growth, Mortality and Population Structure of Silver Croaker *Pennahia argentata* (Houttuyn, 1782) and Red Bigeye *Priacanthus macracanthus* Cuvier, 1829 in the North-Central Taiwan Strait. *J. Appl. Ichthyol.* **2016**, *32*, 652–660. [CrossRef]
31. Jeon, B.S.; Choi, J.H.; Kim, D.N.; Im, Y.J.; Lee, H.W. Age and Growth of White Croaker *Pennahia argentata* in the Southern Sea of Korea by Otolith Analysis. *Korean J. Fish. Aquat. Sci.* **2021**, *54*, 53–63.
32. Attaqi, A.N. Age Structure and Growth of Big-Head Pannah Croaker, *Pennahia macrocephalus* in the Southwestern Waters off Taiwan: An Approach Using Both Thin-Section and Weight of Otoliths. Master's Thesis, College of Ocean Science and Resource, National Taiwan Ocean University,, Keelung, Taiwan, 2018.
33. Vanderkooy, S.; Carroll, J.; Elzey, S.; Gilmore, J.; Kipp, J. (Eds.) *A Practical Handbook for Determining the Ages of Gulf of Mexico and Atlantic Coast Fishes*, 3rd ed.; Gulf States Marine Fisheries Commission and Atlantic States Marine Fisheries Commission: Ocean Springs, MS, USA, 2020.
34. Tsai, T.W. Age and Growth of Big Head Pennah Croaker (*Pennahia macrocephalus*) Sampled from the Southwestern Waters off Taiwan. Master's Thesis, Department of Fisheries Production and Management of National Kaohsiung University of Science and Technology, Kaohsiung, Taiwan, 2009.
35. Farley, J.; Krusic-Golub, K.; Eveson, P.; Clear, N.; Roupsard, F.; Sanchez, C.; Nicol, S.; Hampton, J. Age and Growth of Yellowfin and Bigeye Tuna in the Western and Central Pacific Ocean from Otoliths (No. WCPFC-SC16-2020/SA-WP-02). In Proceedings of the WCPFC Scientific Committee 16th Regular Session, Electronic Meeting, 11–20 August 2020.
36. Roff, D.A. A Motion for the Retirement of the Von Bertalanffy Function. *Can. J. Fish. Aquat. Sci.* **1980**, *37*, 127–129. [CrossRef]
37. Katsanevakis, S. Modelling Fish Growth: Model Selection, Multi-Model Inference and Model Selection Uncertainty. *Fish. Res.* **2006**, *81*, 229–235. [CrossRef]
38. Katsanevakis, S.; Maravelias, C.D. Modelling Fish Growth Multi-model Inference as a Better Alternative to a Priori Using von Bertalanffy Equation. *Fish Fish.* **2008**, *9*, 178–187. [CrossRef]
39. Williams, A.J.; Farley, J.H.; Hoyle, S.D.; Davies, C.R.; Nicol, S.J. Spatial and Sex-Specific Variation in Growth of Albacore Tuna (*Thunnus alalunga*) across the South Pacific Ocean. *PLoS ONE* **2012**, *7*, e39318. [CrossRef] [PubMed]
40. Carbonara, P.; Intini, S.; Kolitari, J.; Joksimović, A.; Milone, N.; Lembo, G.; Casciaro, L.; Bitetto, I.; Zupa, W.; Spedicato, M.T.; et al. A Holistic Approach to the Age Validation of *Mullus barbatus* L., 1758 in the Southern Adriatic Sea (Central Mediterranean). *Sci. Rep.* **2018**, *8*, 13219. [CrossRef] [PubMed]
41. Velasco, G.; Oddone, M.C. Growth Parameters and Growth Performance Indexes for Some Populations of Marine Catfishes (Actinopterygii, Siluriformes, Ariidae). *Acta Biol. Leopoldensia* **2004**, *26*, 307–313.
42. Ba, K.; Thiaw, M.; Lazar, N.; Sarr, A.; Brochier, T.; Ndiaye, I.; Faye, A.; Sadio, O.; Panfili, J.; Thiaw, O.T.; et al. Resilience of Key Biological Parameters of the Senegalese Flat Sardinella to Overfishing and Climate Change. *PLoS ONE* **2016**, *11*, e0156143. [CrossRef]
43. Murua, H.; Rodríguez-Marín, E.; Neilson, J.; Farley, J.; Juan-Jordá, M.J. Fast versus Slow Growing Tuna Species: Age, Growth, and Implications for Population Dynamics and Fisheries Management. *Rev. Fish Biol. Fish.* **2017**, *27*, 733–773. [CrossRef]
44. Pauly, D.; Munro, J.L. Once More on the Comparison of Growth in Fish and Invertebrates. *Fishbyte* **1984**, *2*, 21.
45. Juan-Jordá, M.J.; Mosqueira, I.; Dulvy, N. The Conservation and Management of Tunas and Their Relatives: Setting Life History Research Priorities. *PLoS ONE* **2013**, *8*, e70405. [CrossRef] [PubMed]
46. Pauly, D.; Binohlan, C. FishBase and AUXIM as Tools for Comparing the Life-History Patterns, Growth and Natural Mortality of Fish: Applications to Snappers and Groupers. *ICLARM Conf. Proc.* **1996**, *48*, 223–247.
47. Chang, S.-K.; DiNardo, G.; Farley, J.; Brodziak, J.; Yuan, Z.-L. Possible Stock Structure of Dolphinfish (*Coryphaena hippurus*) in Taiwan Coastal Waters and Globally Based on Reviews of Growth Parameters. *Fish. Res.* **2013**, *147*, 127–136. [CrossRef]
48. Chang, S.-K.; Maunder, M. Aging Material Matters in the Estimation of von Bertalanffy Growth Parameters for Dolphinfish (*Coryphaena hippurus*). *Fish. Res.* **2012**, *119–120*, 147–153. [CrossRef]
49. Nelson, J.S. *Fishes of the World*, 4th ed.; John Wiley & Sons Inc: Hoboken, NJ, USA, 2006.
50. Shao, K.T. Taiwan Fish Database. WWW Web Electronic Publication. 2022. Available online: http://fishdb.sinica.edu.tw (accessed on 21 February 2022).
51. Tang, D.-S. A study of sciaenoid fishes of China. *Amoy Mar. Biol. Bull.* **1937**, *2*, 47–88.

52. Jordan, D.S.; Thompson, W.F. A Review of the Sciaenoid Fishes of Japan. *Proc. United States Natl. Mus.* **1911**, *39*, 241–261. [CrossRef]
53. Chang, S.-K.; Yuan, T.L.; Wang, S.P.; Chang, Y.J.; DiNardo, G.; Chen, Y.; Mello, L. Deriving a Statistically Reliable Abundance Index from Landings Data: An Application to the Taiwanese Coastal Dolphinfish Fishery with a Multispecies Feature. *Trans. Am. Fish. Soc.* **2019**, *148*, 106–122. [CrossRef]
54. Froese, R. Cube Law, Condition Factor and Weight–Length Relationships: History, Meta-Analysis and Recommendations. *J. Appl. Ichthyol.* **2006**, *22*, 241–253. [CrossRef]
55. Zakeyudin, M.-S.; Mat, M.; Salmah, M.; Md Sah, A.S.R.; Hassan, A. Assessment of Suitability of Kerian River Tributaries Using Length-Weight Relationship and Relative Condition Factor of Six Freshwater Fish Species. *J. Environ. Earth Sci.* **2012**, *2*, 52–60.
56. Jisr, N.; Younes, G.; Sukhn, C.; El-Dakdouki, M.H. Length-Weight Relationships and Relative Condition Factor of Fish Inhabiting the Marine Area of the Eastern Mediterranean City, Tripoli-Lebanon. *Egypt. J. Aquat. Res.* **2018**, *44*, 299–305. [CrossRef]
57. James, G.; Witten, D.; Hastie, T.; Tibshirani, R. (Eds.) *An Introduction to Statistical Learning: With Applications in R*; Springer Texts in Statistics; Springer: New York, NY, USA, 2013; ISBN 978-1-4614-7137-0.
58. Ashworth, E.C.; Hesp, S.A.; Hall, N.G. A New Proportionality-Based Back-Calculation Approach, Which Employs Traditional Forms of Growth Equations, Improves Estimates of Length at Age. *Can. J. Fish. Aquat. Sci.* **2017**, *74*, 1088–1099. [CrossRef]
59. Ogle, D.H. *Introductory Fisheries Analyses with R*, 1st ed.; CRC Press: New York, NY, USA, 2016; ISBN 978-1-315-37198-6.
60. Karlson, S.; Michalsen, K.; Folkvord, A.; Karlson, S.; Age, A. Age Determination of Atlantic Halibut (*Hippoglossus hippoglossus* L.) along the Coast of Norway: Status and Improvements. *ICES J. Mar. Sci.* **2012**, *70*, 50–55. [CrossRef]
61. Gompertz, B. On the Nature of the Function Expressive of the Law of Human Mortality, and on a New Mode of Determining the Value of Life Contingencies. *Philos. Trans. R. Soc.* **1825**, *115*, 513–583. [CrossRef]
62. Von Bertalanffy, L. A Quantitative Theory of Organic Growth. *Hum. Biol.* **1938**, *10*, 181–213.
63. Richards, F.J. A Flexible Growth Function for Empirical Use. *J. Exp. Bot.* **1959**, *10*, 290–301. [CrossRef]
64. Ricker, W.E. Computation and Interpretation of Biological Statistics of Fish Populations. *Bull. Fish. Res. Board Can.* **1975**, *191*, 1–382.
65. Flinn, S.A.; Midway, S.R. Trends in Growth Modeling in Fisheries Science. *Fishes* **2021**, *6*, 1. [CrossRef]
66. Quist, M.C.; Pegg, M.A.; Devries, D.R. Age and Growth. Fisheries Techniques. In *Age and Growth. Fisheries Techniques*; Zale, A.V., Parrish, D.L., Sutton, T.M., Eds.; American Fisheries Society: Bethesda, MD, USA, 2012; pp. 677–731.
67. R-Development Core Team. *R: A Language and Environment for Statistical Computing*; R Foundation for Statistical Computing: Vienna, Austria, 2021.
68. Akaike, H. Information Theory and an Extension of the Maximum Likelihood Principle. In Proceedings of the 2nd International Symposium on Information Theory, Tsahkadsor, Armenia, 2–8 September 1973; Volume I. pp. 267–281.
69. Burnham, K.P.; Anderson, D.R. *Model Selection and Multimodel Inference: A Practical Information-Theoretic Approach*, 2nd ed.; Springer: New York, NY, USA, 2002; ISBN 978-0-387-95364-9.
70. Wang, X.; Qiu, Y.; Du, F.; Lin, Z.; Sun, D.; Huang, S. Population Parameters and Dynamic Pool Models of Commercial Fishes in the Beibu Gulf, Northern South China Sea. *Chin. J. Oceanol. Limnol.* **2012**, *30*, 105–117. [CrossRef]
71. Chen, Z.; Huang, Z.; Qiu, Y. 南海北部白姑鱼生长和死亡参数的估算 (Estimation of Growth and Mortality Parameters of *Argyrosomus argentatus* in Northern South China Sea). *Chin. J. Appl. Ecol.* **2005**, *16*, 712–716. (In Chinese). Available online: http://www.cqvip.com/qk/90626a/200504/15404417.html (accessed on 1 October 2021).
72. FishBase. World Wide Web Electronic Publication. 2022. Available online: https://www.fishbase.de/summary/Pennahia-anea (accessed on 1 October 2021).
73. Nhuận, M.C.; Khương, Đ.V. Stock Biological Characteristics of Big-Head Pennah Croacer (*Pennahia macrocephalus*, Tang 1937) in the Shared Fishing Zone in the Gulf of Tonkin, during 2006–2010. *J. Fish. Sci. Technol. Vietnam.* **2013**, 2, 83–88.
74. Liu, H.C.; Tzeng, W.N. Age and growth of the white croaker, *Argyrosomus argentatus* (HOUTTUYN), in the Southern part of the East China Sea and Taiwan Strait. *J. Fish. Soc. Taiwan* **1972**, *1*, 21–38.
75. Lu, Z.; Dai, Q.; Yan, Y. Population dynamics of major demersal fish species offshore Fujian. *J. Oceanogr. Taiwan Strait Chin.* **1999**, *18*, 100–105.
76. Hu, Y.; Qian, S. A study on the age and growth of the white croaker. *Mar. Fish.* **1989**, *4*, 158–162.
77. Higuchi, T.; Yamaguchi, A.; Takita, T. Age and Growth of White Croaker, *Pennahia argentata*, in the Ariake Sound, Japan. *Bull. Fac. Fish. Nagasaki Univ. Jpn.* **2003**, 47–51.
78. Kakuda, S.; Matsumoto, K. On the Age and Growth of the White Croaker *Argyrosomus argentatus*. *Hiroshima Univ.* **1977**, *16*, 115–122.
79. Yan, Y.; Hou, G.; Lu, H.; Yin, Q. Age and Growth of *Pennahia pawak* in the Beibu Gulf. *Chin. Fish. Sci.* **2011**, 145–155.
80. Yi, X.; Qiu, K.; Zhou, X.; Zhao, C.; Deng, Y.; He, X.; Yan, Y. Analysis of Fishery Biology of *Pennahia pawak* in Beibu Gulf. *J. Shanghai Ocean Univ.* **2021**, *30*, 515–524. [CrossRef]
81. Bhuyan, S.K.; Kumar, P.; Jena, J.K.; Pillai, B.R.; Chakraborty, S.K. Studies on the Growth of Otolithes Ruber (Bloch & Schneider, 1801), Johnius Carutta Bloch, 1793 and *Pennahia macrophthalmus* (Bleeker, 1850) from Paradeep Coast, Orissa, India. *Indian J. Fish* **2012**, *59*, 89–93.
82. Ingles, J.; Pauly, D. An Atlas of the Growth, Mortality and Recruitment of Philippine Fishes. Available online: https://www.worldfishcenter.org/publication/atlas-growth-mortality-and-recruitment-philippine-fishes (accessed on 2 May 2022).

83. Ziegler, B. Growth and Mortality Rates of Some Fishes of Manila Bay, Philippines as Estimated from the Analysis of Length Frequencies. Master's Thesis, Kiel University, Kiel, Germany, 1979.
84. Abu Talib, A. *Population Dynamics of Big-Eye Croaker (Pennahia macrophthalmus, Sciaenidae) off Kedah, Penang and Perak States, Malaysia*; FAO Fisheries Report; FAO: Rome, Italy, 1988.
85. Chakraborty, S.K.; Deshmukh, V.D.; Khan, M.Z.; Vidyasagar, K.; Raje, S.G. Estimates of Growth, Mortality, Recruitment Pattern and MSY of Important Resources from the Maharashtra Coast. *J. Indian Fish. Assoc.* **1994**, *24*, 1–39.
86. Chakraborty, S.K. Stock assessment of big-eye croaker, *Pennahia macropthalmus* (Bleeker) (Pisces/Perciformes/Sciaenidae) from Bombay waters. *Indian J. Mar. Sci.* **1996**, *25*, 316–319.
87. Jayasankar, P. Population Dynamics of Big-Eye Croaker *Pennahia macrophthalmus* and Blotched Croaker *Nibea maculata* (Pisces/Perciformes/Sciaenidae) in the Trawling Grounds off Rameswaram Island, East Coast of India. *Indian J. Mar. Sci.* **1995**, *24*, 153–157.
88. Chakraborty, S.K.; Devadoss, P.; Manojkumar, P.P.; Feroz Khan, M.; Jayasankar, P.; Sivakami, S.; Gandi, V.; Appanna Sastry, Y.; Raju, A.; Livingston, P.; et al. Fishery, Biology and Stock Assessment of Jew Fish Resources of India. In *Marine Fisheries Research and Management*; Pillai, V.N., Menon, N.G., Eds.; Central Marine Fisheries Research Institute: Cochin, India, 2000; pp. 604–616.
89. Chakraborty, S.K. Growth Studies of Sciaenids from Mumbai Waters Using Bhattarcharya's Method. *Naga ICLARM Q.* **2001**, *24*, 40–41.
90. Menon, M.; Maheswarudu, G.; Rohit, P.; Laxmilatha, P.; Madhumita, D.; Rao, K.N. Biology and Stock Assessment of the Bigeye Croaker *Pennahia Anea* (Bloch, 1793) Landed along Andhra Pradesh, North-East Coast of India. *Indian J. Fish* **2015**, *62*, 46–51.
91. Wagiyo, K.; Tirtadanu, T.; Chodriyah, U.; Wang, X.; Qiu, Y.; Du, F.; Lin, Z.; Sun, D.; Huang, S. Biology Characteristic, Abundance Index and Fishing Aspect of Donkey Croaker *Pennahia anea* (Bloch, 1793) in the Tangerang Waters. *E3S Web Conf.* **2020**, *30*, 105–117. [CrossRef]
92. Pauly, D. A Preliminary Compilation of Fish Length Growth Parameters. *Ber. Inst. Meereskd. Univ. Kiel* **1978**, *55*, 200.
93. Kao, P. Stock Discrimination of Blackmouth Croaker *Atrobucca Nibe* in Taiwan. Master's Thesis, The Institute of Fisheries Science, National Taiwan University, Taipei, Taiwan, 2019. (In Chinese).
94. Sato, T. Fishery Biology of Black Croaker, *Argyrosomus nibe* (Jordan et Thompson). I. On the Age and Growth of the Black Croaker in the Central and Southern Parts of the East China Sea. *Bull. Seikai Reg. Fish Res. Lab.* **1963**, *29*, 75–96.
95. Salarpouri, A.; Kaymaram, F.; Valinassab, T.; Behzadi, S.; Darvishi, M.; Kamali, E.; Rezwani, S.; Memarzadeh, M.; Karami, N. *A Survey on Black Mouth Croaker (Atrobucca nibe) Resources in the North-West of Oman Sea*; Iranian Fisheries Science Research Institute: Tehran, Iran, 2015.
96. Chang, J.J. Effects of Fishing on Life History Parameters of Atrobucca Nibe from Surrounding Waters of Guei-Shan Island, Northeastern Taiwan. Master's Thesis, College of Ocean Science and Resource, National Taiwan Ocean University, Keelung, Taiwan, 2008.
97. Tsai, C.N. Age and growth of black croaker, *Atrobucca nibe*, in adjacent waters of Tungkang, Taiwan. Master's Thesis, National Taiwan Ocean Univeristy, Keelung, Taiwan, 1993.
98. Hwang, G.M.; Chen, C.T. Fishery Biology of Black Croaker, *Atrobucca Nibe* (Jordan et Thompson) from the Surrounding Waters of Guei Shan Island, Taiwan. *J. Fish. Soc. Taiwan* **1984**, *11*, 35–52.
99. Memon, K.; Liu, Q.; Kalhoro, M.; Nazir, K.; Waryani, B.; Chang, M.S.; Mirjatt, A.N. Estimation of Growth and Mortality Parameters of Croaker *Atrobucca alcocki* in Pakistani Waters. *J. Agric. Sci. Technol.* **2016**, *18*, 669–679.
100. Zhu, L.; Li, L.; Liang, Z. Comparison of Six Statistical Approaches in the Selection of Appropriate Fish Growth Models. *Chin. J. Oceanol. Limnol.* **2009**, *27*, 457–467. [CrossRef]
101. Kimmerer, W.; Avent, S.R.; Bollens, S.M.; Feyrer, F.; Grimaldo, L.F.; Moyle, P.B.; Nobriga, M.; Visintainer, T. Variability in Length–Weight Relationships Used to Estimate Biomass of Estuarine Fish from Survey Data. *Trans. Am. Fish. Soc.* **2005**, *134*, 481–495. [CrossRef]
102. Morey, G.; Moranta, J.; Massutí, E.; Grau, A.; Linde, M.; Riera, F.; Morales-Nin, B. Weight–Length Relationships of Littoral to Lower Slope Fishes from the Western Mediterranean. *Fish. Res.* **2003**, *62*, 89–96. [CrossRef]
103. CMFRI/BOBP-IGO/GoI. *Training Manual on Stock Assessment of Tropical Fishes*; Central Marine Fisheries Research Institute (CMFRI): Kochi, India, 2016; Volume 156.
104. Mredul, M.M.H.; Alam, M.R.; Akkas, A.B.; Sharmin, S.; Pattadar, S.N.; Ali, M.L. Some Reproductive and Biometric Features of the Endangered Gangetic Leaf Fish, *Nandus nandus* (Hamilton, 1822): Implication to the Baor Fisheries Management in Bangladesh. *Aquac. Fish.* **2021**, *6*, 634–641. [CrossRef]
105. Tesch, F.W. Age and growth. In *Methods for Assessment of Fish Production in Freshwaters*; Ricker, W.E., Ed.; Blackwell Scientific Publications: Oxford, UK, 1968; pp. 101–113.
106. Lee, J.J. Reproductive Biology of Big Head Pennah Croaker (*Pennahia macrocephalus*) Sampled from the Southwestern Waters off Taiwan. Master's Thesis, Department of Fisheries Production and Management of National Kaohsiung University of Science and Technology, Kaohsiung, Taiwan, 2010. (In Chinese).
107. Matsui, I. 東海黄海に於ける底曳網漁場と底棲生物群聚との関係に就て (Relation between the Trawling Grounds and the Benthic Association in the East China Sea and the Yellow Sea). *Nippon Suisan Gakkaishi Jpn.* **1951**, *16*, 159–167. [CrossRef]
108. Kao, P.; Rong, J.; Zhang, W.; Li, Y. A Review of the Research on the Biology of *Atrobucca nibe* Fishery. *Fish. Ext. Natl. Taiwan Univ.* **2018**, *29*, 46–57.

109. Liu, F.; Chen, J. A survey of *Nibea nibe* Jordan et Thompson in northern Taiwan. *China Aquat. Prod.* **1954**, *24*, 23–45.
110. Hsiao, L.; Chen, C.; Wu, C.; Chen, Y.; He, J. Research on Reproductive Biology of Heliconia in the Southwest Waters of Taiwan. *J. Taiwan Fish. Res.* **2017**, *25*, 15–25.
111. Schneider, J.C.; Laarman, P.W.; Gowing, H. Length-Weight Relationships. Chapter 17. In *Manual of Fisheries Survey Methods II: With Periodic Updates*; Michigan Department of Natural Resources, Fisheries Division: Lansing, MI, USA, 2000; pp. 1–18.
112. De Giosa, M.; Czerniejewski, P.; Rybczyk, A. Seasonal Changes in Condition Factor and Weight-Length Relationship of Invasive *Carassius gibelio* (Bloch, 1782) from Leszczynskie Lakeland, Poland. *Adv. Zool.* **2014**, *2014*, e678763. [CrossRef]
113. He, J.S.; Huang, X.H.; Wu, Y.S.; Lai, C.C.; Weng, J.S. Reproductive Biology of the Females of the *Pennahia macrocephalus* in the Southwestern Waters of Taiwan. *J. Taiwan Fish. Res.* **2020**, *28*, 13–24.
114. Costa, M.D.P.; Muelbert, J.H.; Moraes, L.E.; Vieira, J.; Castello, J.P. Estuarine Early Life Stage Habitat Occupancy Patterns of Whitemouth Croaker *Micropogonias furnieri* (Desmarest, 1830) from the Patos Lagoon, Brazil. *Fish. Res.* **2014**, *160*, 77–84. [CrossRef]
115. Putnis, I.; Korņilovs, G. Manual for Age Determination of Baltic Herring. In Proceedings of the ICES 2008 Report of the Workshop on Age Reading of Baltic Herring (WKARBH), Riga, Latvia, 9–13 June 2008; p. 37.
116. Newman, S.J.; Dunk, I.J. Age Validation, Growth, Mortality, and Additional Population Parameters of the Goldband Snapper (*Pristipomoides multidens*) off the Kimberley Coast of Northwestern Australia. *Fish. Bull.* **2003**, *101*, 116–128.
117. Hara, K.; Furumitsu, K.; Aoshima, T.; Kanehara, H.; Yamaguchi, A. Age, Growth, and Age at Sexual Maturity of the Commercially Landed Skate Species, *Dipturus chinensis* (Basilewsky, 1855), in the Northern East China Sea. *J. Appl. Ichthyol.* **2018**, *34*, 66–72. [CrossRef]
118. Wang, S.B. *Survey of Fisheries Economic Activities in the Nearby Area of Mai-Liao, Yunlin County, Taiwan*; The Final Report for Capture Fisheries in 2012; Century Engineering Consultant Co., Ltd.: New Taipei, Taiwan, 2011. (In Chinese)
119. Jones, C. Determining Age of Larval Fish with the Otolith Increment Technique. *Fish. Bull.* **1986**, *84*, 91–102.
120. Cailliet, G.M.; Goldman, K.J.; Carrier, J.; Musick, J.A.; Heithaus, M. (Eds.) *Age Determination and Validation in Chondrichthyan Fishes*; CRC Press: Boca Raton, FL, USA, 2004.
121. Chen, Y.; Mello, L.G.S. Growth and Maturation of Cod (*Gadus morhua*) of Different Year-Classes in the Northwest Atlantic, NAFO Subdivision 3Ps. *Fish. Res.* **1999**, *42*, 87–101. [CrossRef]
122. Fisheries Agency. *Fisheries Statistical Yearbook—Taiwan, Kinmen and Matsu Area*; Fisheries Agency, Council of Agriculture, Executive Yuan: Taipei, Taiwan, 2021. (In Chinese)
123. Haimovici, M.; Cavole, L.M.; Cope, J.M.; Cardoso, L.G. Long-Term Changes in Population Dynamics and Life History Contribute to Explain the Resilience of a Stock of *Micropogonias furnieri* (Sciaenidae, Teleostei) in the SW Atlantic. *Fish. Res.* **2021**, *237*, 105878. [CrossRef]
124. Rice, J.; Gislason, H. Patterns of Change in the Size Spectra of Numbers and Diversity of the North Sea Fish Assemblage, as Reflected in Surveys and Models. *ICES J. Mar. Sci.* **1996**, *53*, 1214–1225. [CrossRef]

 MDPI

Article

Timing of Increment Formation in Atlantic Bluefin Tuna (*Thunnus thynnus*) Otoliths

Enrique Rodriguez-Marin [1,*], Dheeraj Busawon [2], Patricia L. Luque [3], Isabel Castillo [1], Nathan Stewart [2], Kyne Krusic-Golub [4], Aida Parejo [1] and Alex Hanke [2]

1 Centro Oceanográfico de Santander (COST-IEO), Instituto Español de Oceanografía (IEO-CSIC), Av. de Severiano Ballesteros, 16, 39004 Santander, Cantabria, Spain
2 Large Pelagics Group, St. Andrews Biological Station, 125 Marine Science Drive, St. Andrews, NB E5B 0E4, Canada
3 AZTI Marine Research, Basque Research and Technology Alliance (BRITA), Herrera Kaia, Portualdea z/g, 20110 Pasaia, Gipuzkoa, Spain
4 Fish Ageing Services, Queenscliff, VIC 3225, Australia
* Correspondence: enrique.rmarin@ieo.csic.es

Abstract: Controversies remain regarding the periodicity, or seasonality, of otolith growth band formation, which directly influences a correct age determination of Atlantic bluefin tuna using this structure. The aim of this work was to apply marginal increment analysis and marginal edge analysis to determine the timing of band deposition. The index of completion was analyzed using general additive models to evaluate the importance of variables, such as month, age/size, and reader. Results indicate that the opaque band formation begins in June and is completed by the end of November. From the end of the year to the beginning of the following year, there is minimal marginal edge growth as the translucent band begins to form. The translucent zone then reaches a maximum development in May. The results obtained in this study provide evidence that the annulus formation in the otoliths of Atlantic bluefin tuna are completed later in the calendar year than previously thought. This would mean it is necessary to delay the date of the current July 1st adjustment criterion to November 30.

Keywords: marginal increment analysis; band deposition; growth; otolith; *Thunnus thynnus*

Citation: Rodriguez-Marin, E.; Busawon, D.; Luque, P.L.; Castillo, I.; Stewart, N.; Krusic-Golub, K.; Parejo, A.; Hanke, A. Timing of Increment Formation in Atlantic Bluefin Tuna (*Thunnus thynnus*) Otoliths. *Fishes* **2022**, *7*, 227. https://doi.org/10.3390/fishes7050227

Academic Editor: Josipa Ferri

Received: 19 July 2022
Accepted: 22 August 2022
Published: 31 August 2022

1. Introduction

The Atlantic bluefin tuna (*Thunnus thynnus*, ABFT) is one of the largest osteichthyans, has a life span of more than 35 years, and has a wide distribution in the Atlantic Ocean and Mediterranean Sea, where it carries out extensive feeding and reproductive migrations. This species is exploited by many countries, and even though the total catch is not very large compared to other tuna species, it is one of the world's most valuable fish species [1–3].

The description of the life cycle and effective management of ABFT requires comprehensive knowledge on the age and growth of this species. One of the most widely used methods for estimating the age of ABFT is based on examination of calcified structures [3]. The estimation of absolute age by reading otoliths is validated by the bomb radiocarbon method [4] and the otolith strontium (Sr):calcium (Ca) profiles revealed periodicity of the formation of the annual increments [5].

Direct age assignment depends not only on the number of annuli, comprising an opaque and a translucent band found in the calcified structure, but also on the time of the year when each type of band is formed. This is so because, to convert the band count into an age estimate, it is necessary to consider the marginal band type found in the structure related to the catch date and the birth date. The study by Siskey et al. [5], shows that compared periodicity of annuli formation in otoliths of ABFT against Sr:Ca oscillations, indicates that opaque bands occurs cyclically, consistent with seasonal cycles in Sr:Ca that serve as a proxy for seasonal temperatures variation (i.e., higher otolith Sr:Ca in opaque bands is presumed to occur during winter months), confirming the assumed inverse relationship between Sr uptake and temperature. In contrast, a higher Sr concentration was detected in the translucent bands of the otolith of southern bluefin tuna (*Thunnus maccoyii*) [6] when applying the same technique. Similarly, Sr concentrations assessed across annulus of the first dorsal fin spine (fin spine, hereafter) of juvenile ABFTs were significantly higher in the translucent bands [7], and were interpreted as having formed in winter. In support of that, edge type and marginal increment analyses of the ABFT fin spines indicated a yearly periodicity of annulus formation, with the translucent bands appearing in winter [8].

The periodicity of annuli formation is commonly determined by marginal increment analysis (MIA), in which the distance from the last completely formed annulus to the edge of the otolith is tracked over time. This method requires a good representation of observations throughout the year to detect any seasonal trend in the formation of growth bands [9,10]. The selected mark must be accurate enough to allow detection of its formation at the extreme edge of the otolith. Edge type identification, either as translucent or opaque, in otoliths of ABFT, is difficult. The thickness of the section and the diffraction of light can influence the perception of the of edge type, making identification of marginal areas in the ABFT otoliths more difficult. Recently, it was accepted that using transmitted light for otoliths direct ageing improves agreement on marginal edge recognition [11,12]. However, long-standing controversies remain regarding the periodicity, or seasonality, of otolith type of band formation, which directly influences the correct age determination of ABFT using this calcified structure. Thereby, the aim of this work was to apply MIA to determine the timing of band deposition. To address this, growth bands (annuli) were measured from otoliths of ABFT collected monthly. We also performed edge type analysis (EA) by analyzing the change in the relative frequency of the opaque or translucent bands over the months.

2. Materials and Methods

To be able to effectively use the marginal increment analysis (MIA) method, sampling is required in all months of the year and, as far as possible, for all ages in each month. The seasonality and size-selectivity of ABFT fisheries, in addition to the migratory behavior of this widely distributed species, makes comprehensive year-round sampling difficult for ABFT. For this study, ABFT otoliths were collected from specimens obtained from both sides of the Atlantic Ocean and from the Mediterranean Sea, thanks to extensive sampling by the Atlantic-Wide Research Programme for Bluefin Tuna of the International Commission for the Conservation of Atlantic Tunas (GBYP, ICCAT), the contributions of St. Andrews Biological Station (SABS) and the Spanish Institute of Oceanography (IEO-CSIC). Sampling was carried out from 2009 to 2021 and used a combination of capture methods (e.g., traps, long-lining, bait boats, purse seine, etc.). No live animals were used for this study. Fish were measured mostly at straight fork length (SFL), but also at curved fork length, snout length and round weight. The latter measurements were converted into SFL, following the biometric relationships established by ICCAT [13,14]. Size range sampled was from 50 to 295 cm SFL.

Otoliths were prepared by three laboratories (Fish Ageing Services (FAS), SABS, and IEO-CSIC) following the standardized methodology described in Rodriguez-Marin et al. [11]. Three research centers were involved in the otolith age reading and analysis (SABS, IEO-CSIC, and AZTI) using the standardized reading criterion [11]. In addition, otoliths read with the previous reading criterion of Busawon et al. [15] were also included in the analysis to ensure a better monthly sampling and size range coverage. These two criteria differ mainly in that the opaque bands at the otolith edge were only counted if it was completely formed and translucent otolith material was visible between the opaque band and the marginal edge, following Rodriguez-Marin et al. [11], and were counted, even if not complete, following Busawon et al. [15]. As such, to use the same band counting criteria to all samples, otolith readings using Busawon et al. [15] criterion were adjusted as follows: if the otolith edge was opaque, then number of bands = number of bands − 1. The number of otoliths read applying the Busawon et al. [15] criterion was 152, and for the Rodriguez-Marin et al. [11] criterion was 2128.

The timing of band formation was assessed by examining the outermost edge of the ventral arm of the otolith section. Following the recommendation from Campana [9], a minimum of two complete cycles needs to be examined to reduce variability. MIA was used to examine the periodicity of otolith band formation. As such, MIA was determined using the index of completion [16]:

$$C = Wn/((Wn\text{-}1 + Wn\text{-}2)/2) \times 100 \quad (1)$$

where Wn is the width of the marginal increment (distance from the end of the last opaque zone to the marginal edge, regardless of edge type); and Wn-1 and Wn-2 are widths of the two previously completed increments (the distance from the end of the second or third most outer opaque zones to the last and penultimate opaque zone). This method only allows analysis of MIA in otoliths of specimens over two years old, while for specimens older than one year, only one complete cycle was used:

$$C = Wn/Wn\text{-}1 \times 100 \quad (2)$$

Measurements were performed with a digital image analysis system. Otoliths were measured along an axis in the area between the sulcus margin and the ventral groove of the ventral arm where the annuli are most distinct, using a standardized "measurement line" [11] (Figure 1).

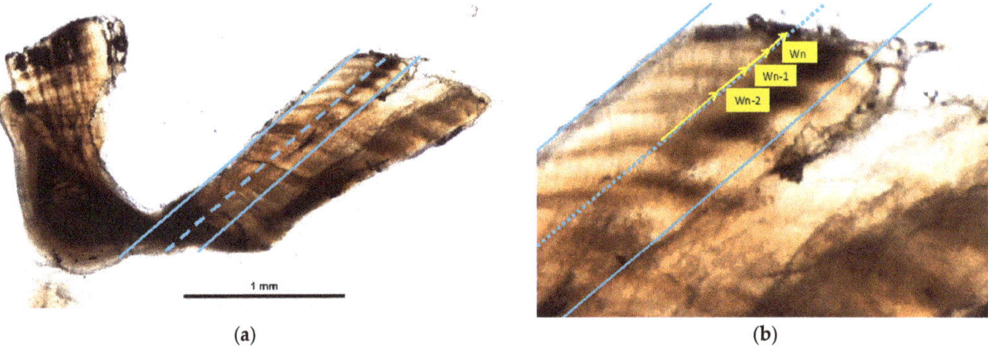

(a) (b)

Figure 1. Bluefin tuna otolith showing measurement lines (**a**) and Wn, Wn-1, and Wn-2 measurements identified with yellow arrows; (**b**) is a 7-year-old individual with a narrow translucent marginal edge.

Five readers participated in the measurements, although one of them read with both types of light, transmitted and reflected. This reader was considered as two different readers according to the type of light used. This resulted in a total of 6 readers. The number of images read by each light type was 1740 for transmitted and 540 for reflected. Readers were requested to provide the following information by sample: Light type (reflected, transmitted), number of annual bands (opaque), reading criterion (1 Busawon et al. [15]; 2 Rodriguez-Marin et al. [11]), ventral arm marginal edge type (wide translucent, WT; narrow translucent, NT or opaque, O), edge type confidence (1 = not confident; 2 = confident in completeness but not with the edge type and 3 = confident), readability code (1 = pattern present—no meaning, 2 = pattern present—unsure with age estimate, 3 = good pattern present—slightly unsure in some areas, and 4 = good pattern—confident with age estimate), Wn, Wn-1, and Wn-2 widths in mm, agreed band count ("Yes" for agreed and "No" for individual decision), agreed edge type ("Yes" for agreed and "No" for individual decision), measuring date, reader coding, and notes with observations about the sample.

The agreement fields were used because all samples were previously read by expert readers as part of several ageing exercise exchanges or read by FAS, however we did not want to restrict new readers to the pre-assigned band counts if they did not agree with the originally assigned band count or marginal edge type. An agreement with previous readings of 49% in the number of bands (were mostly +/− 1 band) and 50% in the type of edge (with the three categories) was reached.

A total of 2280 ABFT otolith images were available for participants to determine yearly periodicity of annulus formation. Approximately 8% of the available samples were not included for further analysis due to poor image quality, missing measurements, age (age 0), or obvious outliers for size-at-age. Furthermore, 18 samples were identified as outliers using the following criteria: if Wn > Wn-1 > Wn-2 and the index of completion C was greater than 120%, they were re-ordered Wn < Wn-1 < Wn-2. This resulted in a total of 2101 samples available for analysis, which included overall good monthly representation by age (number of complete opaque bands), except for months 2, 3, and 4 (Figure 2).

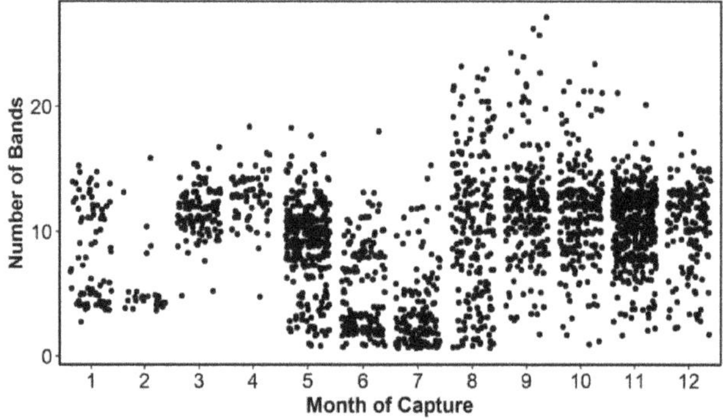

Figure 2. Number of samples identified by number of bands and month of capture.

An alternative measure of annuli completion (CB) was derived using mean annulus width values derived from a recently completed back-calculation study [17], which was used to provide expected values in the following equation: CB = (observed/expected) × 100, where *observed* was an observed annuli width measured in the current study. The *expected* width was calculated as the difference in the mean total annuli width (measured from the origin of the ventral arm to the end of a given opaque band) for each annulus for band numbers 1−28. Due in part to the number of available samples decreasing at older age classes, this calculation was achieved to yield negative values at band numbers 24 and 28. To generalize the expected values, we used a GAM (mean annulus width ~ spline (number of bands), family = Gaussian (link = log) to provide the expected values at each annuli width for the CB equation. Plots of mean annual increment measurements and the monthly mean index of completion (CB) were examined for differences in completion trends.

The index of completion was also analyzed using general additive models (GAMs) to evaluate the importance of variables, such as month, age/size, reading criteria, light type, and reader. The following transformations were applied to the data prior to fitting the models: to account for cases where C was greater than 100, samples with an index of completion of 100 were changed to 99.9. Edge type was not included in the model, since it represents phases of completion and is essentially a derivative of the index of completion, that is the completion index for NT < WT < O. Formulations of the model with the light effect as a random effect showed similar trends that differed slightly in the peak of completion, with rate of completion being the highest in July for reflected light vs. September for transmitted light. Since the trends were similar, and due to the fact that the reflected light trend was not very informative due to the timing/span of the sample collection, light type was excluded from the final model. Furthermore, the final model was fitted using both SFL and age; however, SFL yielded a better fit compared to age. This better fit is likely due to length being a more continuous variable compared to age or number of bands. GAMs were fitted assuming beta-distributed data for the response with a logit link. A cyclic cubic spline basis function was specified for smoothers involving month of capture and was limited to 8 knots. Reader, age, and SFL were included as random effects. The best fitted GAM was:

$$\text{Model: C} \sim \text{s (Month, k = 8)} + \text{s (SFL)} + \text{s (Reader)} \quad (3)$$

3. Results

The index of completion (C) by number of bands showed an increase in C with the number of bands, although there was a clear change in the slope of this line, forming three groups: from 1 to 6 bands (juveniles), from 7 to 16 bands (adults) and more than 16 bands (old adults) (Figure 3). These three slope sections made us explore MIA and EA with all annual/age bands grouped together and by the band groups or bins described above.

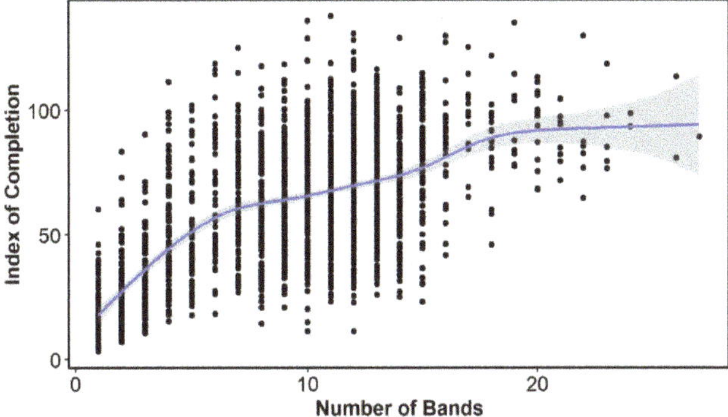

Figure 3. Index of completion by estimated age (i.e., number of opaque bands).

The trend in C appeared to be bimodal, indicating that the highest rate of completion occurred during August to October and in April (Figure 4a). When C was analyzed by age group, this bimodality continued to be observed, with June and July again showing the lowest values, indicating that the otolith margin grew very slowly in these months. The maximum C values of the first months were supported by few samples, while the maximum values of the months from mid-summer to mid-fall were well sampled and indicated higher C values for the months of August–September, through November. This was especially evident in the adults group, which was the best sampled group (Figure 4b). Mean annual band increment measurements from the back-calculation study [17] and the present study (AI1 = Wn-1 and AI2 = Wn-2) were the same, except in annual bands 0, 1 and 4 to 6, where there were small differences, but with a similar trend of decreasing size. The alternative completion index (CB) also showed a bimodal pattern, with greater growth in August to October compared to the rest of the year (Figure 5).

When the type of edge was analyzed, the month of greatest change in the ratio of translucent to opaque bands indicated that the formation of opaque bands occurred between July and August in the overall marginal status (Figure 6a) and from June to July in the marginal estate of adult groups (7–16 bands, Figure 6b). This indicated that the opaque bands began to form in June–July and continued to form up to and including October. The translucent band started to form in November with a narrow translucent margin until February, and the incidence of otoliths with a wide translucent margin from March to June in the best sampled group of adults (7–16 bands group, Figure 6b).

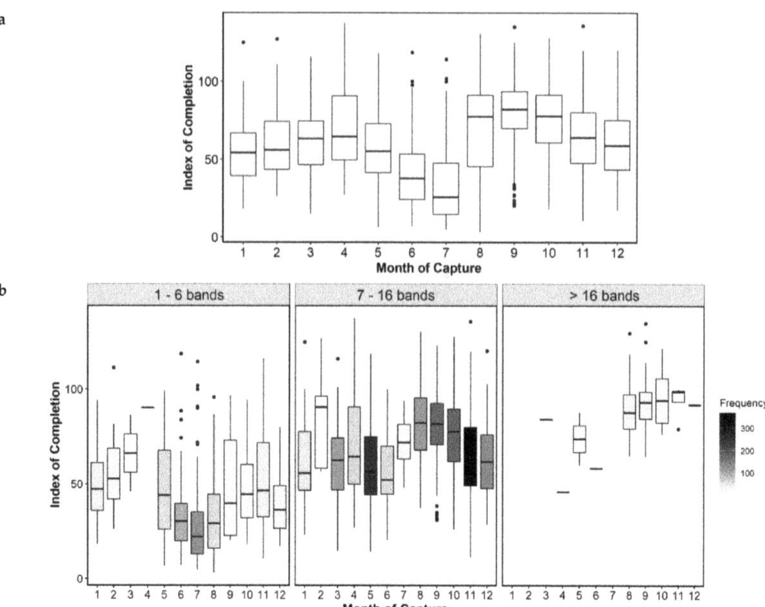

Figure 4. Mean index of completion (C) by month, all age classes combined (**a**), and by month for each age group (number of annual bands binned) (**b**). The gray scale indicates the number of samples analyzed. Months 3–4 for bin 1, month 2 for bin 2, and months 11–7 for bin 3 have a low number of samples (5 or less).

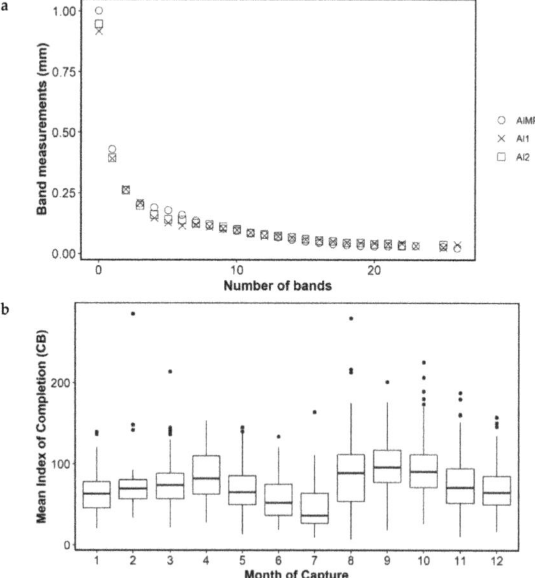

Figure 5. (**a**) plot comparing mean annual band increment measurements from back calculation study (AIMP, [17]), and present study (AI1 = Wn-1 and AI2 = Wn-2). (**b**), mean index of completion (CB) by month.

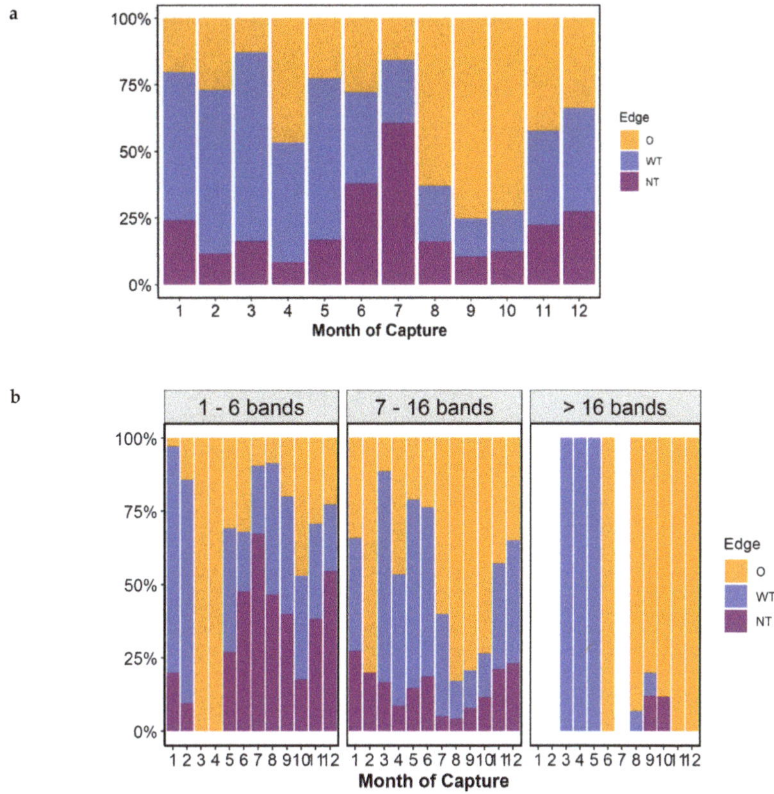

Figure 6. Percent marginal edge type by month, (**a**), and by month and age group (number of annual bands binned), (**b**). Months 3–4 for bin 1, month 2 for bin 2, and months 11–7 for bin 3 have a low number of samples (5 or less). Marginal edge type (O = opaque, WT = wide translucent, and NT = narrow translucent).

Considering the group of samples as a whole, MIA and EA showed consistency in the results among readers and among quality values in the identification of marginal edge type. The type of light used for reading (reflected or transmitted) also had no influence, except in the month of August, where with transmitted light, the opaque marginal edge represented nearly 20%, while with reflected light, it represented 70%.

The functional relationship between month of capture and C obtained from the GAM model showed that the maximum C was obtained in the month of September, and from that month onwards, it decreased until it reached a minimum value in May (Figure 7). The trend represents the population-level predicted values and their corresponding confidence intervals (i.e., the uncertainty related to the variance parameters for the mean random effects associated with reader and SFL are not included). Variability in C of the marginal increment was large. The variability associated with the month of capture was relatively small compared to sources such as the reader (Table 1). Nevertheless, there is evidence for the level of completion to be highest in September. The factors of month, age, SFL, and reader accounted for a significant amount of the variability in C, with the model explaining 47.3% of the variation in the dataset, with an R^2 of 0.46. Diagnostic plots for the GAM model indicated no issues (Figures 8 and 9).

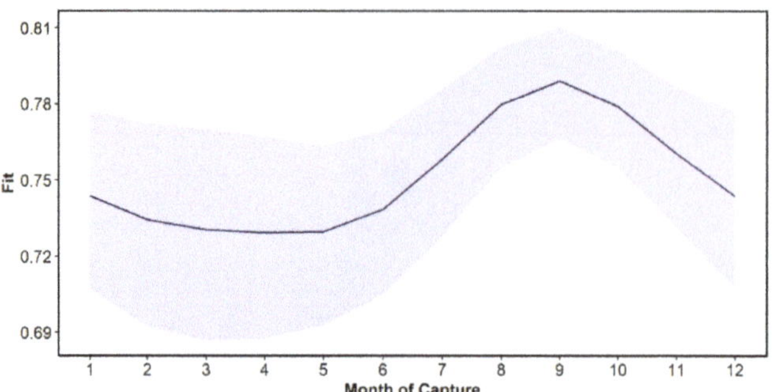

Figure 7. Prediction interval for the month of capture effect from the GAM model.

Table 1. The variance and standard deviation for each smooth in the final GAM model.

Component	Variance	Std_dev	Lower_ci	Upper_ci
Month	0.0108	0.104	0.0378	0.287
SFL	0.00000936	0.00306	0.00143	0.00654
Reader	0.266	0.516	0.250	1.06

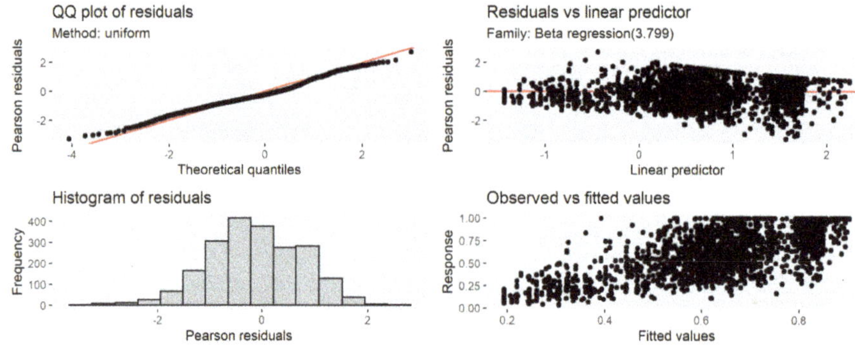

Figure 8. Diagnostic plots for the final GAM model. The QQ-plot and the histogram of the residuals are used to verify normality. The plot of standardized residuals against fitted values assesses homogeneity.

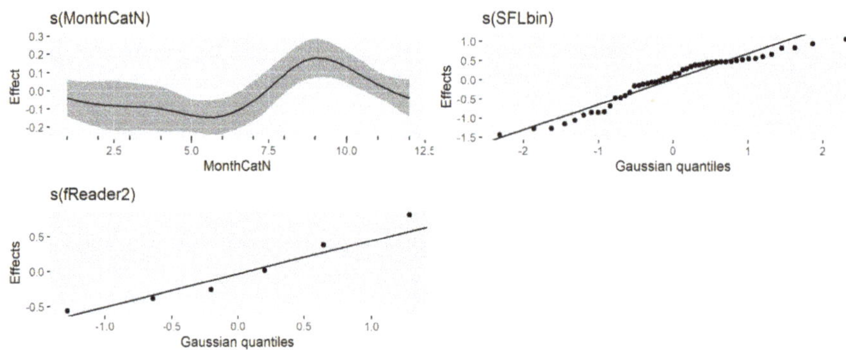

Figure 9. Estimated smoothing effects obtained by the GAM model.

4. Discussion

In this study, MIA, and EA were used to determine the annual periodicity in the annuli formation of ABFT otoliths. Despite the difficulty in accurately interpreting the increment structure in ABFT otoliths (where opaque bands are not easily distinguished from translucent bands), and the difficulty of sampling this species in entirety throughout the year and from all size classes, we identified when and how the growth of the otolith marginal edge occurs throughout the year.

Both MIA and EA analyses are not recommended with pooled age data because there may be different patterns of growth band formation at different life stages (e.g., at sexual maturity) [9,18]. However, grouping the data by age groups, such as young and old fish, can reduce this age effect and improve the reliability of the results [18,19]. In this study, the change in slope in C by number of bands identified three main groups: juveniles (1 to 6 bands), adults (7 to16 bands) and old adults (>16 bands). Despite each of these age groups being unbalanced in terms of samples available for all months, MIA results show similar trends, whereas EA results between age groups did not. While the differences between groups could be largely attributed to the difficulty in identifying the type of marginal edge and further complicated by the low number of specimens sampled, we cannot rule out the effect due to changes in the seasonal timing of the marginal increment with age or capture location.

The band formation process is related to age and associated with energy destined towards somatic growth in relation to biological processes, such as spawning and food availability [20–22]. A graph of MIA thus follows a sawtooth shape, where growth rates are sensitive to environmental conditions and individual ontogeny [9,23]. The expected pattern in an annual MIA cycle is that it presents a single maximum, however, we found two periods of maximum growth in C, although the peak in March and April is lower than the one in September and is based on an inadequate number of samples. This first peak of the year could be related to the development of the wide translucent margin beginning in March. It is necessary to rely on modeling using GAM in order to clearly identify the period of greatest growth in C. A modeling approach is necessary to identify the temporal annual periodicity of growth increments, as it is not always so obvious and is influenced by differences between years and geographic areas [24–26].

The concordance between the MIA and EA results in the adults group (7–16 bands), which was the best sampled group, the similarity between the results of the band measurements and C and CB [17], and the fit of the GAM analysis provide us with a marginal growth pattern with a yearly sinusoidal cycle. We can confirm that this marginal edge deposition is annual and consists of an opaque and a translucent zone because the annulus formation was previously validated through the bomb radiocarbon method [4], and the annual periodicity of its formation was confirmed thanks to Sr:Ca oscillations [5]. The greatest growth of the marginal edge occurs from June onwards and peaks in September. From this month forward, it starts to decrease and slows down from November to May.

Since the completion index is based on the measurement between the end of consecutive opaque bands, the minimum value of C from MIA and GAM models after the maximum values in the period from August to October indicate that the opaque band is being completed during November. This is also supported by EA, which indicates that the deposition of the opaque band finishes forming in November. From the end of the year to the first months of the following year, there is minimal marginal edge growth, and this is when the translucent band begins to form and reaches its maximum development in May, with the highest proportion of wide translucent zones from March to June (particularly evident from adults group interpretation in Figure 6b).

Individual ages are estimated using the count of opaque bands, otolith edge classification, catch date, assumed spawning month or birth date, and timing of opaque band formation. The last opaque band is only counted when translucent material can be observed between the last opaque and the margin, i.e., a narrow or wide translucent edge [11]. Therefore, the time of year when the otolith edges are changing from opaque into narrow translucent needs to be the period for the initial edge type zone adjustment. Using a date of November 30th as a date of opaque completion, a two-step age adjustment protocol is proposed: firstly, a zone formation adjustment. If this initial step is not completed, it is possible that individuals with the same birth year, even when captured on the same date, could be assigned to different age classes due to either reader differences in interpreting edge type or individual variability in when fish are completing their opaque band formation on their otoliths (Table 2a). Secondly, a biological age adjustment using June 1st as the birth date, assumed for both ABFT management units based on the ABFT reproductive cycle, where spawning occurs from May to June in the western Atlantic (Gulf of Mexico) and eastern Mediterranean (Levantine Sea), or from June to July in the western and central Mediterranean (Balearic Islands waters, South of Tyrrhenian Sea and Sea of Sicily) [1] (Table 2b); and a second option, instead of adjusting for biological age, is to perform a calendar year adjustment to properly identify each year class (Table 2c). To convert opaque band counts (N) into age estimates following the ageing protocol of Busawon et al. [15], which states that opaque bands are counted even if not complete, the following steps need to be used. (1) Subtract 1 from the zone count for each sample assigned with an opaque (O) edge type, (2) leave unchanged the number of bands of the samples with a translucent (T) edge type, and (3) covert zone count into age using June 1st for the biological age adjustment (2b) or January 1st for the calendar year adjustment (2c).

Table 2. (**a**–**c**) To convert opaque band counts (N) into age estimates (A) following the ageing protocol of Rodriguez-Marin et al. [11] the following two-step age adjustment should be used: (1) Based on otolith marginal edge type (NT = narrow translucent, WT = wide translucent or O = opaque), the adjustment date of 30th November and the date of capture (**a**). And (2) June 1st for the biological adjustment (**b**) or January 1st for the calendar year adjustment (**c**).

a. Band Formation Adjustment.			
		Month of Capture	
Edge type	**Dec–May**	**Jun–July**	**Aug–Nov**
NT	N	N	N − 1
WT	N	N	N
O	N + 1	N	N

b. Biological Adjustment (ensuring that the ages align to the biological management units June 1st).		
	Month of Capture	
	Dec–May	**Jun–Nov**
Biological year	A	A + 1

c. Calendar year adjustment.		
	Month of Capture	
	Jan–Nov	**Dec**
Calendar year	A + 1	A

To ensure that the age adjustment tables were providing correct age assignment, i.e., incrementing the age by a count of one each year while still ensuring that individuals hatched in the same year remained in the correct age cohort, a conceptual age adjustment table for a fictional individual ABFT was created for comparison. This table shows (1) the progression of monthly and yearly (decimal) age and (2) the band count or age for each combination of band counts and edge types for a given month of life of an individual hatched in June 2020 and caught 3.6 years later (see Supplementary Table S1).

The proposed new adjustment criterion involves delaying the date of the current adjustment criterion from 1 July [11] to November 30. The 1 July criterion was based on a preliminary study with a smaller number of samples, especially in the first 4 months of the year, than those used in the present study, and used a different band measurement methodology. This study also showed an increase in the completion rate in the last months of the year, although this increase is delayed by two months with respect to the present study. To evaluate the influence of applying the existing age adjustment criterion [11] and the new one proposed in this study, we used the current ICCAT length at age database to see how the cohorts were reflected using the otolith readings available in that database [27], and in particular to see how it influenced the location of the strong 2003 year class [28,29]. The change in the date of the otolith fitting criterion allows for a better outline of the strong 2003 year class, despite the fact that the otolith samples in this database do not proportionally cover all fisheries in every year (Figure 10).

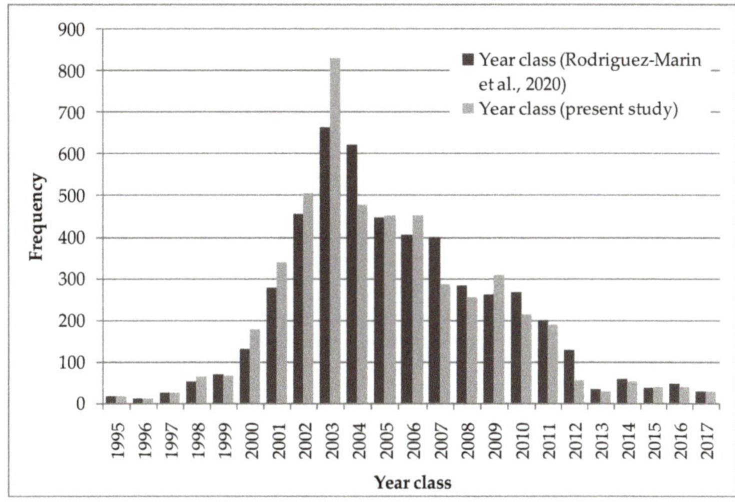

Figure 10. Number of Atlantic bluefin tuna by year class by applying the current [11] and present study age adjustment criteria to otolith band counts.

The occurrence of translucent bands in ABFT otoliths were described in studies over 40 years ago, describing their appearance from October to December to May of the following year [30–32]. The findings of the present study, in relation to the formation of the otolith edge type, would be in agreement with these studies and also with the seasonal deposition patterns found in ABFT fin spines [8]. It would also agree with the results of the higher Sr:Ca ratio concentration found in the translucent bands of the same species and in southern bluefin tuna, Luque et al. [7] and Clear et al. [6], respectively. In contrast, Siskey et al. [5] found higher Sr:Ca in the opaque bands in western ABFT adults over 10 years of age, which would imply that the bands formed during the winter months, due to the inverse relationship between Sr uptake and temperature. A possible explanation for this discrepancy, between the present study and that of Siskey et al. [5], is that this difference in the "presumed" timing of the opaque band deposition could be due to the migrations of western ABFT. These individuals migrate from the spawning grounds in the Gulf of Mexico in spring (warm waters) to the Mid-Atlantic Bight and Canadian shelf areas in summer (cold waters); the type of signal given by Sr:Ca could be influenced by this temperature change, where it is precisely the colder waters that the western ABFT occupy in summer [33,34].

Stock assessments used to provide management advice for ABFT rely on age-structured population analysis. The method applied for transforming length-structured data to age structured data is age slicing using a deterministic growth model [35]. This method divides size distributions into age classes under the assumption that there are distinct lengths separating adjacent age classes. Given the variability in growth between individuals, which increases as fish grow, this technique can assign an incorrect age when one or more age classes overlap in length. Growth variability over time and strong cohort signals can also be affected by this deterministic method [36]. An alternative approach to estimating age composition, including variability in growth, is the use of age–length keys (ALKs). It is precisely in the construction of these ALKs that our study is essential to convert band counts in calcified structures, such as otoliths, into age estimates. The age adjustment we propose is suitable for both ABFT management units or stocks, the western stock, and the eastern stock and Mediterranean Sea; in fact, we used samples from both management units. Being able to apply the same adjustment, including the biological adjustment (in relation to the date of birth), explained above, facilitates its application. Moreover, rather than differences in growth between the two management units, it is the type of fitting made to the length at age data (e.g., Richards' or von Bertalanffy's growth models) and the size range covered in the sampling that has the greatest influence [3,37]. Furthermore, it was shown that there is no difference in the growth of both management units [17].

5. Conclusions

We used MIA and EA to determine when the opaque and translucent band of each growth annuli forms at the marginal edge of the ABFT otolith throughout the year. We used GAM to clearly identify marginal edge growth. This modeling was necessary to interpret the results, due to the difficulty of identifying translucent and opaque bands and the difficulty of sampling this species in its entirety throughout the year and from all size classes. The use of GAM also allowed the evaluation of some of the variables affecting edge growth, such as month, age/size, and reader.

Opaque bands are present at the marginal edge of the otolith from June 1st to November 30th, with a maximum in September (opaque bands in January–May are late forming or late identified from the previous year). Translucent bands are present on the otolith from December 1st to May 31st, going from narrow to wide in March and reaching the maximum growth of the translucent band in May. Each type of band, opaque or translucent, having a six-month time period. The results obtained in this study provide evidence that the annulus formation in the otoliths of Atlantic bluefin tuna are completed later in the calendar year than previously thought. This would mean that it is necessary to delay the date of the current July 1st adjustment criterion to November 30.

The discrepancies found among the references for this species and the results of the present study regarding the season in which the translucent and opaque bands form may be due to migrations of western adult specimens from the warmer waters of the Gulf of Mexico during the spring spawning season to the cooler waters of the Mid-Atlantic Bight and Canadian shelf areas in summer to feed. The type of signal given by the Sr:Ca concentration in the opaque bands, the source of the discrepancy, could be influenced by this temperature shift.

Our study is essential for converting band counts in calcified structures, such as otoliths, into age estimates. This allows using an alternative approach to estimate catch at age composition that includes growth variability. This variability is not currently accounted for in age-structured population analyses used to provide management advice in ABFT stock assessments, which use deterministic growth models.

Supplementary Materials: The following are available online at https://www.mdpi.com/article/10.3390/fishes7050227/s1, Table S1: Conceptual age adjustment table.

Author Contributions: Conceptualization, methodology, visualization, formal analysis, E.R.-M., D.B., P.L.L., N.S., A.H., K.K.-G.; resources, data curation, I.C., A.P., D.B., P.L.L., N.S., K.K.-G.; software, D.B., N.S., A.H.; writing—review and editing, E.R.-M., D.B., P.L.L., N.S., K.K.-G., A.H.; project administration, funding acquisition, E.R.-M. All authors have read and agreed to the published version of the manuscript.

Funding: This work was funded by the Atlantic Bluefin Tuna Research Programme (GBYP) of the International Commission for the Conservation of Atlantic Tunas (ICCAT). Time spent by two participants has been co-funded by the European Union through the European Maritime Fisheries and Aquaculture Fund (EMFF) within the National Program of collection, management and use of data in the fisheries sector and support for scientific advice regarding the Common Fisheries Policy.

Data Availability Statement: The raw data which support this study are available from the corresponding author at reasonable request.

Acknowledgments: This work was carried out under the provision of the International Commission for the Conservation of Atlantic Tunas (ICCAT) Atlantic Wide Research Programme for Bluefin Tuna (GBYP) Phase 11.

Conflicts of Interest: The authors declare no conflict of interest.

References

1. Rooker, J.R.; Bremer, J.R.A.; Block, B.A.; Dewar, H.; Metrio, G.D.; Corriero, A.; Kraus, R.T.; Prince, E.D.; Rodríguez-Marín, E.; Secor, D.H. Life history and stock structure of Atlantic bluefin tuna (*Thunnus thynnus*). *Rev. Fish. Sci.* **2007**, *15*, 265–310. [CrossRef]
2. Fromentin, J.M.; Bonhommeau, S.; Arrizabalaga, H.; Kell, L.T. The spectre of uncertainty in management of exploited fish stocks: The illustrative case of Atlantic bluefin tuna. *Mar. Policy.* **2014**, *47*, 8–14. [CrossRef]
3. Murua, H.; Rodriguez-Marin, E.; Neilson, J.D.; Farley, J.H.; Juan-Jordá, M.J. Fast versus slow growing tuna species: Age, growth, and implications for population dynamics and fisheries management. *Rev. Fish. Sci.* **2017**, *27*, 733–773. [CrossRef]
4. Neilson, J.D.; Campana, S.E. A validated description of age and growth of western Atlantic bluefin tuna (*Thunnus thynnus*). *Can. J. Fish. Aquat. Sci.* **2008**, *65*, 1523–1527. [CrossRef]
5. Siskey, M.; Lyubchich, V.; Liang, D.; Piccoli, P.; Secor, D. Periodicity of strontium: Calcium across annuli further validates otolith-ageing for Atlantic bluefin tuna (*Thunnus thynnus*). *Fish. Res.* **2016**, *177*, 13–17. [CrossRef]
6. Clear, N.P.; Gunn, J.S.; Rees, A.J. Direct validation of annual increments in the otoliths of juvenile southern bluefin tuna, *Thunnus maccoyii*, by means of a large-scale mark-recapture experiment with strontium chloride. *Fish. Bull.* **2000**, *98*, 25–40.
7. Luque, P.L.; Zhang, S.; Rooker, J.R.; Bidegain, G.; Rodriguez-Marin, E. Dorsal fin spines as a non-invasive alternative calcified structure for microelemental studies in Atlantic bluefin tuna. *J. Exp. Mar. Biol. Ecol.* **2016**, *486*, 127–136. [CrossRef]
8. Luque, P.; Rodriguez-Marin, E.; Landa, J.; Ruiz, M.; Quelle, P.; Macias, D.; Ortiz de Urbina, J.M. Direct ageing of *Thunnus thynnus* from the eastern Atlantic Ocean and western Mediterranean Sea using dorsal fin spines. *J. Fish. Biol.* **2014**, *84*, 1876–1903. [CrossRef]
9. Campana, S.E. Accuracy, precision and quality control in age determination including a review of the use and abuse of age validation methods. *J. Fish. Biol.* **2001**, *59*, 197–242. [CrossRef]
10. Panfili, J.; de Pontual, H.; Troadec, H.; Wright, P.J. *Manual of Fish. Sclerochronology*; IFREMER-IRD Co-Edition: Brest, France, 2002; pp. 1–464.
11. Rodriguez-Marin, E.; Quelle, P.; Addis, P.; Alemany, F.; Bellodi, A.; Busawon, D.; Carnevali, O.; Cort, J.L.; Di Natale, A.; Farley, J.; et al. Report of the ICCAT GBYP international workshop on Atlantic bluefin tuna growth. *Collect. Vol. Sci. Pap. ICCAT.* **2020**, *76*, 616–649.
12. Rodriguez-Marin, E.; Busawon, D.; Addis, P.; Allman, R.; Bellodi, A.; Castillo, I.; Garibaldi, F.; Karakulak, S.; Luque, P.L.; Parejo, A.; et al. Calibration of Atlantic bluefin tuna otolith reading conducted by an independent fish ageing laboratory contracted by the ICCAT Research Programme GBYP. *Collect. Vol. Sci. Pap. ICCAT* **2021**, *78*, 938–952.
13. Secor, D.H.; Busawon, D.; Gahagan, B.; Golet, W.; Koob, E.; Neilson, J.; Siskey, M. Conversion factors for Atlantic bluefin tuna fork length from measures of snout length and otolith mass. *Collect. Vol. Sci. Pap. ICCAT* **2014**, *70*, 364–367.
14. Rodriguez-Marin, E.; Ortiz, M.; Ortiz de Urbina, J.M.; Quelle, P.; Walter, J.; Abid, N.; Addis, P.; Alot, E.; Andrushchenko, I.; Deguara, S.; et al. Atlantic bluefin tuna (*Thunnus thynnus*) Biometrics and Condition. *PLoS ONE.* **2015**, *10*, e0141478. [CrossRef] [PubMed]
15. Busawon, D.S.; Rodriguez-Marin, E.; Luque, P.L.; Allman, R.; Gahagan, B.; Golet, W.; Koob, E.; Siskey, M.; Ruiz, M.; Quelle, P.; et al. Evaluation of an Atlantic bluefin tuna otolith reference collection. *Collect. Vol. Sci. Pap. ICCAT.* **2015**, *71*, 960–982.
16. Tanabe, T.; Kayama, S.; Ogura, M.; Tanaka, S. Daily increment formation in otoliths of juvenile skipjack tuna *Katsuwonus pelamis*. *Fish. Sci.* **2003**, *69*, 731–737. [CrossRef]
17. Stewart, N.D.; Busawon, D.S.; Rodriguez-Marin, E.; Siskey, M.; Hanke, A.R. Applying mixed-effects growth models to back-calculated size-at-age data for Atlantic bluefin tuna (*Thunnus thynnus*). *Fish. Res.* **2022**, *250*, 106260. [CrossRef]
18. Hyndes, G.A.; Loneragan, N.R.; Potter, I.C. Influence of sectioned otoliths on marginal increment trends and age and growth estimates for the flathead *Platycephalus speculator*. *Fish. Bull.* **1992**, *90*, 276–284.
19. Hesp, S.A.; Potter, I.A.; Hall, N.G. Age and size composition, growth rate, reproductive biology, and habitats of the West Australian dhufish (*Glaucosoma hebraicum*) and their relevance to management of this species. *Fish. Bull.* **2002**, *100*, 214–227.
20. Morales-Nin, B. Review of the growth regulation processes of otolith daily increment formation. *Fish. Res.* **2000**, *46*, 53–67. [CrossRef]
21. Ashworth, E.C.; Hall, N.G.; Hesp, S.A.; Coulson, P.G.; Potter, I.C. Age and growth rate variation influence the functional relationship between somatic and otolith size. *Can. J. Fish. Aquat Sci.* **2017**, *74*, 680–692. [CrossRef]
22. Katayama, S.A. Description of four types of otolith opaque zone. *Fish. Sci.* **2018**, *84*, 735–745. [CrossRef]

23. Mercier, L.; Panfili, J.; Paillon, C.; N'diaye, A.; Mouillot, D.; Darnaude, A.M. Otolith reading and multi-model inference for improved estimation of age and growth in the gilthead seabream *Sparus aurata* (L.). *Estuar Coast Shelf Sci.* **2011**, *92*, 534–545. [CrossRef]
24. Farley, J.H.; Williams, A.J.; Clear, N.P.; Davies, C.R.; Nicol, S.J. Age estimation and validation for South Pacific albacore *Thunnus alalunga*. *J. Fish. Biol.* **2013**, *82*, 1523–1544. [CrossRef] [PubMed]
25. Hidalgo-de-la-Toba, J.A.; Morales-BojoÂrquez, E.; González-Peláez, S.S.; Bautista-Romero, J.J.; Lluch-Cota, D.B. Modeling the temporal periodicity of growth increments based on harmonic functions. *PLoS ONE* **2018**, *13*, e0196189. [CrossRef]
26. Salgado-Cruz, L.; Quiñonez-Velázquez, C.; García-Domínguez, F.A.; Pérez-Quiñonez, C.I. Detecting *Mugil curema* (Perciformes: Mugilidae) phenotypic stocks in La Paz Bay, Baja California Sur, Mexico, using geometric morphometrics of otolith shape, growth, and reproductive parameters. *Rev. Mex. Biodivers.* **2020**, *91*, e913273. [CrossRef]
27. Rodriguez-Marin, E.; Quelle, P.; Busawon, D. Description of the ICCAT length at age data base for bluefin tuna from the eastern Atlantic, including the Mediterranean Sea. *Collect. Vol. Sci. Pap. ICCAT* **2022**, *79*, 275–280.
28. Suzuki, Z.; Kimoto, A.; Sakai, O. Note on the strong 2003 year-class that appeared in the Atlantic bluefin fisheries. *Collect. Vol. Sci. Pap. ICCAT* **2013**, *69*, 229–234.
29. Ailloud, L.E.; Rouyer, T.; Kimoto, A.; Sharma, R.; Butterworth, D.S. Detailed analysis of the catch-at-length and age composition data to check results for recent year recruitment estimates for eastern Atlantic bluefin tuna. *Collect. Vol. Sci. Pap. ICCAT.* **2018**, *74*, 3472–3482.
30. Butler, M.J.A.; Accy, J.F.; Dickson, C.A.; Hunt, J.J.; Burnett, C.D. Apparent age and growth, based on otoliths analysis, of giant bluefin tuna (*Thunnus thynnus thynnus*) in the 1975–76 Canadian catch. *Col. Vol. Sci. Pap. ICCAT* **1977**, *6*, 318–330.
31. Hurley, P.C.F.; Iles, T.D. Age and growth estimation of Atlantic Bluefin tuna, *Thunnus thynnus*, using otoliths. *NOAA Tech. Rep. NMFS* **1983**, *8*, 71–75.
32. Lee, D.W.; Prince, E.D.; Crow, E. Interpretation of growth bands on vertebrae and otoliths of Atlantic bluefin tuna, *Thunnus thynnus*. *NOAA Tech. Rep. NMFS* **1983**, *8*, 61–70.
33. Galuardi, B.; Royer, F.; Golet, W.; Logan, J.; Neilson, J.; Lutcavage, M. Complex migration routes of Atlantic bluefin tuna (*Thunnus thynnus*) question current population structure paradigm. *Can. J. Fish. Aquat. Sci.* **2010**, *67*, 966–976. [CrossRef]
34. Richardson, D.E.; Marancik, K.E.; Guyon, J.R.; Lutcavage, M.E.; Galuardi, B.; Lam, C.H.; Walsh, H.J.; Wildes, S.; Yates, D.A.; Hare, J.A. Discovery of a spawning ground reveals diverse migration strategies in Atlantic bluefin tuna (*Thunnus thynnus*). *Proc. Natl. Acad. Sci. USA* **2016**, *113*, 3299–3304. [CrossRef] [PubMed]
35. Anonymous. Report of the 2017 ICCAT bluefin Stock Assessment Meeting. *Col. Vol. Sci. Pap. ICCAT* **2018**, *74*, 2372–2535.
36. Kell, L.T.; Ortiz, M. A comparison of statistical age estimation and age slicing for Atlantic Bluefin tuna (*Thunnus thynnus*). *Col. Vol. Sci. Pap. ICCAT* **2011**, *66*, 948–955.
37. Ailloud, L.E.; Lauretta, M.V.; Hanke, A.R.; Golet, W.J.; Allman, R.J.; Siskey, M.R.; Secor, D.H.; Hoenig, J.M. Improving growth estimates for Western Atlantic bluefin tuna using an integrated modeling approach. *Fish. Res.* **2017**, *191*, 17–24. [CrossRef]

 MDPI

Article

The Use of Daily Growth to Analyze Individual Spawning Dynamics in an Asynchronous Population: The Case of the European Hake from the Southern Stock †

Cristina García-Fernández [1,*], Rosario Domínguez-Petit [2] and Fran Saborido-Rey [1]

1 Instituto de Investigaciones Marinas (IIM-CSIC), 36208 Vigo, Spain
2 Centro Oceanográfico de Vigo (IEO-CSIC), 36390 Vigo, Spain
* Correspondence: cgarcia@iim.csic.es
† The paper was rewritten/reproduced from author's works in PhD's thesis and has not been published.

Abstract: Daily growth patterns and their relationship with reproduction was analyzed in the European hake from the Galician Shelf, where it shows a very protracted spawning with three spawning peaks. The daily growth analysis was performed in otoliths of adult females on the transversal section of the sagittae otolith. Daily increments were measured from the border to the nucleus in females until they were discernible. Results show that daily growth of females decreases during the spawning period because they allocate less energy to somatic growth in favor of the production of gametes, with an increase in growth in July. Lastly, daily growth individual trends showed a "spawning pattern" in 28% of medium and large females, suggesting an individual spawning period of one to two months, with 4–5 valleys of narrow daily increments, likely associated to batch release: individual spawning frequency would be 4–5 days. This is the first time that individual spawning frequency in hake is estimated based on individual data. Finally, the spawning pattern is detected only once per year, indicating that a single female participates only in one spawning peak per year, supporting the hypothesis of the existence of two or more spawning components in the stock.

Keywords: trade-off; daily growth; spawning; otolith; European hake

Citation: García-Fernández, C.; Domínguez-Petit, R.; Saborido-Rey, F. The Use of Daily Growth to Analyze Individual Spawning Dynamics in an Asynchronous Population: The Case of the European Hake from the Southern Stock. *Fishes* **2022**, *7*, 208. https://doi.org/10.3390/fishes7040208

Academic Editor: Josipa Ferri

Received: 20 July 2022
Accepted: 17 August 2022
Published: 18 August 2022

1. Introduction

Energy devoted to growth and reproduction is balanced along the whole lifetime of an individual hake [1]. Energetic resources are used for maintenance and somatic growth when the fish is immature. However, when it reaches sexual maturity, the energy allocation changes: the surplus energy, which was previously only used for growing, is redistributed between growth and reproduction [2,3]. Part of this energy is transferred for reproduction processes (gamete production and reproductive behaviour) with the consequent growth rate decrease. The seasonality on growth (intra-annual variability) in fish is also triggered by physiological events like spawning, but also by environmental conditions [4].

This trade-off is strongly related to the energy acquisition strategy of the species. In capital breeders, the growth rate is expected to be low during spawning season because of feeding reduction during this period. If this is so, a reduction in growth rate should be traceable in the otolith daily growth pattern [5]. On the other hand, in income breeder species, there is not an interruption of feeding while spawning that could mask the influence of reproduction on growth [6–8].

Based on this trade-off, it could be interesting to study the daily growth to validate whether the individual reproductive activity can be monitored on daily otolith growth patterns. In fact, it would be very useful in exploited marine species with protracted spawning seasons and high population asynchrony in spawning due to the associated temporal dispersion on the estimation of reproductive parameters (spawning fraction, spawning frequency, etc.).

Thus, analyzing daily growth in the otolith could provide individual information on species where behavior at the population level hinders the estimation of parameters. The target of the study is the European hake, *Merluccius merluccius*. This species can be found in temperate areas of the Eastern Atlantic, principally throughout the Northeast Atlantic Ocean and the Mediterranean Sea, from Norway to Iceland in the north and to Mauritania in the south [9,10]. Along its distribution, the species presents geographical variability in different aspects, such as reproductive traits [11], i.e., size at maturity [12], diet [13,14], or spawning season [12]. This study is focused on the population located in the Galician shelf (NW Iberian Peninsula), the main spawning area in the southern stock [15–17]. It is a very plastic species that adapts its reproductive potential to environmental conditions, energy availability, and fishing pressure [18,19]. This capacity to adapt well is even more evident in highly productive environments such as the Galician shelf, where important upwelling events take place [20]. Due to this, in the study area, the species presents a protracted spawning season with high-population reproductive asynchrony and an income breeding strategy. Despite the protracted spawning season, three spawning peaks are clearly identified: the main one in winter−spring, the secondary one in summer and occasionally another one in autumn. However, there is great interannual variability in both intensity and timing [10,21,22], likely because of the influence of several internal and external factors such as environment, food availability, or even growth.

Based on its population asynchrony, key individual parameters such as spawning duration, spawning fraction, or spawning frequency are not known for this species. Therefore, the main objective of this study is to test the use of daily growth patterns to extract reproductive parameters at the individual level in one species with high population asynchrony. In relation to this objective, the daily growth of the adult females was analyzed to elucidate if there are changes in the female physiology due to the reproductive cycle, and specifically spawning. If the trade-off between spawning and growth is detectable, it would be possible to estimate reproductive parameters at individual scale in the spawning stock of *M. merluccius* in the Galician Shelf. To do so, the daily growth of mature females was studied. Thus, the objectives of the study are to test if (i) the daily rings formed in the otoliths during the spawning peaks are narrower than those rings formed in the rest of the year, (ii) there are differences in the magnitude of this narrowing between spawning peaks (i.e., there are differences in energy investment in each spawning peak), and (iii) it is possible to estimate individual reproductive parameters based on this method.

2. Materials and Methods

The approach of the present study was created on the analysis of daily growth in otoliths of adult females. Based on the assumption that actively spawning females feed less and devote more energy to reproduction than growth, a decrease in daily growth during the spawning peak is expected, i.e., narrow daily increments.

Based on the reproductive season, and assuming that during the spawning period, the daily growth of an active female decreases, each female would present a low daily growth pattern once perceived in each female's otoliths if it participates in one spawning peak. On the contrary, if each female spawns in both of the two main spawning periods, that pattern should appear two or more times in each female's otoliths. Finally, if spawning activity does not impact daily growth, no detectable pattern in the otoliths would be expected (Figure 1).

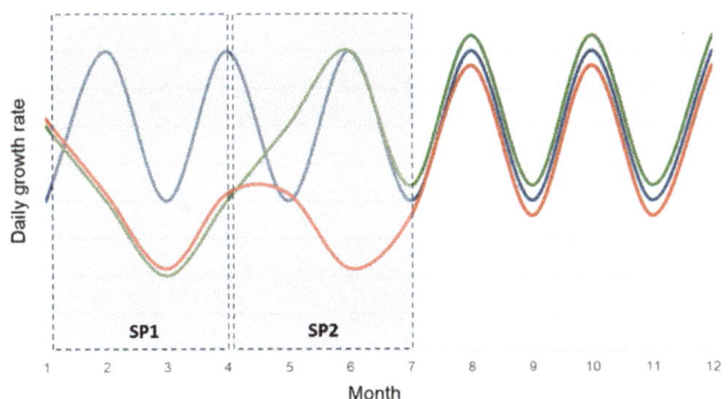

Figure 1. Plot for describing the three potential patterns of the daily growth of adult females of European hake, assuming lower growth when active spawning takes place (i.e., narrow daily rings in otoliths). Grey boxes: each spawning peak (SP1: winter-spring and SP2: summer). Red line: assuming that each female spawns in the two main spawning periods; Green line: assuming that each female participates only in one spawning peak (in this plot a female actively spawning only in SP1); and blue line: assuming that daily growth is not related to spawning activity.

To carry out this study, 60 adult females were selected from the monthly sampling of the commercial gillnet fleets on the Galician shelf during the whole of 2017. This sampling effort was higher during the first half of the year to cover the two main spawning peaks. Data recorded from sampled females included the total length (TL) (cm), the total, gutted, gut, liver, and gonad weights (g), as well as sex and macroscopic ovary stages. Selected females were sampled between the two main spawning peaks (April–May), and after the second spawning peak (July), and the total length ranged from 34 to 80 cm.

The daily growth increments analyses were performed on the otolith edges of developing and post-spawner females which were previously identified by analyzing the ovary microscopic maturity stages using histological sections (according to Brown-Peterson et al. [23]). Larger females were not considered for this analysis because otolith daily increments in large individuals are incomplete or compressed at the outer edges [24]. From the total, 35 specimens were post-spawning females (regressing-RG and regenerating-RN) and 25 specimens were pre-spawning ones (developing-D) that were used as a control group, assuming that D females capitalize all energy for gonadal growth and gamete development prior to the spawning [23].

From the selected females, six were discarded due to over-polishing of the otolith during sample processing. Accordingly, 54 females were analyzed: 23 in D, 18 in RG, and 13 in RN. Additionally, females were classified in two categories in relation to the spawning peak, based on the capture date and the ovary microscopic phase: females which present evidences of having spawn in the winter-spring spawning peak (SP1) and females with evidences of having spawn in the summer spawning peak (SP2) (Table 1).

Table 1. Number of sampled females (n) in developing (D), regressing (RG), and regenerating (RN) for the daily growth analysis considering the two first spawning peaks (SP1, SP2) and the total length range (TL range, cm).

Ovary Phase	SP1		SP2		Total n
	n	TL Range (cm)	n	TL Range (cm)	
D	4	34–73	19	49–79	23
RG	9	45–72	9	46–56	18
RN	6	47–60	7	49–58	13
Total general	19	34–73	35	46–79	54

For daily increments analysis, right sagittae otoliths were selected (Figure 2A). Then, otoliths were embedded in epoxy resin, cut in cross-sections (transversal section) and assembled with Crystalbond adhesive on a slide for subsequent polishing. Finally, using the cross-sectional preparation, sequential images of the entire ventral axis were taken at 20× magnification (Figure 2B).

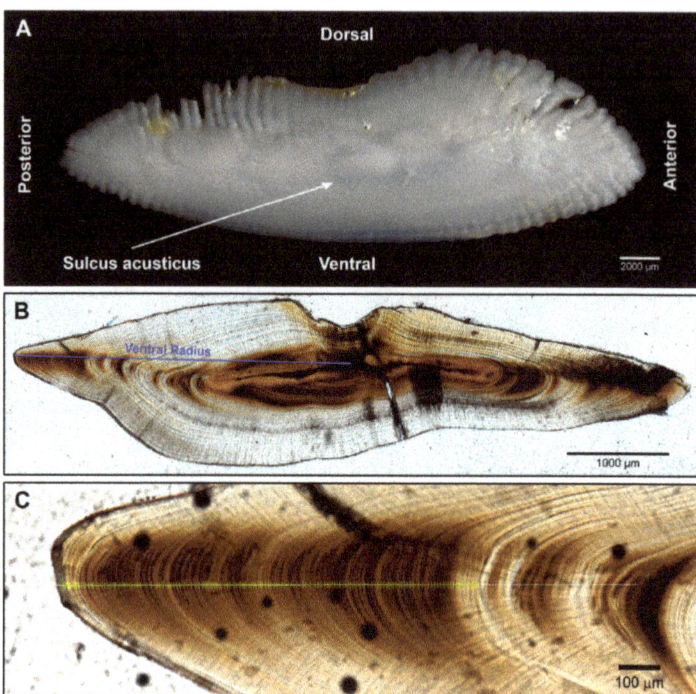

Figure 2. (**A**) Image of European hake right sagittae otolith (5× of magnification) before its processing. (**B**) Mosaic of the cross-section of right sagittae otolith (20× of magnification), with the ventral radius (blue line). (**C**) Axis (white line) and marks (yellow crosses) of the ventral lecture of the cross-section of the right otolith (20× magnification).

The radius was measured from the core along the ventral axis, and daily rings were counted following the same axis, from border to the nucleus of the otolith, until daily growth rings were discernible (Figure 2C).

Daily ring increments were standardized based on the otolith radius for posterior analyses to remove the female's size effect. An ANOVA was performed to compare standard distance mean between months (from January to July). As the duration of each ovarian development phase in European hake remains unknown due to its reproductive cycle, comparisons between phases were performed using only the last 15 days before the capture, assuming that females were in the same ovarian stage within the 15 days prior to capture as on the day they were caught, i.e., each ovary phase lasts at least 15 days. Additionally, an analysis of covariance (ANCOVA) was used to test for seasonal differences of the standard distance mean between otoliths of females at different ovarian development phases. Then, a post-hoc (Tukey's test) test was performed to compare between all of the levels for each factor in these analyses.

A generalized additive mixed model (GAMM) was performed to identify temporal differences in daily increments (standard distance) of the last 15 days before the capture of the female considering the capture day, ovary phases (D, RG and RN), and SP (SP1 and SP2). Capture day was included as an explanatory fixed variable but with a double interaction: ovarian phases and SP. Moreover, female ID was included as random effect to take into account the autocorrelation inherent in the daily rings in each female. GAMM models were performed using "mgcv" R package (version 1.8.31 Simon Wood, Edinburgh, United Kingdom [25]). The model had the following structure:

$$\begin{aligned}\text{Standard distance} \\ &= \beta + f_1(\text{Capture day, interaction} = \text{SP} - \text{Ovarian phase}) \\ &+ \text{Random effect (ID)} + \varepsilon\end{aligned} \quad (1)$$

where β is the intercept and ε the residual error vector. The optimal model was selected based on the minimization of the Akaike information criterion (AIC [26]), and pairwise comparisons among factors were performed using Tukey's test.

Likewise, and because of the asynchrony of *M. merluccius*, the time series of the standard distance five-days rolling mean of each female's otoliths was examined by the naked eye in order to detect patterns in the daily growth that may be related to the reproductive cycle, and specifically to with spawning.

All statistical analyses were conducted using R software version 3.4.3 [27], with a significance level of $\alpha = 0.05$.

3. Results

Daily growth analysis was performed in order to identify changes in the growth pattern that could be related to active spawning with the last aim of trying to obtain individual reproductive parameters in the population. A total of 54 females with a total length range of 34–80 cm, at D, RG, and RN ovarian phases and collected both in the SP1 and SP2 were analyzed (Table S1). Otolith daily increment counts ranged from 18 to 288 by individual (with a mean of 75 read increments). Dorsal radius of the otoliths was measured in all the otoliths and ranged from 2 730 to 4 517 μm. Both inter- and intraindividual data presented high dispersion, with strong fluctuations throughout the analyzed time series.

From January to March, the mean daily increment increased 4%, from 9.91×10^{-4} to 1.02×10^{-3} μm. In April, growth dropped by 8% until the minimum value (9.49×10^{-4} μm) with similar daily mean distance in May (9.55×10^{-4} μm). Afterwards, the daily mean increment increased steadily almost 10% until July, when it reached the maximum value (1.05×10^{-3} μm) (Figure 3). Hence, monthly significant differences were detected in the daily increment ($p < 0.001$; Tukey test results in Table 2).

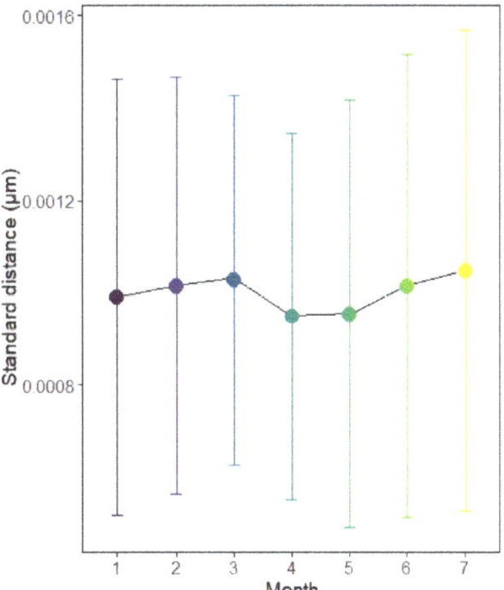

Figure 3. Temporal trend of the standard distance between daily rings in otoliths of European hake females captured in 2017. Points represent the mean and whiskers the standard deviation.

Table 2. p-values for Tukey multicomparison tests of the ANOVA analysis standard distance~month. Significant p-values are shown in boldface type.

	Jan.	Feb.	Mar.	Apr.	May	Jun.	Jul.
Jan.	1.00						
Feb.	1.00	1.00					
Mar.	0.94	1.00	1.00				
Apr.	0.86	0.35	**<0.05**	1.00			
May	0.92	0.45	0.05	1.00	1.00		
Jun.	0.99	1.00	1.00	0.07	0.11	1.00	
Jul.	0.81	0.99	1.00	**<0.05**	0.07	0.96	1.00

Additionally, mean standard distance (std dist) was analyzed by SP and ovarian development phases in females, considering only the last 15 days before capture. Females that spawned in SP1 presented lower daily increments than females that spawned in SP2. RG females in SP1 presented the lowest daily increments (std dist mean = 8.52×10^{-4} µm), followed by D females of SP1 (std dist mean = 9.59×10^{-4} µm) and then RN of the same SP (std dist mean = 9.85×10^{-4} µm).

Wider rings were estimated in RN females (std dist = 1.03×10^{-3} µm) and D females of SP2 (std dist mean = 1.09×10^{-3} µm). Lastly, RG females from SP2 had the highest daily growth rates of all sampled females, with a standard distance mean of 1.10×10^{-3} µm (Figure 4). Hence, the SP-ovary phase interaction showed significant influence on standard distance ($p < 0.001$; Tukey test results in Table 3).

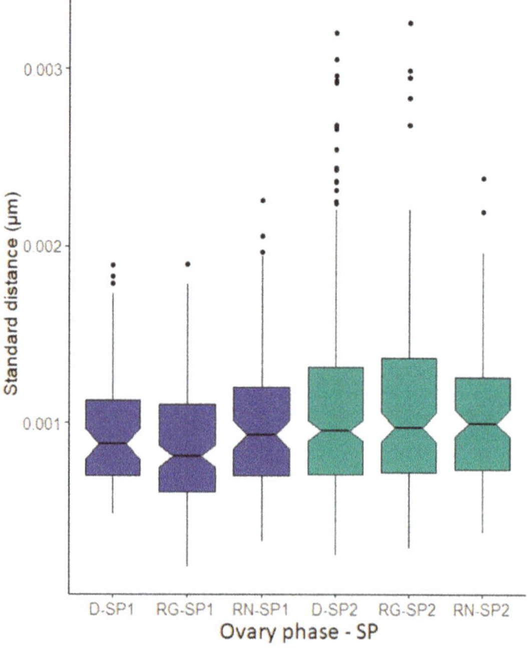

Figure 4. Standard distances (µm) of daily increments in the last 15 days before capture in European hake females captured in 2017 by ovarian development phase (D: developing, RG: regressing, RN: regenerating) and spawning peak (SP1, SP2). The box represents the interquartile range (25th to the 75th percentile), the line in the box represents the mean value, the solid vertical lines represents the standard error, and the notch is the 95% confidence interval of the median.

Table 3. *p*-values for Tukey multicomparison tests of the ANCOVA analysis standard distance~SP-ovarian development phases in the last 15 days before capture in European hake females. Significant *p*-values are shown in boldface type.

	SP1-D	SP1-RG	SP1-RN	SP2-D	SP2-RG	SP2-RN
SP1-D	1.00					
SP1-RG	0.66	1.00				
SP1-RN	1.00	0.27	1.00			
SP2-D	0.32	**<0.001**	0.39	1.00		
SP2-RG	0.31	**<0.001**	0.38	1.00	1.00	
SP2-RN	0.92	**<0.05**	0.98	0.87	0.83	1.00

The GAMM performed with the standardized daily increments of last 15 days of life in sampled females showed different effects of ovarian development phases and SP on the standard distance. Not all SP-ovary phase combinations had a significant effect on the daily increment: in RG and RN, females from both SP standard distances significantly increased as days progress ($p_{SP1\text{-}RG} < 0.05$, $p_{SP1\text{-}RN} < 0.05$, $p_{SP2\text{-}RG} < 0.001$, $p_{SP2\text{-}RN} < 0.05$) while in D, females from both SP1 and SP2 did not show significant variation of standard distance with time ($p_{SP1\text{-}D} = 0.24$, $p_{SP2\text{-}D} = 0.15$) (Table 4). Figure 5 shows the additive effect of the capture day on the standard distance: D females did not present any pattern in either of the two spawning peaks. Conversely, post-spawning females (RG and RN) of SP1 and SP2 showed an upward trend, which appears more intense in the SP2-RG group.

Table 4. Parameter estimates by the fixed effects of the generalized additive mixed model (GAMM) of daily increments in the last 15 days before capture in developing (D), regressing (RG) and regenerating (RN) European hake females captured in 2017. Parametric coefficients present: estimate, standard error (Std. error), z value, and *p*-value; smooth terms present: estimated degrees of freedom (edf), standard error (Std. error), z value, and *p*-value. Moreover, total Akaike Information Criterion (AIC) was calculated.

Parametric Coefficients	Estimate	Std. Error	*t* Value	*p*-Value
Intercept	−6.94	0.05	−152.50	<0.001
Smooth Terms	**edf**	**Ref.df**	**Chi.sq**	***p*-Value**
SP1–D	1.00	1.00	1.37	0.24
SP1–RG	1.00	1.00	4.95	<0.05
SP1–RN	1.00	1.00	4.30	<0.05
SP2–D	1.00	1.00	2.12	0.15
SP2–RG	1.99	1.99	7.44	<0.001
SP2–RN	1.00	1.00	3.93	<0.05
		AIC = 531.40		

GAMM was associated with population asynchrony, thus, individual trends were analyzed by the naked eye to identify possible patterns associated with spawning: a relationship between spawning and the daily increment. In some females was detected a period of one to two months during which 4–6 valleys of narrow daily increments appear. The described pattern was not repeated over time—it appears once in each female sampled—and did not overlap on the same dates in females due to population asynchrony. Only 15 females—28% of the sampled females—presented this pattern of daily increment variability that may be related with spawning, more specifically 7 females from SP1 (5 RG and 2 RN) and 8 females from SP2 (3 RG, 4RN and 1 D) (Figure S1 represents the time series of the sampled females with the detected pattern). In general, the expected pattern of daily increments that could be linked to spawning activity was not observed in the otoliths of the largest females (>60 cm), excepting one female of 72 cm, despite sampled females reaching sizes of 80 cm.

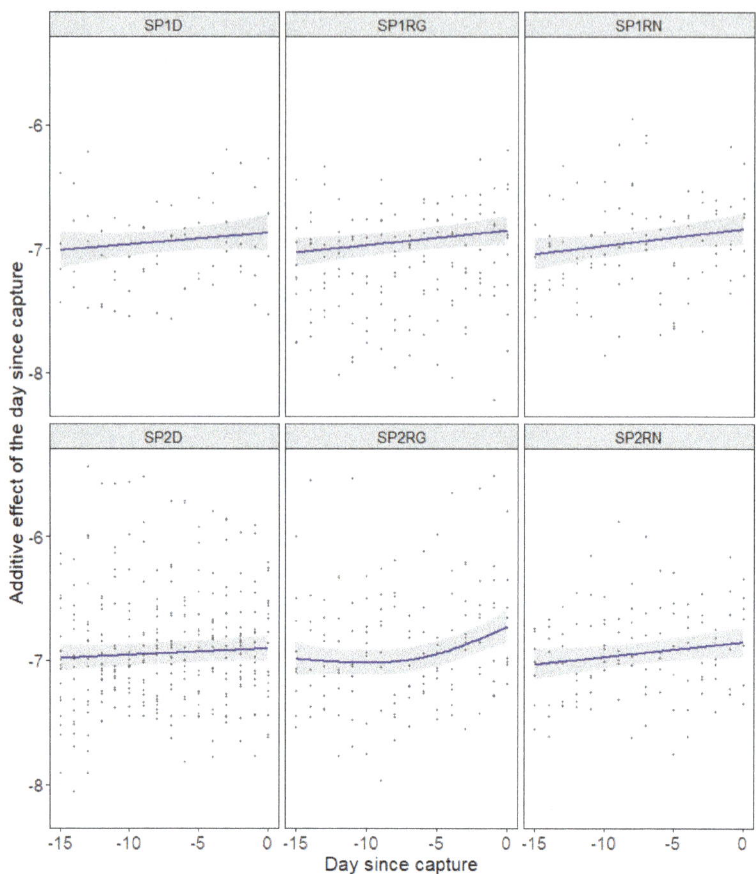

Figure 5. Response plots on the partial effects on the generalized additive mixed model (GAMM) of daily increments in the last 15 days before capture in European hake females captured in 2017. Plot includes the additive effect spawning peaks (SP1, SP2) and ovarian development phase (D, RG, RN). Solid lines denote smoothed values and shaded areas represent 95% confidence intervals.

4. Discussion

Spawning dynamics, both in terms of space and time, is a critical component in reproductive success because of its effect on development and the survival of the offspring and hence on recruitment [28–30]. Specially, spawning time has long been revealed as a one of the most relevant factors [31]. Compared with other areas, European hake in the Iberian shelf has a very protracted spawning season, likely an adaptation to the highly productive but also highly variable environmental conditions in this area [10,22]. Indeed, the intra-annual variability in spawning timing has been also evidenced in other species of *Merluccius*, and the existence of numerous peaks of spawning was associated to different stocks or sub-stocks [32–34]. Hake in the study area presents several spawning peaks, but it is difficult to elucidate its reproductive dynamics because the analysis of maternal attributes does not provide clear evidences. Likewise, the impossibility of captive breeding hinders the estimation of individual reproductive parameters. Therefore, given that otoliths are records of fish life and daily growth is expected to be linked to reproductive activity, we tested whether patterns in daily growth can be identified that can be related to individual spawning (onset, duration, frequency, or subcomponents).

Considering the trade-off between growth and reproduction, the oocyte dynamics may impact growth performance in European hake [35]. The initial assumption considers a lower growth in females when they are spawning, from the oocyte recruitment until the egg release, because gamete production and reproductive behaviour have a cost in terms of energy [2]. Several studies have detected the presence of discontinuities in the width pattern of annual otolith increments which are associated to reproductive processes, such as the onset of sexual maturity (only once in the lifetime of the individual [36–38]) or spawning activity [39]. This last study was focused on the relationship between translucent and opaque zones in the otoliths, but results were not conclusive. However, as active reproduction is expected to lead to low growth rate, the hypothesis of the present study is that daily increments in otoliths will be narrower during the spawning peak than outside it. The analysis is based on two assumptions. First, pre-spawner females capitalize most of their energy intake in oocyte development for spawning, and second, females that have spawn (RG and RN) devote most of their surplus energy to growth. The result is that in RN, females the daily increments at the edge of the otolith should reflect an accelerated growth pattern after a period of low daily growth, during spawning.

Temporal differences in daily growth were detected in our analyses, with narrow rings in April–May and wide ones in July–August, already detected in juveniles, i.e., immature fish, of the same sampling area [40]. This result associated to the temporal dynamics showed that daily growth may be more influenced by other factors than the impact of reproduction, such as environmental conditions. Seasonal differences in the Galician Shelf with cooler waters on winter—spring than the rest of the year in association to the other oceanographic conditions of the area [41,42] probably influenced the growth result of slower daily growth rates [43].

Regarding mature females, the individuals with the lowest daily growth increments were those in regressing phase after the winter—spring spawning peak. In addition to the environmental conditions, narrower daily rings may be associated with a spawning event. In concordance with previous studies, spawners from winter—spring present higher reproductive effort than spawners from the rest of the years, i.e., fecundity, egg diameter, or dry weight are higher in that spawning peak than in the summer and autumn [10,22,44,45]. So, the acquired energy may be devoted to ensuring maintenance and recovery of body mass, rather than length, after an exhausted spawning period.

However, since then, daily growth steadily increased in all the post-spawning females and the highest daily growth rates are recorded in the regressing females from the summer spawning peak. In contrast to the spring—winter peak, the summer spawning peak with less reproductive effort allows for a greater growth of the females. As well, upwelling events during summer enhances food availability due to higher primary production, and consequently, higher energy acquisition for growing [46,47].

This indicates a complex and intricate trade-off between the feeding, growth, and reproductive dynamics. However, some limitations in this analysis may hinder the temporal variation between ovarian development phases in the daily increments, such as the high dispersion of the data and the inability to determine the duration of each ovary phase. Nevertheless, our modelling approach that includes temporal variability may shed light into the matter of growth dynamics related with spawning. Developing females caught both in the winter—spring and summer spawning peak did not show any trend in daily growth, likely because these females maintain basal growth and invest the entire surplus in reproduction [23]. On the other hand, post-spawning females in both peaks had an accelerated growth, evidencing the recovery after the energy expenditure in spawning. Nevertheless, the daily increments showed large inter- and intraindividual variability, making it difficult to reach robust conclusions. As previously mentioned during the study, the large variability, the uncertainty about the duration of each ovary phase throughout the year, and the high asynchrony of the population hamper the analysis of the hake spawning dynamics in determining the individual reproductive behaviour exactly, but it provides information that allows us to observe certain individual spawning patterns.

One particular and interesting pattern observed in the daily ring increments time series is the existence of regular intervals of decreasing growth rate every four to five days in several females which as a "spawning pattern" could be related to the spawning events (batch release) and thus be a reflection of the spawning frequency. In fact, this is the first report of an approximation of individual spawning frequency in European hake.

This "spawning pattern" in the otolith was discernible only in 15 females (28%), all of them with medium and large size. Small females likely invest less in reproductive effort and still require a major investment in growth. Rollefsen (in 1933) [48] also found spawning-related growth patterns in 60% of the analyzed otoliths of cod in Lofoten based on annual rings and suggested that younger females showed minor changes in daily growth during spawning because the energy invested in reproduction was lower. In other words, if the energy investment does not compromise female's growth, a spawning pattern cannot be detected in the otolith because of the absence of spawning checks. A similar pattern, but on a daily basis, can be expected in young female hake, explaining the absence of the "spawning pattern" in the otoliths.

Daily growth dynamics analysis in European hake showed a relationship with the spawning dynamics, but is not sufficiently conclusive about the number of spawning components. This study has shed light on the study of the population dynamics of this stock: the pattern is detected only once per year in the otolith, so a female would participate in only one spawning peak. The presence or two or more spawning components seems to better explain the observed pattern in growth and spawning dynamics of European hake and the involved trade-offs, as was hypothesized in previous works of this stock [22,44]. Thus, the next step may be the analysis of the contribution of each spawning subcomponent to the annual recruitment in the Southern stock of hake.

Several spawning-related factors may impact daily growth, such as spawning vertical migrations, seasonal habitat changes for spawning (different environmental conditions), or the reduction of food intake during the egg release. Other factors, as fluctuations in prey availability or reduction in metabolism (e.g., parasites or diseases) will certainly cause alteration in growth rate [48–50], but not in a rhythmic pattern as observed.

A part of the knowledge gap existing on the otolith formation relates to energy trade-off [50]; one constraint on using otolith daily growth to correlate with spawning activity is the fact that the detection of daily rings in adult females is laborious and complex. It has been shown that the number of daily rings decreases and becomes less discernible with age [24,51,52]. Moreover, the detection was made with the underlying assumption of a constancy in the daily increment periodicity. However, the existence of sub-daily and even the absence of daily rings make the otolith increments interpretation difficult [24,53]. Therefore, a validation of daily increments and aging criteria in adult hake is necessary for a better interpretation of data.

5. Conclusions

In spite of the large variability recorded in the daily growth patterns of European hake in the Galician waters, an evident daily growth pattern coincident with the expected spawning pattern is observed in some females, indicating that individual spawning season lasts 1.5–2 months and that spawning frequency is 4–5 days. In fact, this is the first report of an approximation of individual spawning frequency in European hake. Thus, this study highlights the usefulness and potential of daily growth analysis in the estimation of individual reproductive parameters. In combination with otolith microchemistry or isotopes, the knowledge of the reproductive parameters of exploited species may improve notably, especially when the species presents population asynchrony.

Supplementary Materials: The following supporting information can be downloaded at: https://www.mdpi.com/article/10.3390/fishes7040208/s1, Table S1: Data from the daily growth analysis of the adult females of European hake captured in 2017. A total of 54 females were analysed: 23 developing (D) females, 18 regressing (RG) females and 13 regenerating (RN) females. Each row corresponds to one sampled female, SP to the assumed spawning peak in which each female spawn (SP1 or SP2), capture date, TL to the total length of each female (in cm), otolith radius to the ventral radius (in µm), *n* rings the number of rings that were able to be read and Std distance, the standard distance mean between daily increments (and the standard deviation) (in µm). Figure S1: Rolling mean standard distance (µm) of daily increment time series (of the last 150 days of life) in European hake females captured in 2017 with a detected spawning pattern: period with 4–6 valleys of reduced daily growth extending over 30–60 days (period marked in the figure between vertical dashed grey lines). Dark blue represents females which spawn in SP1 (Dec–Mar) and dark green represents females which spawn in SP2 (Apr–Jul). Ovary phase (D: developing, RG: regressing, RN: regenerating) and total length (cm) are showed in the upper part of each plot. Note the difference in the y axis.

Author Contributions: Conceptualization, formal analysis and methodology, C.G.-F.; validation, R.D.-P.; investigation, C.G.-F., R.D.-P. and F.S.-R.; resources, F.S.-R.; writing—original draft preparation, C.G.-F.; writing—review and editing, R.D.-P. and F.S.-R.; supervision, R.D.-P. and F.S.-R.; project administration, F.S.-R.; funding acquisition, F.S.-R. All authors have read and agreed to the published version of the manuscript.

Funding: C.G.-F. was funded by Predoctoral Fellowship from the Fundación Tatiana Pérez de Guzmán el Bueno and the study was carried out with financial support from the DREAMER project (CTM2015-66676-C2-1-R) funded by the Spanish National Research Program.

Institutional Review Board Statement: Ethical review and approval were waived for this study, because the described scientific sampling did not require ethical permission according to the applicable international, EU and national laws.

Informed Consent Statement: Not applicable.

Data Availability Statement: The data that support the findings of this study are available from the corresponding author upon reasonable request.

Acknowledgments: Special thanks to Lorena Rodríguez, Carmen Piñeiro, Maria Sainza, Santiago Cerviño and Antonio Gomez from Centro Oceanográfico de Vigo (IEO-CSIC) for providing samples and for suggestions for improving the study.

Conflicts of Interest: The authors declare no conflict of interest.

References

1. Heino, M.; Kaitala, V. Evolution of resource allocation between growth and reproduction in animals with indeterminate growth. *J. Evol. Biol.* **1999**, *12*, 423–429. [CrossRef]
2. Roff, D.A. An allocation model of growth and reproduction in fish. *Can. J. Fish. Aquat. Sci.* **1983**, *40*, 1395–1404. [CrossRef]
3. Ware, D.M. Bioenergetics of stock and recruitment. *Can. J. Fish. Aquat. Sci.* **1980**, *37*, 1012–1024. [CrossRef]
4. Campana, S.E. Otolith science entering the 21st century. *Mar. Freshw. Res.* **2005**, *56*, 485–495. [CrossRef]
5. Morales-Nin, B.; Moranta, J. Recruitment and post-settlement growth of juvenile *Merluccius merluccius* on the western Mediterranean shelf. *Sci. Mar.* **2004**, *68*, 399–409. [CrossRef]
6. Lucio, P.; Murua, H.; Santurtún, M. Growth and reproduction of hake (*Merluccius merluccius*) in the Bay of Biscay during the period 1996–1997. *Ozeanografika* **2000**, *3*, 325–354.
7. Piñeiro, C.; Morgado, C.; Saínza, M.; McCurdy, W.J. *Hake Age Estimation: State of the Art and Progress towards a Solution*; ICES Cooperative Research Report No. 294; ICES/CIEM: Copenhagen, Denmark, 2009.
8. Otxotorena, U.; Díez, G.; de Abechuco, E.L.; Santurtún, M.; Lucio, P. Estimation of age and growth of juvenile hakes (*Merluccius merluccius* Linneaus, 1758) of the Bay of Biscay and Great Sole by means of the analysis of macro and microstructure of the otoliths. *Fish. Res.* **2010**, *106*, 337–343. [CrossRef]
9. Casey, J.; Pereiro, F.J. European hake (*M. merluccius*) in the North-East Atlantic. In *Hake: Fisheries, Ecology and Markets*; Alheit, J., Pitcher, T.J., Eds.; Chapman and Hall: London, UK, 1995; pp. 125–147.
10. Domínguez-Petit, R. *Study of Reproductive Potencial of Merluccius merluccius in the Galician Shelf*; University of Vigo: Vigo, Spain, 2007.

11. Mellon-Duval, C.; de Pontual, H.; Métral, L.; Quemener, L. Growth of European hake (*Merluccius merluccius*) in the Gulf of Lions based on conventional tagging. *ICES J. Mar. Sci.* **2010**, *67*, 62–70. [CrossRef]
12. Werner, K.M.; Staby, A.; Geffen, A.J. Temporal and spatial patterns of reproductive indices of European hake (*Merluccius merluccius)* in the northern North Sea and Norwegian coastal areas. *Fish. Res.* **2016**, *183*, 200–209. [CrossRef]
13. Velasco, F.; Olaso, I. European hake *Merluccius merluccius* (L., 1758) feeding in the Cantabrian Sea: Seasonal, bathymetric and length variations. *Fish. Res.* **1998**, *38*, 33–44. [CrossRef]
14. Cartes, J.E.; Hidalgo, M.; Papiol, V.; Massutí, E.; Moranta, J. Changes in the diet and feeding of the hake *Merluccius merluccius* at the shelf-break of the Balearic Islands: Influence of the mesopelagic-boundary community. *Deep. Res. Part I Oceanogr. Res. Pap.* **2009**, *56*, 344–365. [CrossRef]
15. Pérez, N.; Pereiro, F.J. Reproductive aspects of hake (*Merluccius merluccius* L.) on the Galician and Cantabrian shelves. *Bol. Inst. Esp. Oceanogr.* **1985**, *2*, 39–47.
16. Sánchez, F.; Gil, J. Hydrographic mesoscale structures and Poleward Current as a determinant of hake (*Merluccius merluccius*) recruitment in southern Bay of Biscay. *ICES J. Mar. Sci.* **2000**, *57*, 152–170. [CrossRef]
17. Izquierdo, F.; Paradinas, I.; Cerviño, S.; Conesa, D.; Alonso-Fernández, A.; Velasco, F.; Preciado, I.; Punzón, A.; Saborido-Rey, F.; Pennino, M.G. Spatio-Temporal Assessment of the European Hake (*Merluccius merluccius*) Recruits in the Northern Iberian Peninsula. *Front. Mar. Sci.* **2021**, *8*, 614675. [CrossRef]
18. Domínguez-Petit, R.; Korta, M.; Saborido-Rey, F.; Murua, H.; Sainza, M.; Piñeiro, C. Changes in size at maturity of European hake Atlantic populations in relation with stock structure and environmental regimes. *J. Mar. Syst.* **2008**, *71*, 260–278. [CrossRef]
19. Domínguez-Petit, R.; Saborido-Rey, F. New bioenergetic perspective of European hake (*Merluccius merluccius* L.) reproductive ecology. *Fish. Res.* **2010**, *104*, 83–88. [CrossRef]
20. Fraga, F. Upwelling off the Galacian Coast, northwest Spain. In *Coastal Upwelling*; Richards, F.A., Ed.; American Geophysical Union: Washington, DC, USA, 1981; Volume 1, pp. 176–182.
21. Mehault, S.; Domínguez-Petit, R.; Cerviño, S.; Saborido-Rey, F. Variability in total egg production and implications for management of the southern stock of European hake. *Fish. Res.* **2010**, *104*, 111–122. [CrossRef]
22. García-Fernández, C.; Domínguez-Petit, R.; Aldanondo, N.; Saborido-Rey, F. Seasonal variability of maternal effects in European hake *Merluccius merluccius*. *Mar. Ecol. Prog. Ser.* **2020**, *650*, 125–140. [CrossRef]
23. NBrown-Peterson, J.; Wyanski, D.M.; Saborido-Rey, F.; Macewicz, B.J.; Lowerre-Barbieri, S.K. A standardized terminology for describing reproductive development in fishes. *Mar. Coast. Fish.* **2011**, *3*, 52–70. [CrossRef]
24. Campana, S.E.; Neilson, J.D. Microstructure of fish otoliths. *Can. J. Fish. Aquat. Sci.* **1985**, *42*, 1014–1032. [CrossRef]
25. Wood, S.N. Fast stable restricted maximum likelihood and marginal likelihood estimation of semiparametric generalized linear models. *J. R. Stat. Soc. Ser. B (Stat. Methodol.)* **2011**, *73*, 3–36. [CrossRef]
26. Akaike, H. Information theory as an extension of the maximum likelihood principle. In *Second International Symposium on Information Theory*; Petrov, B.N., Csaki, F., Eds.; Akademiai Kiado: Budapest, Hungary, 1973; pp. 267–281.
27. R Core Team. *R: A Language and Environment for Statistical Computing*; R Core Team: Vienna, Austria, 2019.
28. Myers, R.A.; Barrowman, N.J. Is fish recruitment related to spawner abundance? *Fish. Bull.* **1996**, *94*, 707–724.
29. Gilbert, D.J. Towards a new recruitment paradigm for fish stocks. *Can. J. Fish. Aquat. Sci.* **1997**, *54*, 969–977. [CrossRef]
30. Saborido-Rey, F. Fish Reproduction. In *Encyclopedia of Ocean Sciences*, 3rd ed.; Cochran, J., Bokuniewicz, H., Yager, P., Eds.; Elsevier: Cambridge, UK, 2016; pp. 232–245.
31. Wright, P.J.; Trippel, E.A. Fishery-induced demographic changes in the timing of spawning: Consequences for reproductive success. *Fish Fish* **2009**, *10*, 283–304. [CrossRef]
32. Helser, T.E.; Alade, L. A retrospective of the hake stocks off the Atlantic and Pacific coasts of the United States: Uncertainties and challenges facing assessment and management in a complex environment. *Fish. Res.* **2012**, *114*, 2–18. [CrossRef]
33. Jansen, T.; Gislason, H. Population Structure of Atlantic Mackerel (*Scomber scombrus*). *PLoS ONE* **2013**, *8*, e64744. [CrossRef]
34. Jansen, T.; Kainge, P.; Singh, L.; Wilhelm, M.; Durholtz, D.; Strømme, T.; Kathena, J.; Erasmus, V. Spawning patterns of shallow-water hake (*Merluccius capensis*) and deep-water hake (*M. paradoxus*) in the Benguela Current Large Marine Ecosystem inferred from gonadosomatic indices. *Fish. Res.* **2015**, *172*, 168–180. [CrossRef]
35. Saborido-Rey, F.; Kjesbu, O.S. *Growth and Maturation Dynamics*; 2005; p. 26. Available online: https://digital.csic.es/bitstream/10261/47150/3/Growth%20and%20maturation%20dynamics%20%281%29.pdf (accessed on 19 July 2022).
36. Rijnsdorp, A.D.; Storbeck, F. Determining the onset ofsexual maturity from otoliths of individual female North Sea plaice, *Pleuronectes platessa* L. In *Recent Developments in Fish Otolith Research*; Secor, D.H., Dean, J.M., Campana, S., Eds.; University of South Carolina Press: Columbia, SC, USA, 1995; pp. 581–598.
37. Francis, R.I.C.C.; Horn, P.L. Transition zone in otoliths of orange roughy (*Hoplostethus atlanticas*) and its relationship to the onset of maturity. *Mar. Biol.* **1997**, *129*, 681–687. [CrossRef]
38. Reglero, P.; Mosegaard, H. Onset of maturity and cohort composition at spawning of Baltic sprat *Sprattus sprattus* on the basis of otolith macrostructure analysis. *J. Fish. Biol.* **2006**, *68*, 1091–1106. [CrossRef]
39. Irgens, C. Size Distributions, Growth and Settling of Northeast Arctic cod Juveniles (*Gadus morhua*) and the Use of Otolith Shape for Age Group Discrimination. Master's Thesis, University of Bergen, Bergen, Norway, 2010.
40. Piñeiro, C.; Rey, J.; de Pontual, H.; García, A. Growth of Northwest Iberian juvenile hake estimated by combining sagittal and transversal otolith microstructure analyses. *Fish. Res.* **2008**, *93*, 173–178. [CrossRef]

41. Pingree, R.D.; le Cann, B. Three anticyclonic slope water oceanic eDDIES (SWODDIES) in the Southern Bay of Biscay in 1990. *Deep Sea Res. Part A Oceanogr. Res. Pap.* **1992**, *39*, 1147–1175. [CrossRef]
42. Torres, R.; Barton, E.D. Onset of the Iberian upwelling along the Galician coast. *Cont. Shelf Res.* **2007**, *27*, 1759–1778. [CrossRef]
43. Houde, E.D. *Fish Reproductive Biology*; Wiley-Blackwell: Oxford, UK, 2009.
44. Serrat, A.; Saborido-Rey, F.; Garcia-Fernandez, C.; Muñoz, M.; Lloret, J.; Thorsen, A.; Kjesbu, O.S. New insights in oocyte dynamics shed light on the complexities associated with fish reproductive strategies. *Sci. Rep.* **2019**, *9*, 18411. [CrossRef] [PubMed]
45. Korta, M.; Domínguez-Petit, R.; Murua, H.; Saborido-Rey, F. Regional variability in reproductive traits of European hake *Merluccius merluccius* L. populations. *Fish. Res.* **2010**, *104*, 64–72. [CrossRef]
46. Bode, A.; Casas, B.; Fernández, E.; Marañón, E.; Serret, P.; Varela, M. Phytoplankton biomass and production in shelf waters off NW Spain: Spatial and seasonal variability in relation to upwelling. *Hydrobiologia* **1996**, *341*, 225–234. [CrossRef]
47. Reboreda, R.; Cordeiro, N.G.F.; Nolasco, R.; Carmen, G. Modelling the seasonal and interannual variability (2001–2010) of chlorophyll-a in the Iberian margin. *J. Sea Res.* **2014**, *93*, 133–149. [CrossRef]
48. Rollefsen, G. *The Otoliths of the Cod. Preliminary Report*; Report on Norwegian Fishery und Marine Investigations No. 3; Fiskeridirektoratets Skrifter Serie Havundersøkelser: Bergen, Norway, 1933; Volume 4, pp. 3–14.
49. Irgens, C. *Otolith Structure as Indicator of Key Life History Events in Atlantic Cod (Gadus morhua)*; University of Bergen: Bergen, Norway, 2018.
50. Irgens, C.; Folkvord, A.; Otterå, H.; Kjesbu, O.S. Otolith growth and zone formation during first maturity and spawning of Atlantic cod (*Gadus morhua*). *Can. J. Fish. Aquat. Sci.* **2020**, *77*, 113–123. [CrossRef]
51. Pannella, G. Fish Otoliths: Daily Growth Layers and Periodical Patterns. *Science* **1971**, *173*, 1124–1127. [CrossRef]
52. Morales-Nin, B.; Torres, G.J.; Lombarte, A.; Recasens, L. Otolith growth and age estimation in the European hake. *J. Fish Biol.* **1998**, *53*, 1155–1168. [CrossRef]
53. Dougherty, A.B. Daily and sub-daily otolith increments of larval and juvenile walleye pollock, *Theragra chalcogramma* (Pallas), as validated by alizarin complexone experiments. *Fish. Res.* **2008**, *90*, 271–278. [CrossRef]

MDPI

Article

Abundance and Growth of the European Eels (*Anguilla anguilla* Linnaeus, 1758) in Small Estuarine Habitats from the Eastern English Channel

Jérémy Denis [1,*], Kélig Mahé [2] and Rachid Amara [1]

[1] Université du Littoral Côte d'Opale, Université Lille, CNRS, IRD, UMR 8187, LOG, Laboratoire d'Océanologie et de Géosciences, F-62930 Wimereux, France
[2] IFREMER, Fisheries Laboratory, 150 Quai Gambetta, BP 699, F-62321 Boulogne-sur-Mer, France
* Correspondence: jeremy.denis@univ-littoral.fr; Tel.: +33-321-996-427

Abstract: Abundance and growth of the European eel from six small northern French estuaries during their growth phase were examined to explore variations according to the local habitat characteristics. The length–weight relationships and growth models fitted to length-at-age back-calculated otolith growth increments were used to compare the growth. Higher abundances were observed in the smaller estuaries (2.4 to 10.5 ind. fyke nets 24 h^{-1}). The eel length ranged from 215–924 mm with an age range of 4–21 years. There was no significant difference in fish eel lengths or age except in the Liane estuary where the individuals were larger. The length–weight relationships showed an isometric or positive allometric growth in most estuaries. The Gompertz growth models, which best fits the growth, showed no significant differences between estuaries except for female eels from the Liane and the Somme estuaries where the growth performance index was higher. The estimated annual growth rate varied from 2.7 to 115.0 mm·yr^{-1} for female and from 4.4 to 90.5 mm·yr^{-1} for male. The present study shows that eels in the six estuaries had CPUE and growth rates similar to those previously reported in larger habitats. These results reinforce the idea that small estuaries are important habitats that contribute significantly to the eel population and, therefore, play an essential role in conservation strategies for European eel.

Keywords: length–frequency distribution; age; sex-ratio; hydro-morpho-sedimentary characteristics

Citation: Denis, J.; Mahé, K.; Amara, R. Abundance and Growth of the European Eels (*Anguilla anguilla* Linnaeus, 1758) in Small Estuarine Habitats from the Eastern English Channel. *Fishes* **2022**, *7*, 213. https://doi.org/10.3390/fishes7050213

Academic Editor: Josipa Ferri

Received: 20 July 2022
Accepted: 23 August 2022
Published: 23 August 2022

1. Introduction

The European eel (*Anguilla anguilla* L. 1758) is an emblematic species, considered critically endangered by the IUCN and has a complex life cycle. During its continental life phase (i.e., growth phase), the European eel has a large geographic distribution from Northern Europe to North Africa [1]. The growth phase of eels is essential for transoceanic migration, fecundity and reproductive success [2–4]. Variations of growth depend of longer distances from the spawning site [5,6], latitudes and/or temperature range [7,8], habitat productivity as well as differences in fish density [9]. It has also been reported that eel growth depends on habitat use [7,10]. The European eel is a facultative catadromous species [11] that occupies a variety of freshwater, brackish and marine habitats during the growth phase [5,10,12]. Three main migratory tactics have been identified, with the marine and brackish resident, the freshwater resident, and the "migrant" eel that moves from one habitat to another [13–15]. Marine and brackish resident eels generally have a higher growth rate than freshwater residents [7,10,16], due probably to the high potential of available prey (mainly marine macrozoobenthos species) and the length of the growing season (i.e., a temperature above 12 °C corresponding to the growth limit temperature of eel [17]) compared to freshwater habitats.

Knowledge of the growth phase of eels is still limited in brackish and shallow coastal habitats [18], despite the fact that these habitats can be considered as important for the

European eel, especially in brackish systems such as estuaries, which host higher eel densities than freshwater habitats [19]. Studies of individual fish and population growth are important for understanding life-history strategies [20], and are necessary for analyzing eel population dynamics, stock assessment and management across the range of habitats used. The few studies on eel population growth have focused on differences between habitat use (i.e., marine, brackish and freshwater; [10,21–26]), temperature ranges and/or latitudes [7,27], sexual dimorphism and/or silvering stages [7,16,22,28,29], as well as on habitat quality (i.e., contaminant concentrations) [30]. Habitat characteristics, such as water temperature and/or the geographical location (i.e., latitude), appear to be the most influential external factors on eel growth. Eels living in higher temperature habitats, particularly in southern Europe, grow faster than northern individuals, where temperatures are lower [21]. However, growth differences have also been attributed in some cases to local habitat characteristics such as hydro-morpho-sedimentary features (e.g., intertidal seascape [31], river or estuary [32]) or food-web and productivity [33–35].

To our knowledge, the variation of European eel growth in estuary habitats according to the local habitat characteristics has never been examined. In addition, investigations on eels focused almost exclusively on large estuary (e.g., in the Severn Estuary [30] and in the Gironde Estuary [10,25]), neglecting the role of small estuaries, which are the most numerous. Conservation and implementation of management measures for eel populations require an understanding of the importance of small estuaries for eels, particularly in terms of their carrying capacity and growth potential [36,37]. In the present study, we analyzed the European eel abundance and growth attributes in six small estuaries along the French coast of the eastern English Channel and explored whether local habitat characteristics influence them. Specifically, we estimated and compared eel abundances, sex ratio, length structure, length–weight relationship and growth models based on length-at-age back-calculated otolith growth measurements. Finally, we assessed the influence of an estuary's hydro-morpho-sedimentary and anthropogenic characteristics on both female and male eel's growth.

2. Materials and Methods

2.1. Eels Sampling

The European eels were sampled in six estuaries (i.e., Slack, Wimereux, Liane, Canche, Authie and Somme) along the French coast of the eastern English Channel (Figure 1 a). The authorization to collect fish in the estuaries and access to the field site was issued by the Interregional Directorate for the Eastern English Channel-North Sea (dram-npe@equipement.gouv.fr; Decision n°196/2019). This study was conducted in accordance with European Commission recommendation 2010/63/EU, on revised guidelines for the accommodation and care of animals used for experimental and other scientific purposes. Eels were captured in 2019 and 2020 during four sampling periods each year (February–March, May–June, July–August and October–November). Eels were sampled using two fyke nets (16 m long with a mesh size of 15 mm at the beginning, 10 mm in the middle and 8 mm at the cod end) deployed in each estuary at three or four stations (separated by 0.5 to 8 km depending on the estuary size; Figure 1b) along the salinity gradients. The fyke nets were installed along the shoreline at low tide for a period of 2 × 24 h.

2.2. Sampling Areas Description

The six estuaries are located at very close latitudes (i.e., between 50°13′31.78″ N and 50°48′18.79″ N) and have similarities in terms of water temperature and salinity ranges [38] but have their own hydro-morpho-sedimentary characteristics. The Slack, Wimereux and Liane estuaries are the smallest estuaries. The Slack, Wimereux and Liane estuaries have a mean annual flows of 0.64, 1.04 and 4.35 $m^3 \cdot s^{-1}$ (water agency hydro. eaufrance.fr) and a surface area of 21.8, 22.1 and 38.2 km^2 (IGN-F maps), respectively. These estuaries have a narrow mouth width of about 0.1 km, sheltering them from tidal action, are and characterised by a dominant substrate of mud/sand for the Slack and Wimereux, and mud for the Liane [38]. The water quality of these estuaries is considered to be in medium ecological status but in good chemical status (SDAGE 2016-2021). The Slack and Liane estuaries include the existence of dams that delimit the lower and upper parts, leading to higher exposure to freshwater inputs. The Canche, Authie and Somme estuaries have higher mean annual flows of 12.1, 7.3 and 35.2 $m^3 \cdot s^{-1}$ and a surface area of 5.3, 11.9 and 41.7 km^2, respectively. They are composed mainly of sand and gravel sediments [32]. These medium estuaries have mouth widths of 4.6, 2.9 and 5 km, respectively, and are more exposed to tidal action [32]. Their ecological and chemical status is similar to that of the smaller estuaries, with the exception of the Somme estuary, which has a poor chemical status. Human activities are much more important in the medium estuaries, notably due to the presence of agricultural activities, industry and dams in the upstream (Naïades database: naiades.eaufrance.fr).

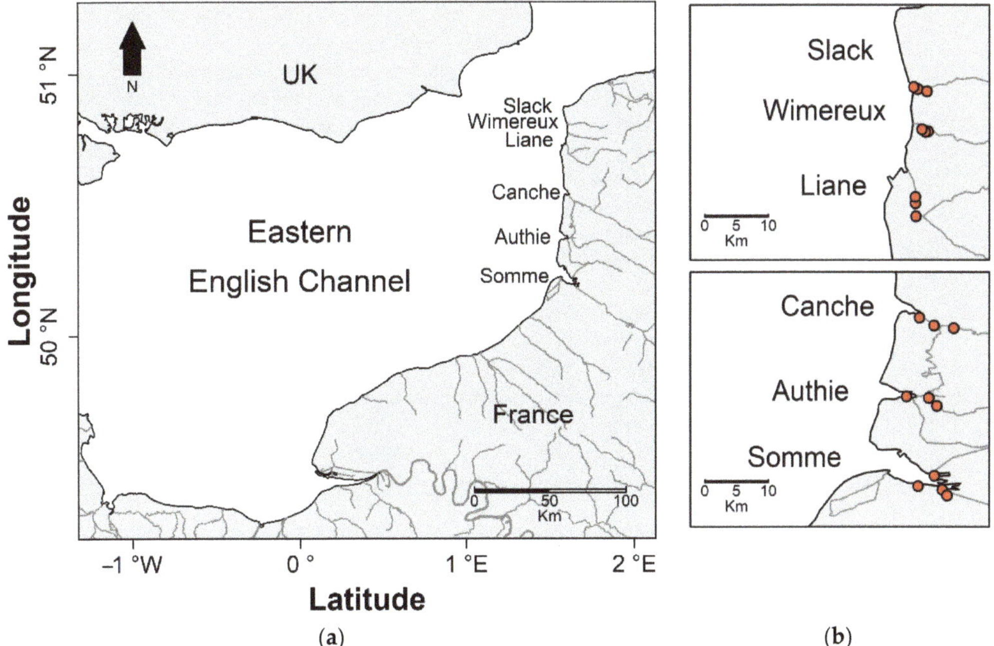

Figure 1. Location of (**a**) the six sampling estuaries along the French coast of the Eastern English Channel and (**b**) location of the three or four sampling stations per estuary (red dot).

2.3. Sample Processing

The eels captured were anaesthetized with eugenol solution (0.04 $mL \cdot L^{-1}$; Thermo Scientific^{TM}) before being individually weighed (total Weight, W ± 1 g) and measured (Total Length, TL ± 0.1 cm). The sex of all eels caught was determined using the "silvering index" [39] based on measurements of body length, body weight, horizontal and vertical eye diameter (mm), and pectoral fin length (mm), and classified as undifferentiated growth phase (OH), female growth phase (FII), female pre-migrant phase (FIII), female migrating phases (FIV and FV) and male migrating phase (MII). A total of 226 eels (i.e., four to six eels for each sampling period, year and estuary) were then euthanized with a saturated eugenol solution and stored at −20 °C for otolith analysis. The other fish captured were released alive in the vicinity of the sampling station. Sex determination was confirmed for stored individuals by visual examination of morphological criteria [18]. Since female eels are larger than males at the same length due to sexual dimorphism [40], females and males were analyzed separately. Undifferentiated individuals for which sex determination had not yet taken place were included in the growth in both female and male models.

The otolith radius (R_o), age (year) and growth of eels were determined on the right sagittal otoliths due to an asymmetry between the right and left otoliths [41] and to compare the results with other studies. The otoliths were extracted, cleaned with demineralized water, dried and embedded in Crystal bond® before polished with 200–800 µm micro-abrasive discs (LP Unalon®). The otoliths were examined under a stereomicroscope (oil-immersion, Olympus BX51) according to the international reading method [42]. Age, otolith radius and annual increments width (µm) were measured along the longest axis on the ventral side, from the elver mark corresponding to the beginning of the growth phase [42] to the edge. Annual increments in eel otoliths have been validated by considering a complete annual increment with a hyaline zone (i.e., cold period) and an opaque zone (i.e., warm period) [43,44].

2.4. Eel Abundance, Total Length and Weight

The abundance of eels was related to the fishing effort called Catch Per Unit Effort (CPUE), expressed in number of eels per gear and per unit of time (ind. fyke nets 24 h^{-1}). Since the data did not comply with the parametric assumption of normality (Shapiro–Wilk test) and homoscedasticity of variance (Levene's F test), the female and male eels' CPUE, total length and weight (combined and separately) were compared between estuaries each year (i.e., 2019 and 2020) using the non-parametric Kruskal–Wallis test and Dunn's multiple comparison test. The Kruskal–Wallis, Dunn test, Shapiro–Wilk test and Levene's F test were performed using the *stats* package in R 4.0.2.

2.5. Length–Weight Relationship

The eels' length–weight relationship (LWR) was fitted separately for female and male eels in each estuary. The length–weight relationship was calculated using the following Equation (1) and $\log_{10}$-transformed data were fitted using a least squared linear model following Equation (2):

$$W = aTL^b \tag{1}$$

$$\log_{10}(W) = \log_{10}(a) + b\log_{10}(TL) \tag{2}$$

where W is the total weight (g), TL is the total length (mm), a is the intercept or body shape coefficient and b is the slope and the growth coefficient (b value indicating an isometry growth when equal to 3.0; [45–47]).

2.6. Growth Model Selection

The linear relationship between total length and otolith radius was tested using a simple linear regression (i.e., $TL = a + b \times R_o$ [48]) for females and males separately.

Length-at-age (mm) was back-calculated using total length and otolith increment measurements and following the modified Fraser–Lee back-calculation procedure [49]:

$$TL_t = TL_c + (TL_c - TL_{bi}) \times \frac{(R_t - R_o)}{(R_o - R_{bi})} \tag{3}$$

where TL_t is the length-at-age at age t, TL_c is total length at capture, TL_{bi} is length at the biological intercept, R_t is otolith radius at age t, R_o is otolith radius at capture, and R_{bi} is otolith radius at the biological intercept.

The non-linear growth models were fitted to length-at-age data obtained from the back-calculated total length of female and male eels separately. Five most commonly used non-linear growth models were tested:

Von Bertalanffy model (vbp) [50]:

$$TL_t = TL_\infty\left(1 - e^{-k(t-t0)}\right) \tag{4}$$

Von Bertalanffy model forced at *t0* = 0 (vbt0p):

$$TL_t = TL_\infty - \left(TL_\infty.e^{-k.t}\right) \tag{5}$$

Von Bertalanffy model forced at TL_{1i} (vbL1p):

$$TL_t = TL_\infty - (TL_\infty - TL_1).e^{-k(t-1)} \tag{6}$$

Gompertz model (gp.p) [51]:

$$TL_t = TL_\infty.e^{\ln\left(\frac{TL_1}{TL_\infty}\right).e^{-k(t-1)}} \tag{7}$$

Logistic model (log.p) [52]:

$$TL_t = \frac{TL_\infty}{1 + \left(\left(\frac{TL_\infty}{TL1}\right) - 1\right).e^{-k.t}} \tag{8}$$

where TL_∞ is the asymptotic length, *k* is the rate at which the asymptote is reached and *t0* is the theoretical age (in years) at zero length. The value of *t0* has no biological significance [53]. The optimal growth model was selected based on the Akaike Information Criterion (AIC; [54,55]). The TL_∞ and *k* parameters of the selected growth model were used to characterize the female and male eels' growth [56,57] separately in each estuary.

In order to compare the selected growth model between estuaries, the growth performance index ($mm{\cdot}yr^{-1}$, [58]) for female and males, was calculated separately:

$$\Phi = \log_{10}k + 2.31\ log_{10}TL_\infty \tag{9}$$

where Φ is the growth performance, *k* and TL_∞ are the growth parameters of the selected model equation.

2.7. Growth Increments

Back-calculation of annual growth increments ($mm{\cdot}yr^{-1}$) was calculated, separately for female and male, using the formula:

$$I_t = TL_t - TL_{t-1} = (TL_\infty - TL_{t-1})\left(1 - e^{-k}\right) \tag{10}$$

where *It* is the growth increments at age *t*, and TL_{t-1} is the length-at-age $t - 1$. *k* and TL_∞ are the growth parameters of the selected model equation for female and male by estuary.

Normality and homoscedasticity of female and male eels' growth rates were assessed using a Shapiro–Wilk test ($p < 0.05$) and a Levene's F test ($p < 0.05$), respectively. Since the data did comply with the parametric assumption of normality and homoscedasticity of variance, two-way analysis of variance (ANOVA) and multiple Tukey (HSD Tukey) tests were used to compare growth rates between estuaries. To compare annual growth increments of eels of the same age between estuaries with a minimum of five individuals per estuary, annual growth increments from 1 to 10 years-old for females and from 1 to 7 years-old for males were considered in the tests. The ANOVA and HSD Tukey tests were performed using the *stats* package in R 4.0.2.

2.8. PCA Analysis

The variation of eel growth in estuary habitats according to the local habitat characteristics was analyzed using a Principal Component Analysis (PCA). The PCA was performed to determine how the spatial difference between female and male eels' growth parameters (*b*, TL_∞, *k*, TL_1, Φ, and I_t) could be explained by hydro-morpho-sedimentary and anthropogenic parameters (surface area, mean annual flow, wave exposure, mouth depth, substrate, mouth width, number of dams and chemical status) in the six estuaries.

The growth parameters were used as input variables and the hydro-morpho-sedimentary parameters were considered as quantitative supplementary variables. The PCA was used to summarize the variables into principal components and the relationship between variables was measured with the linear correlation coefficient (Pearson). The distance between observations (i.e., six estuaries) was based on the Euclidean metric. The growth, hydro-morpho-sedimentary and anthropogenic variables were log-transformed (log + 1) to reduce the skewness of the distribution, then centered and reduced before analyses. The PCA was performed using *FactoMineR* [59] package in R 4.0.2.

3. Results

3.1. Eel Population Characteristics

The eel development stages were mainly undifferentiated (50%), followed by females (40%) and males (10%) (Figure 2a). The sex-ratio was favorable for females in all estuaries and particularly in the Canche (23%), Authie (36%) and Somme (45%) estuaries. Males were more abundant in the Slack (23%), Wimereux (21%) and Liane (25%) estuaries. Eel lengths ranged from 215 to 924 mm and the majority of individuals (73%) were small, ranging from 250 to 500 mm (Figure 3), except for the Liane estuary, where most of the individuals were larger, with a majority of individuals (85%) between 300 and 650 mm. These differences were similar between the two years (Figure 3).

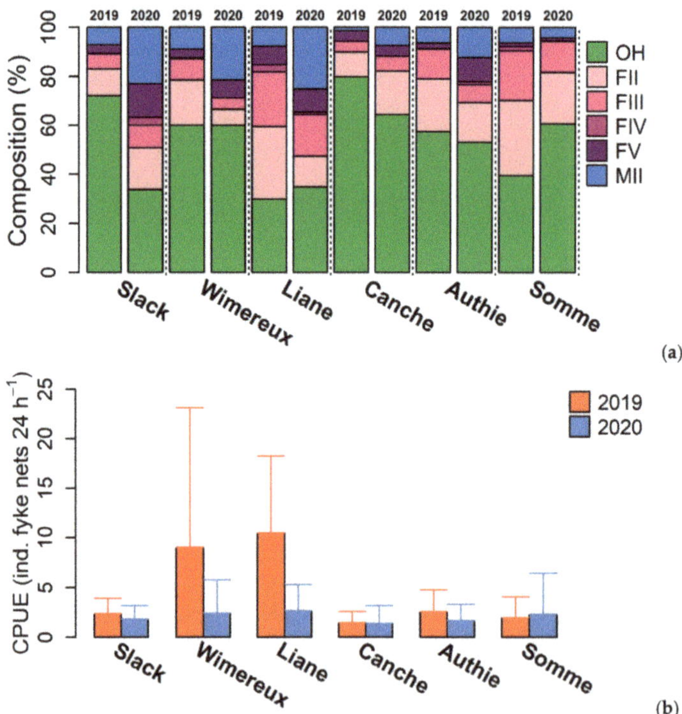

Figure 2. Variation of (**a**) the percentage of individuals per silvering stage (OH undifferentiated growth phase, FII female growth phase, FIII female pre-migration phase, FIV and FV female migration phases and MII male migration phase) and (**b**) means ± standard deviations in CPUE (ind. fyke nets 24 h^{-1}) of the eels in the six estuaries during 2019 and 2020.

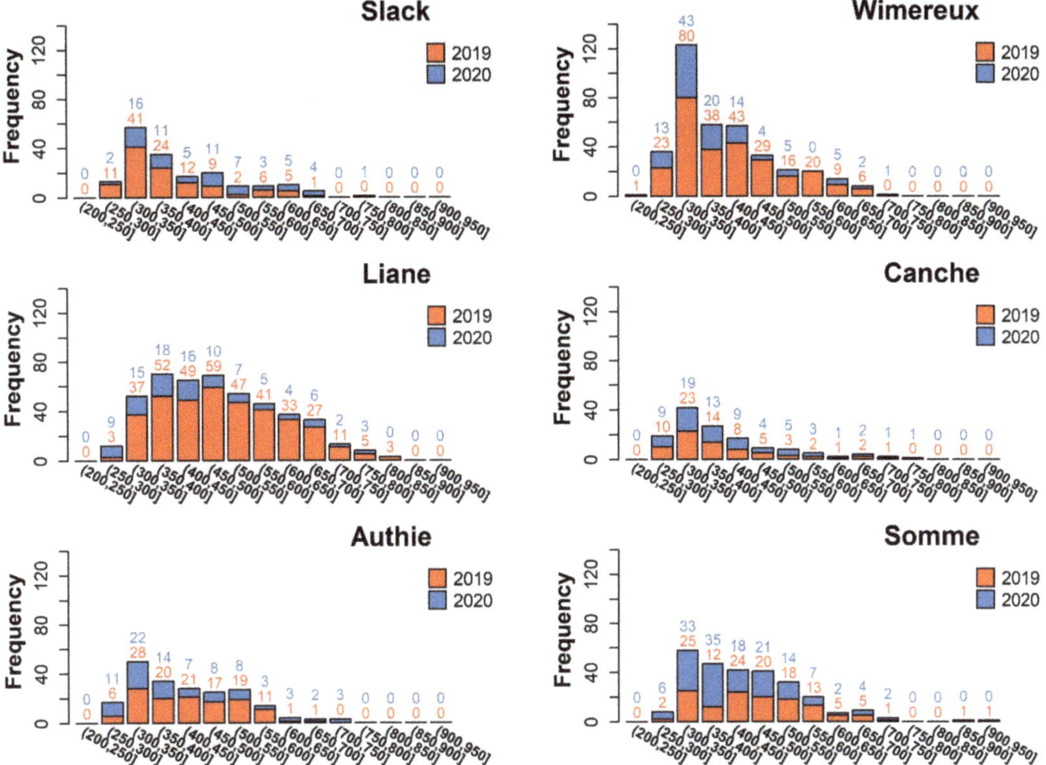

Figure 3. Total length–frequency distribution (TL in mm) of eels measured in the six estuaries during 2019 and 2020. The number of individuals measured is shown above each bar for 2019 (red) and 2020 (blue).

The CPUE means were significantly different across estuaries (Kruska–Wallis test, $p = 0.017$) with higher abundances in the Wimereux (2.4 $\pm$ 3.4 to 9.0 $\pm$ 14.1 ind. fyke nets 24 h^{-1}) and Liane estuaries (2.7 $\pm$ 2.6 to 10.5 $\pm$ 7.6 ind. fyke nets 24 h^{-1}) for both years (Figure 2b). Mean abundances were significantly different between estuaries for both females (Kruskal–Wallis test, $p = 0.0252$) and males (Kruskal–Wallis test, $p = 0.0335$). The abundances were significantly higher in 2019 than in 2020 (Kruskal–Wallis test, $p = 0.009$).

The total length of the eels analyzed ranged from 260 to 924 mm for females (mean: 463.6 $\pm$ 135.8 mm) and from 260 to 494 mm for males (mean: 361.8 $\pm$ 51.4 mm), with no significant differences in mean length between the six estuaries (Kruskal–Wallis test, $p = 0.05$ and 0.16, respectively; Table S1). The female and male eels' total weight also showed no significant differences between estuaries (Kruskal–Wallis test, $p = 0.07$ and 0.38, respectively). Values ranged from 32 to 987 g for females (mean: 238.2 $\pm$ 237.1 g) and from 32 to 247 g for males (mean: 88.3 $\pm$ 41.1 g) (Table S1). There was no difference in age between estuaries for either females (Kruskal–Wallis test, $p = 0.3583$) or males (Kruskal–Wallis test, $p = 0.9014$), with a mean age of 8.4 $\pm$ 2.9 years and 6.8 $\pm$ 1.5 years, respectively (Table S1).The length–weight relationship indicated that the initial body shape coefficient a varied from -17.2 to -12.8 for females and from -13.5 to -11.7 for males, while the growth coefficient b ranged from 2.9 to 3.6 and 2.7 to 3.0, respectively (Table 1). The eels from each estuary showed a significant relationship between total length and weight (Table 1). Almost all the analyzed female eels had a positive allometric growth with b higher than 3.0, except in the Slack estuary with a slight negative allometry (2.9), indicating a tendency to be thinner than individuals in other estuaries. Male eels in the Wimereux, Liane and Canche estuaries exhibited an isometric growth with *b* values equal to 3.0, while eels in the Slack, Authie and Somme estuaries showed a negative allometry ($b = 2.8$), suggesting more elongated fish (Table 1).

Table 1. Length–weight (TL-W) and length–otolith radius (TL-R_o) relationship parameters for female and male eels collected in six estuaries. The values of coefficient *a*, the intercept or initial growth coefficient and *b*, the slope i.e., the growth coefficient and 95% confidence limits (%95 CL). R^2 is the correlation coefficient.

Tested Parameters	Estuary	Female *a*	Female *b*	Female R^2	Male *a*	Male *b*	Male R^2
		%95 CL	%95 CL		%95 CL	%95 CL	
TL-W	Slack	−12.8 −19.4–6.3	2.9 1.9–4.0	0.82	−11.7 −19.4–6.3	2.7 1.9–4.0	0.93
	Wimereux	−17.2 −21.8–12.5	3.6 2.9–4.4	0.89	−12.9 −21.8–12.5	3.0 2.9–4.4	0.91
	Liane	−17.0 −20.3–13.7	3.6 3.1–4.1	0.89	−13.4 −20.3–13.7	3.0 3.1–4.1	0.92
	Canche	−16.7 −20.4–13.0	3.5 3.0–4.1	0.91	−13.5 −20.4–13.0	3.0 3.0–4.1	0.87
	Authie	−14.3 −16.4–12.2	3.2 2.8–3.5	0.95	−13.0 −16.4–12.2	2.8 2.8–3.5	0.94
	Somme	−15.3 −17.6–12.9	3.3 2.9–3.7	0.94	−11.9 −17.6–12.9	2.8 2.9–3.7	0.87
TL-R_o	Slack	15.0 −92.0–121.9	0.27 0.21–0.34	0.76	132.0 68.7–195.4	0.17 0.12–0.22	0.73
	Wimereux	41.9 −42.6–126.4	0.25 0.20–0.31	0.75	209.0 128.6–289.3	0.11 0.05–0.17	0.32
	Liane	−45.6 −180.5–89.4	0.33 0.25–0.41	0.67	205.5 65.2–345.8	0.12 0.02–0.22	0.23
	Canche	17.5 −86.1–121.2	0.30 0.22–0.37	0.70	174.1 54.6–293.5	0.14 0.04–0.24	0.30
	Authie	27.0 −78.2–132.1	0.28 0.21–0.35	0.64	47.5 −52.4–147.4	0.24 0.17–0.31	0.69
	Somme	22.4 −94.8–139.5	0.30 0.23–0.37	0.71	198.6 88.6–308.5	0.13 0.06–0.21	0.43

3.2. Growth Model

A significant linear relationship was found between the female and male eels' total length and otolith radius in each tested estuary (Table 1).

The Gompertz growth model was chosen for both female and male eels, as it was the model that best fitted the eels' total length-at-age data (model with lowest AIC, 18185.89 for female and 8357.17 for male; Table S2 and Figure S1). The asymptotic length (TL_∞) values were higher for female in the Liane and Somme estuaries (751 and 1047 mm, respectively) (Table 2; Figure 4a) and for male in the Slack and Wimereux estuaries (500 and 516 mm, respectively) (Table 2; Figure 4b). The rate at which the asymptote was reached (*k*) showed higher values in the Canche and the Authie for females (0.20 and 0.22, respectively), whereas for the male eels, *k* was higher in the Liane and the Somme (0.29 and 0.30, respectively). The growth performance index Φ of female eels was greater in the Liane and the Somme (59 and 61 $mm \cdot yr^{-1}$, respectively), and similar for other estuaries (58 $mm \cdot yr^{-1}$). The Φ varied little between estuaries for the male eels (55 to 56 $mm \cdot yr^{-1}$) (Table 2).

Table 2. Parameters of the Gompertz growth model and 95% confidence limits (%95 CL) for female and male eels collected in the six estuaries with TL_{∞} the asymptotic length, TL_1 the length in the first year, k the rate at which the asymptote is reached, and Φ the growth performance index (mm·yr^{-1}).

Estuary	Female				Male			
	TL_{∞} %95 CL	TL_1% 95 CL	k %95 CL	Φ	TL_{∞} %95 CL	TL_1 %95 CL	k %95 CL	Φ
Slack	641	109	0.19	57.6	500	107	0.23	55.9
	574–708	97–122	0.15–0.22		436–564	99–114	0.18–0.27	
Wimereux	701	111	0.16	57.9	516	106	0.20	55.7
	601–800	99–123	0.13–0.20		416–616	98–115	0.15–0.25	
Liane	751	112	0.19	59.2	442	110	0.29	55.7
	647–855	98–126	0.15–0.23		367–518	96–125	0.20–0.38	
Canche	621	107	0.20	57.5	437	104	0.27	55.2
	576–667	96–118	0.17–0.22		391–482	97–111	0.22–0.31	
Authie	596	109	0.22	57.5	437	103	0.27	55.3
	523–669	96–122	0.17–0.26		387–488	90–115	0.21–0.33	
Somme	1047	127	0.12	60.6	433	102	0.30	55.7
	873–1221	114–141	0.10–0.14		370–497	88–116	0.22–0.39	

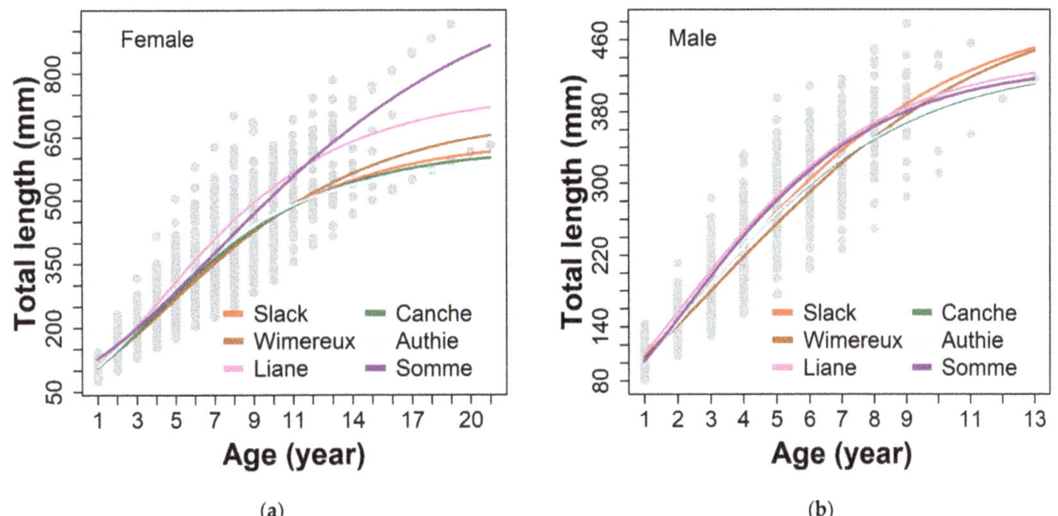

Figure 4. Gompertz growth model (colored lines) fitted to eels' total length-at-age data collected in the six estuaries for (**a**) female and (**b**) male eels (grey dots).

3.3. Annual Growth Rate

The estimated annual growth rate varied from 2.7 to 115.0 mm·yr^{-1} for females and from 4.4 to 90.5 mm·yr^{-1} for males. Mean growth rate decreased rapidly with age; it was around 98.6 ± 8.1 mm·yr^{-1} at one year-old and reached 38.5 ± 18.0 mm·yr^{-1} at 10 years-old for the females (Figure 5a) and 79.8 ± 4.3 mm·yr^{-1} at one year-old and reached 29.5 ± 10.3 mm·yr^{-1} at 7 years-old for the males (Figure 5b). The ANOVA and HSD Tukey tests of female and male eels indicated that female and male eels' annual growth increments were significantly higher in the Liane and the Somme than in any other estuary (HSD Tukey test, $p < 0.001$; Table 3, Figure 5a). Female growth rates were significantly higher in the

Liane estuary (between 39.3 ± 15.4 and 110.9 ± 2.6 mm·yr^{-1}) and the Somme (between 64.2 ± 10.9 and 106.6 ± 1.3 mm·yr^{-1}) than in the other estuaries (between 22.5 ± 16.4 and 95.4 ± 2.5 mm·yr^{-1}).

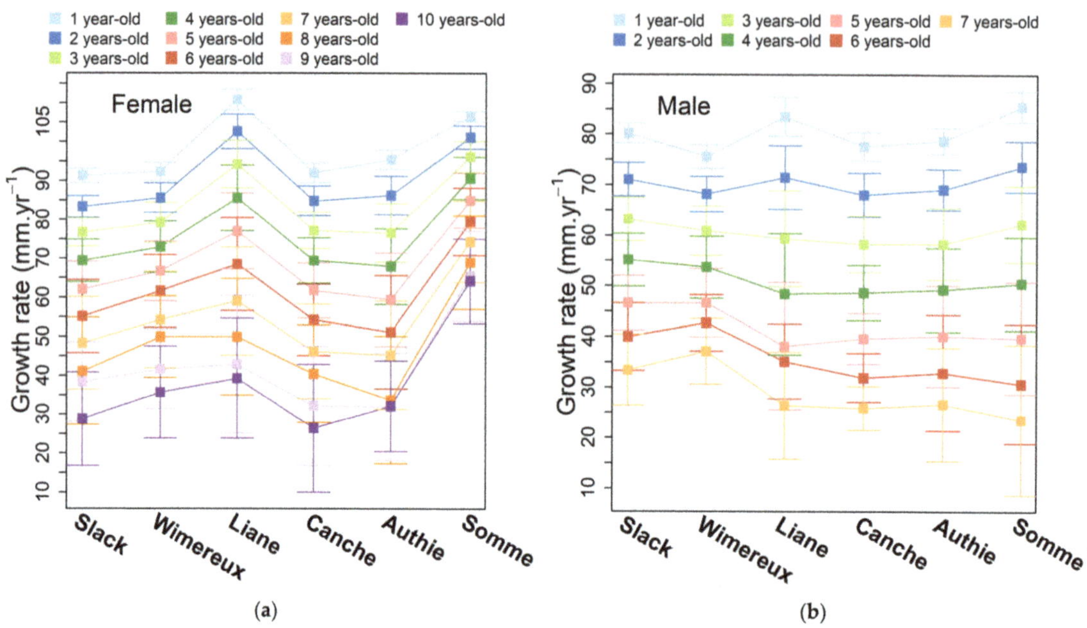

Figure 5. Annual growth increments (mm·yr^{-1}) ± standard deviation from (**a**) 1 to 10 years-old female eels and (**b**) from 1 to 7 years-old male eels collected in the six estuaries.

Table 3. *p*-values of Tukey multiple comparisons tests on the estuary effects on annual growth increments of female and male eels collected in the six estuaries.

Sex	Estuary	Slack	Wimereux	Liane	Canche	Authie
Female	Wimereux	<0.001				
	Liane	<0.001	<0.001			
	Canche	0.999	<0.001	<0.001		
	Authie	0.721	<0.001	<0.001	0.906	
	Somme	<0.001	<0.001	<0.001	<0.001	<0.001
Male	Wimereux	0.706				
	Liane	<0.001	0.056			
	Canche	<0.001	<0.001	0.315		
	Authie	<0.001	<0.001	0.746	0.974	
	Somme	<0.05	0.458	0.947	<0.05	0.195

Mean 1 to 7 years-old male eels' growth increments presented a similar pattern with higher values in the Liane and the Somme; however, their mean growth increments decreased more rapidly from 3 years-old and then reached the same values (59.3 ± 9.5 and 62.3 ± 7.6 mm·yr^{-1}, respectively) as the other estuaries (between 58.4 ± 5.6 and 63.2 ± 4.3 mm·yr^{-1}, respectively) (Figure 5b and Table 3). Their mean growth increments became lower from 5 years-old (38.1 ± 12.6 and 39.7 ± 11.0 mm·yr^{-1}, respectively) compared with those in the Slack, Wimereux and Authie estuaries (46.6 ± 5.4, 46.6 ± 6.7 and 40.1 ± 10.0 mm·yr^{-1}, respectively).

3.4. Local Habitat Characteristics Influence on Eel Growth

The first two axes of the PCA explained 93.5% and 87.6% of the total female and male eels' growth variance, respectively (Figure 6). The PCA showed a clear separation between the variables *k*, *b*, and the variables TL_∞, TL_1, Φ and l_t for both female and male eels. All female eels' growth parameters, except for *b*, showed a positive correlation with the first axis (Figure 6a). TL_∞, TL_1, Φ and l_t from 1 to 10 years-old showed the highest positive correlations (between 0.81 and 1.00) and were associated with the estuary's chemical status (0.84), number of dams (0.86) and wave exposure (0.84). *k* was negatively correlated with the first axis, whereas the *b* (0.64) was positively correlated with the second axis and was associated with the mouth depth (0.84).

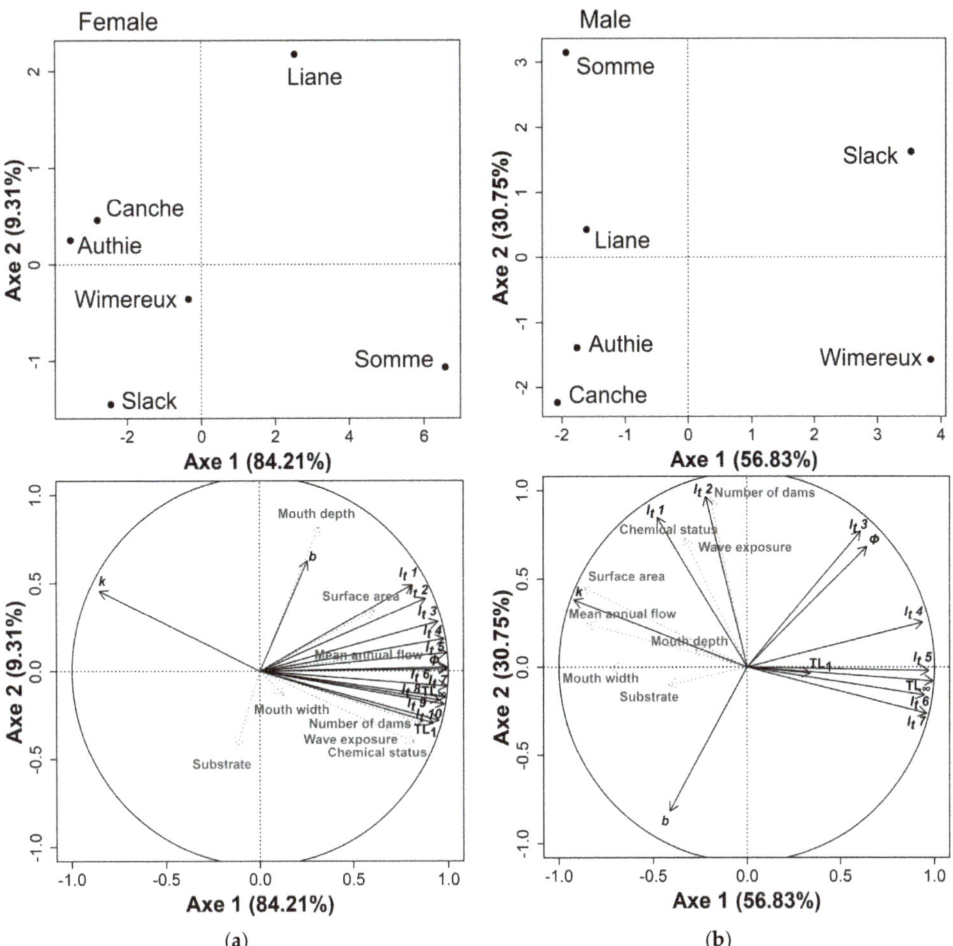

Figure 6. Observations (**top**) and correlation circle (**bottom**) of the first two axes of the Principal Component Analyses (PCA) on the growth of (**a**) female and (**b**) male eels (*b*, TL_∞, *k*, TL_1, Φ, and I_t; black arrows) in the six estuaries as a function of the hydro-morpho-sedimentary (surface area, mean river annual flow, wave exposure, mouth depth and width and substrate; grey arrows) and anthropogenic (number of dams and chemical status; grey arrows) variables.

Male eels' growth parameters (Figure 6b), Φ (0.64), TL_∞ (0.99), and l_t from 4 to 7 years-old (0.94–0.97), were positively correlated with the first axis, and while *k* (−0.92) was negatively correlated with the area surface (−0.87), mouth width (−0.73) and mean annual flow (−0.85). 1 to 3 years-old l_t were positively correlated with the second axis (0.77–0.97) and were associated with chemical status (0.73), number of dams (0.96) and wave exposure (0.73), while *b* was negatively correlated with the second axis (−0.81).

4. Discussion

The stock of the European eel is currently at its historical minimum. Effective conservation management of eels is hampered in particular by an incomplete understanding of the contribution of small estuaries. The present study highlights the importance of small estuaries located along the French coast of the eastern English Channel for European eel. Eels were most abundant in the smaller estuaries of the Wimereux and the Liane. CPUE values for the eels were in range with the highest densities reported in the ICES WGEEL database for larger rivers of Great Britain, France and Spain in the Interreg Atlantic and North Sea areas [60]. This suggests that small estuaries can support large eel populations. The sex ratio of eels caught in the six estuaries was in favor of females, with more than 80% of individuals compared to males. Generally, females are caught more upstream in freshwater, while males are significantly more important downstream in estuaries [61–63]. A sex ratio in favor of females in estuaries could be explained by the decrease in densities in recent years, which would have resulted in a higher proportion of females, since sex determination depends on density [18,64,65]. A possible change in the sex ratio may have occurred in recent years as a result of declining eel stocks [66,67]. The sex ratio of eels can be used as an indicator of population density, when environmental conditions are similar [68]. A higher proportion of males would indicate a higher population density at the time of sex determination [69,70]. However, sex determination may be affected by other parameters such as habitat quality and food availability [68]. In the six estuaries studied, we found a relatively higher proportion of males in the smaller estuaries of the Slack, Wimereux and Liane (23–53%) compared to the other estuaries (<25%). Sex ratios calculated from sex determination based on the silvering index [39], which allows a quick and non-invasive estimation of the sex of the eel, require more accurate verification by macroscopic observation of the gonads [71]. However, the results of sex ratios calculated with the same methodology between estuaries allow comparison of inter-estuary variations and highlights a significant contribution of small estuaries to eel populations. In terms of length structure, except for the Liane estuary which has larger eels, the eel lengths in other small estuaries do not vary compared to the other estuaries.

The length–weight relationship of eels showed that differences in body condition can be observed in both sexes, and that growth values are high in the six estuaries located along the French coast of the eastern English Channel. For the slope of the length–weight relationship, *b* represents isometric growth with a value of 3.0 [45–47]. When *b* is less than 3.0, growth will be negatively allometric and fish become thinner with increasing length, whereas when *b* is greater than 3.0, growth will be positively allometric with larger fish reflecting optimal growth conditions. The *b* measured in this study indicated that the growth of eels is optimal with values between 3.0-3.6 for females and 3.0 for males in most of the estuaries, except for the Slack estuary where values were below 3.0 for both males and females and for the Authie and Somme estuary only the males were below. These *b* results are similar to those described for European eels in brackish and freshwater habitats further south in Europe [24,72,73], and higher than in high latitude [72,74]. These differences reflect the effect of latitude on eel growth, through a decrease with latitude with decreasing water temperature range and time of the growing season [8,18,22,75]. Other more local factors may also impact on the length–weight relations, such as density, food availability or pollution [76,77]. Nevertheless, these results suggest that the six northern French estuaries can be considered as offering optimal conditions for eel growth, close to that found in warmer and larger systems.

The growth model parameters are used to characterize growth [56,57]. The values of the parameter *k* and TL_∞ were highly variable between estuaries (0.12–0.30 and

433–1047 mm, respectively), and similar to other habitats in both brackish and freshwater conditions [21]. These variations in growth parameters are highly dependent on environmental conditions and demographic factors such as the range of sizes of individuals used to fit the growth models [78,79]. The growth performance index (Φ) based on growth model parameters allows for better comparisons of growth between populations and/or habitats [57,79,80]. The growth performance index Φ of eels showed higher values between 57.5 to 60.6 mm·yr^{-1} for the females and 55.2 to 55.9 mm·yr^{-1} for the males, compared to other estuarine habitats with values between 46.4 to 51.4 mm·yr^{-1} (i.e., in the Severn [30] and Guadalquivir estuaries [22]; the estuaries along the Southern Baltic [26] and the German Baltic coast [81]). In comparison with other brackish habitats such as Mediterranean lagoons, eels have growth values between 50.2 to 53.3 mm·yr^{-1} for the females, 47.9 to 53.6 mm·yr^{-1} for the males and 48.4 to 51.5 mm·yr^{-1} for both (i.e., the Valli di Comacchio [29], Vaccarés-Impérieux, Fumemorte [24], Aveiro [82] and Valle Nuova lagoons [29]). Only eels from the Santo André lagoon in Portugal have higher growth, with 62.1 mm·yr^{-1} for females and 59.5 mm·yr^{-1} for males [21]. This highlights that the eels living in the six northern France small estuaries studied have much better growth performance than those living in brackish habitats in warmer latitude (i.e., further south in Europe), suggesting that parameters other than temperature may influence eel growth. Eels from freshwater habitats also presented lower growth values compared to our studied estuaries, with values between 46.6 to 48.5 mm·yr^{-1} for the females and 31.6 to 46.1 mm·yr^{-1} for the males (i.e., the Severn [9], Shannon [83], Frome [84], Koge Lellinge [85], Barrow [78] and Imsa rivers [86] and the Fertö lake [87]). This confirm that brackish habitats such as estuaries support higher growth rates than in freshwater habitats and offer probably more favorable conditions to support eel growth [10,24–26]. Estuarine habitats in temperate ecosystems are much more productive than rivers and allow for a longer period of higher temperatures and longer time of year, which underlines the importance of estuarine habitats for the growth of the eels [10].

The length–weight relations, growth models and growth rates were significantly different between the six estuaries located along the French coast of the eastern English Channel. In general, eels in the smaller estuaries (i.e., the Slack, Wimereux and Liane estuaries) exhibited rather high growth compared to the larger estuaries, except for the Somme that showed an exception for female eels with high growth. The water temperature range (or latitude) and distance to spawning sites are the two major factors that most explain differences in eel growth rate [7,8]. However, at a local scale, these factors do not vary significantly between the six local estuaries that were studied. Other factors may explain these growth variations, such as fish density and/or habitat productivity [9]. In terms of eel density, the Wimereux and Liane estuaries showed higher eel densities (9 $\pm$ 14 and 11 $\pm$ 8 ind. fyke nets 24 h^{-1}, respectively) compared to the other estuaries (between 1.4 $\pm$ 1.1 to 2.6 $\pm$ 2.2 ind. fyke nets 24 h^{-1}), which could lead to strong competition for food and thus limit optimal growth. Our results do not support such a hypothesis since growth rates were relatively higher in these smaller estuaries. Food limitation is one of the major factors influencing the growth of fish, especially juvenile fish in coastal nurseries [88]. Eels have opportunistic feeding plasticity with a feeding preference for macro-crustaceans, capable of feeding on other prey such as fish, mainly when benthic invertebrates are less available [63–65]. This dietary plasticity of eels leads to spatial differences in dietary, in particular in the six estuaries [89], with a diet more focused on fish in smaller estuaries and on macro-crustaceans in larger estuaries (unpublished data). These differences suggest that feeding on more energetic prey, such as fish, would help maintain optimal growth. A previous study also showed that the trophic ecology of eels in these estuaries may be influenced by the availability of macrozoobenthos species, which in turn depends on the hydro-morpho-sedimentary characteristics of the estuary [89]. On a local scale, the characteristics of the estuary may be a likely mechanism for indirect regulation of eel growth through food availability.

Several causes have been suggested to explain the decline in eel stocks, including habitat loss and dams that reduce accessibility to growth habitats, pathogens and pollution [76,90–93]. The presence of barriers to migration and alteration of water quality and their habitats have been suggested as the main causes which have led to their decline [60,94–96]. The six estuaries studied in the present study are less affected by anthropogenic pressures and considered to be of good chemical status and medium ecological status, except for the Somme estuary, which is rated as poor. In addition, only the Liane and Somme estuaries have dams that likely impact on the free movement of eel populations. No negative correlation was found between female eel's growth parameters and the hydro-morpho-sedimentary and anthropogenic variables analyzed. However, the growth of male eels seems to be impacted by some parameters such as the river flow, the surface area and the mouth depth, especially for the older ones (i.e., 3 to 7 years old). The rate (k) at which male eel's approach the population asymptotic size (TL_{∞}) is positively correlated with these later parameters. Male eels being smaller than females may be more sensitive to habitat quality and to obstacles limiting accessibility to growth habitats. It has been shown that the alteration of the banks is likely to hinder the lateral ecological continuity between the main watercourse and the submerged areas (e.g., silt, intertidal flats, etc.) [97,98]. However, the small sample size of male eel's (representing only 10% of the catches) does not allow us to conclude anything regarding the effect of the hydro-morpho-sedimentary and anthropogenic variables analyzed on the growth of the males. Other anthropogenic factors, such as lateral habitat loss due to channelization and reinforcement of the riverbanks need to be to be considered in future studies. Nevertheless, the relative high growth performance measured in the six estuaries compared to other brackish water and estuarine habitats [21,22,24,26,30,81,99] supports the idea that the environmental conditions of these estuaries are favorable to eel growth.

5. Conclusions

CPUE and estimated growth of European eels in the estuarine habitats have highlighted the importance of small estuaries as a crucial habitat for supporting higher growth during their growth phase. The present study shows that eels in the six estuaries located along the French coast of the eastern English Channel had growth rates similar to those measured in brackish habitats further south and higher than those measured in freshwater habitats. In addition, CPUE values were in range with the highest densities previously reported in larger habitats. These results reinforce the idea that small estuaries are important habitats that contribute significantly to the eel population and, therefore, must be considered in both population status assessments, conservation and management strategies for the European eel [36].

Supplementary Materials: The following supporting information can be downloaded at: https://www.mdpi.com/article/10.3390/fishes7050213/s1, Figure S1: Non-linear growth models of female and male eels collected in the six estuaries; Table S1: Number of individuals (N) analyzed of female and male eels for the six estuaries from the North to the South and their mean ± standard deviation (sd) and range (min–max) of total length (mm), total weight (g) and age-at-capture (year); Table S2: Parameters of non-linear growth models for female and male eels collected in the six estuaries with $TL\infty$: the asymptotic length, k: the rate at which the asymptote is reached, TL1: the length in the first year, t0 no biological significance and AIC the Akaike Information Criterion to select the optimal growth model.

Author Contributions: Conceptualization, J.D. and R.A.; Methodology, J.D. and K.M.; Validation, J.D., K.M. and R.A.; Formal analysis, J.D., K.M. and R.A.; Investigation, J.D.; Writing—original draft preparation, J.D., K.M. and R.A.; Supervision, R.A.; Funding acquisition, J.D. and R.A. All authors have read and agreed to the published version of the manuscript.

Funding: This research was funded by "Parc Naturel Marins des Estuaires Picards et de la Mer d'Opale" (DECISION N°2018-28 9 March 2018), and "European Maritime Fisheries Fund" and "Région Hauts de France" (PFEA621220CR0310022). This work also has been partially financially supported by the European Union (ERDF), the French Government, the Région Hauts-de-France and IFREMER, in the framework of the project CPER MARCO 2015–2020 (https://marco.univ-littoral.fr/). The funders had no role in study design, data collection and analysis, decision to publish, or preparation of the manuscript.

Institutional Review Board Statement: The permission to collect fish in the estuaries and field site access was issued by the "Préfète de la region Normandie, préfète de la Seine Maritime, Direction interrégionale de la mer Manche Est-mer du Nord, Service Régulation des Activités et des Emplois Maritimes, Unité Réglementation des Ressources Marines (dram-npe@equipement.gouv.fr): Decision n°196/2019". This study was conducted in accordance with European Commission recommendation 2007/526/EC, on revised guidelines for the accommodation and care of animals used for experimental and other scientific purposes.

Data Availability Statement: The data presented in this study are available on request from the corresponding author.

Acknowledgments: The authors would like to thank K. Rabhi, M. Diop, M. Kazour and K. Boutin for their participation in the collection of the samples and A. Lheriau, R. Elleboode and A. Dussuel for their participation in the preparation of the otoliths. Finally, the authors are very grateful to the three anonymous referees and editor, who helped to greatly improve the initial manuscript with their positive and constructive comments.

Conflicts of Interest: The authors declare no conflict of interest.

References

1. Dekker, W. On the Distribution of the European Eel (*Anguilla anguilla*) and Its Fisheries. *Can. J. Fish. Aquat. Sci.* **2003**, *60*, 787–799. [CrossRef]
2. Belpaire, C.G.J.; Goemans, G.; Geeraerts, C.; Quataert, P.; Parmentier, K.; Hagel, P.; De Boer, J. Decreasing Eel Stocks: Survival of the Fattest? *Ecol. Freshw. Fish* **2009**, *18*, 197–214. [CrossRef]
3. Couillard, C.M.; Verreault, G.; Dumont, P.; Stanley, D.; Threader, R.W. Assessment of Fat Reserves Adequacy in the First Migrant Silver American Eels of a Large-Scale Stocking Experiment. *N. Am. J. Fish. Manag.* **2014**, *34*, 802–813. [CrossRef]
4. van Ginneken, V.J.T.; van den Thillart, G.E.E.J.M. Eel Fat Stores Are Enough to Reach the Sargasso. *Nature* **2000**, *403*, 156–157. [CrossRef] [PubMed]
5. Jessop, B.M. Geographic Effects on American Eel (*Anguilla Rostrata*) Life History Characteristics and Strategies. *Can. J. Fish. Aquat. Sci.* **2010**, *67*, 326–346. [CrossRef]
6. Vélez-Espino, L.A.; Koops, M.A. A Synthesis of the Ecological Processes Influencing Variation in Life History and Movement Patterns of American Eel: Towards a Global Assessment. *Rev. Fish Biol. Fish.* **2010**, *20*, 163–186. [CrossRef]
7. Daverat, F.; Beaulaton, L.; Poole, R.; Lambert, P.; Wickström, H.; Andersson, J.; Aprahamian, M.; Hizem, B.; Elie, P.; Yalçın-Özdilek, S.; et al. One Century of Eel Growth: Changes and Implications: One Century of Eel Growth. *Ecol. Freshw. Fish* **2012**, *21*, 325–336. [CrossRef]
8. Panfili, J.; Ximénès, M.-C.; Crivelli, A.J. Sources of Variation in Growth of the European Eel (*Anguilla anguilla*) Estimated from Otoliths. *Can. J. Fish. Aquat. Sci.* **1994**, *51*, 506–515. [CrossRef]
9. Aprahamian, M.W.; Walker, A.M.; Williams, B.; Bark, A.; Knights, B. On the Application of Models of European Eel (*Anguilla anguilla*) Production and Escapement to the Development of Eel Management Plans: The River Severn. *ICES J. Mar. Sci.* **2007**, *64*, 1472–1482. [CrossRef]
10. Daverat, F.; Tomás, J. Tactics and Demographic Attributes in the European Eel *Anguilla anguilla* in the Gironde Watershed, SW France. *Mar. Ecol. Prog. Ser.* **2006**, *307*, 247–257. [CrossRef]
11. Tesch, F.W.; Rohlf, N. Migration from Continental Waters to the Spawning Grounds. In *Eel Biology*; Aida, K., Tsukamoto, K., Yamauchi, K., Eds.; Springer: Tokyo, Japan, 2003; pp. 223–234. ISBN 978-4-431-65907-5.
12. Thibault, I.; Dodson, J.; Caron, F.; Tzeng, W.; Iizuka, Y.; Shiao, J. Facultative Catadromy in American Eels: Testing the Conditional Strategy Hypothesis. *Mar. Ecol. Prog. Ser.* **2007**, *344*, 219–229. [CrossRef]
13. Daverat, F.; Tomas, J.; Lahaye, M.; Palmer, M.; Elie, P. Tracking Continental Habitat Shifts of Eels Using Otolith Sr/Ca Ratios: Validation and Application to the Coastal, Estuarine and Riverine Eels of the Gironde–Garonne–Dordogne Watershed. *Mar. Freshw. Res.* **2005**, *56*, 619–627. [CrossRef]
14. Shiao, J.C.; Ložys, L.; Iizuka, Y.; Tzeng, W.N. Migratory Patterns and Contribution of Stocking to the Population of European Eel in Lithuanian Waters as Indicated by Otolith Sr:Ca Ratios. *J. Fish Biol.* **2006**, *69*, 749–769. [CrossRef]
15. Tabouret, H.; Bareille, G.; Claverie, F.; Pécheyran, C.; Prouzet, P.; Donard, O.F.X. Simultaneous Use of Strontium:Calcium and Barium:Calcium Ratios in Otoliths as Markers of Habitat: Application to the European Eel (*Anguilla anguilla*) in the Adour Basin, South West France. *Mar. Environ. Res.* **2010**, *70*, 35–45. [CrossRef] [PubMed]

16. Melià, P.; Bevacqua, D.; Crivelli, A.J.; Panfili, J.; De Leo, G.A.; Gatto, M. Sex Differentiation of the European Eel in Brackish and Freshwater Environments: A Comparative Analysis. *J. Fish Biol.* **2006**, *69*, 1228–1235. [CrossRef]
17. Sadler, K. Effects of Temperature on the Growth and Survival of the European Eel, *Anguilla anguilla* L. *J. Fish Biol.* **1979**, *15*, 499–507. [CrossRef]
18. Tesch, F.W. *The Eel*; Blackwell Science: Oxford, UK, 2003; Volume 15, ISBN 0-632-06389-0.
19. Costa, J.L.; Domingos, I.; Assis, C.A.; Almeida, P.R.; Moreira, F.; Feunteun, E.; Costa, M.J. Comparative Ecology of the European Eel, *Anguilla anguilla* (L., 1758), in a Large Iberian River. *Environ. Biol. Fishes* **2008**, *81*, 421–434. [CrossRef]
20. Brown, J.H.; Gillooly, J.F.; Allen, A.P.; Savage, V.M.; West, G.B. Toward a Metabolic Theory of Ecology. *Ecology* **2004**, *85*, 1771–1789. [CrossRef]
21. Correia, M.J.; Domingos, I.; De Leo, G.A.; Costa, J.L. A Comparative Analysis of European Eel's Somatic Growth in the Coastal Lagoon Santo André (Portugal) with Growth in Other Estuaries and Freshwater Habitats. *Environ. Biol. Fishes* **2021**, *104*, 837–850. [CrossRef]
22. Fernéndez-Delgado, C.; Hernando, J.A.; Herrera, M.; Bellido, M. Age and Growth of Yellow Eels, *Anguilla anguilla*, in the Estuary of the Guadalquivir River (South-West Spain). *J. Fish Biol.* **1989**, *34*, 561–570. [CrossRef]
23. Kullmann, B.; Thiel, R. Bigger Is Better in Eel Stocking Measures? Comparison of Growth Performance, Body Condition, and Benefit-Cost Ratio of Simultaneously Stocked Glass and Farmed Eels in a Brackish Fjord. *Fish. Res.* **2018**, *205*, 132–140. [CrossRef]
24. Melia, P.; Bevacqua, D.; Crivelli, A.J.; De Leo, G.A.; Panfili, J.; Gatto, M. Age and Growth of *Anguilla anguilla* in the Camargue Lagoons. *J. Fish Biol.* **2006**, *68*, 876–890. [CrossRef]
25. Patey, G.; Couillard, C.M.; Drouineau, H.; Verreault, G.; Pierron, F.; Lambert, P.; Baudrimont, M.; Couture, P. Early Back-Calculated Size-at-Age of Atlantic Yellow Eels Sampled along Ecological Gradients in the Gironde and St. Lawrence Hydrographical Systems. *Can. J. Fish. Aquat. Sci.* **2018**, *75*, 1270–1279. [CrossRef]
26. Simon, J.; Ubl, C.; Dorow, M. Growth of European Eel *Anguilla anguilla* along the Southern Baltic Coast of Germany and Implication for the Eel Management. *Environ. Biol. Fishes* **2013**, *96*, 1073–1086. [CrossRef]
27. Vaughan, L.; Brophy, D.; O'Toole, C.; Graham, C.; Ó Maoiléidigh, N.; Poole, R. Growth Rates in a European Eel *Anguilla anguilla* (L., 1758) Population Show a Complex Relationship with Temperature over a Seven-Decade Otolith Biochronology. *ICES J. Mar. Sci.* **2021**, *78*, 994–1009. [CrossRef]
28. De Leo, G.A.; Gatto, M. A Size and Age-Structured Model of the European Eel (*Anguilla anguilla* L.). *Can. J. Fish. Aquat. Sci.* **1995**, *52*, 1351–1367. [CrossRef]
29. Rossi, R.; Colombo, G. Sex Ratio, Age and Growth of Silver Eels, in Two Brackish Lagoons in the Northern Adriatic (Valli of Comacchio and Valle Nuova). *Arch. Oceanogr. Limnol.* **1976**, *18*, 227–310.
30. Bird, D.J.; Rotchell, J.M.; Hesp, S.A.; Newton, L.C.; Hall, N.G.; Potter, I.C. To What Extent Are Hepatic Concentrations of Heavy Metals in *Anguilla anguilla* at a Site in a Contaminated Estuary Related to Body Size and Age and Reflected in the Metallothionein Concentrations? *Environ. Pollut.* **2008**, *151*, 641–651. [CrossRef]
31. Teichert, N.; Carassou, L.; Sahraoui, Y.; Lobry, J.; Lepage, M. Influence of Intertidal Seascape on the Functional Structure of Fish Assemblages: Implications for Habitat Conservation in Estuarine Ecosystems. *Aquat. Conserv. Mar. Freshw. Ecosyst.* **2018**, *28*, 798–809. [CrossRef]
32. Nicolas, D.; Lobry, J.; Le Pape, O.; Boët, P. Functional Diversity in European Estuaries: Relating the Composition of Fish Assemblages to the Abiotic Environment. *Estuar. Coast. Shelf Sci.* **2010**, *88*, 329–338. [CrossRef]
33. Basilone, G.; Guisande, C.; Patti, B.; Mazzola, S.; Cuttitta, A.; Bonanno, A.; Kallianiotis, A. Linking Habitat Conditions and Growth in the European Anchovy (*Engraulis encrasicolus*). *Fish. Res.* **2004**, *68*, 9–19. [CrossRef]
34. Hayes, D.B.; Ferreri, C.P.; Taylor, W.W. Linking Fish Habitat to Their Population Dynamics. *Can. J. Fish. Aquat. Sci.* **1996**, *53*, 383–390. [CrossRef]
35. Sogard, S. Variability in Growth Rates of Juvenile Fishes in Different Estuarine Habitats. *Mar. Ecol. Prog. Ser.* **1992**, *85*, 35–53. [CrossRef]
36. Copp, G.H.; Daverat, F.; Bašić, T. The Potential Contribution of Small Coastal Streams to the Conservation of Declining and Threatened Diadromous Fishes, Especially the European Eel. *River Res. Appl.* **2021**, *37*, 111–115. [CrossRef]
37. Cucherousset, J.; Paillisson, J.-M.; Carpentier, A.; Thoby, V.; Damien, J.-P.; Eybert, M.-C.; Feunteun, E.; Robinet, T. Freshwater Protected Areas: An Effective Measure to Reconcile Conservation and Exploitation of the Threatened European Eels (*Anguilla anguilla*)? *Ecol. Freshw. Fish* **2007**, *16*, 528–538. [CrossRef]
38. Selleslagh, J.; Lesourd, S.; Amara, R. Comparison of Macrobenthic Assemblages of Three Fish Estuarine Nurseries and Their Importance as Foraging Grounds. *J. Mar. Biol. Assoc. UK* **2011**, *92*, 85–97. [CrossRef]
39. Durif, C.; Dufour, S.; Elie, P. The Silvering Process of *Anguilla anguilla*: A New Classification from the Yellow Resident to the Silver Migrating Stage. *J. Fish Biol.* **2005**, *66*, 1025–1043. [CrossRef]
40. Kushnirov, D.; Degani, G. Sexual Dimorphism in Yellow European Eels, *Anguilla anguilla* (L.). *Aquac. Res.* **1995**, *26*, 409–414. [CrossRef]
41. Sahyoun, R.; Claudet, J.; Fazio, G.; Silva, C.D.; Lecomte-Finiger, R. The Otolith as Stress Indicator of Parasitism on European Eel. *Vie Milieu* **2007**, *57*, 193–200.
42. ICES. Workshop on Age Reading of European and American Eel (WKAREA). In Proceedings of the Workshop on Age Reading of European and American Eel (WKAREA), Bordeaux, France, 20–24 April 2009.

43. Meunier, F.J. Données sur la croissance de l'anguille (*Anguilla anguilla* L.) dans le cours moyen du Rhin, région alsacienne. *Bull. Fr. Pêche Piscic.* **1994**, *335*, 133–144. [CrossRef]
44. Poole, W.R.; Reynolds, J.D. Age and Growth of Yellow Eel, *Anguilla anguilla* (L.), Determined by Two Different Methods. *Ecol. Freshw. Fish* **1996**, *5*, 86–95. [CrossRef]
45. Froese, R. Cube Law, Condition Factor and Weight–Length Relationships: History, Meta-Analysis and Recommendations. *J. Appl. Ichthyol.* **2006**, *22*, 241–253. [CrossRef]
46. Le Cren, E.D. The Length-Weight Relationship and Seasonal Cycle in Gonad Weight and Condition in the Perch (*Perca fluviatilis*). *J. Anim. Ecol.* **1951**, *20*, 201–219. [CrossRef]
47. Ricker, W. Computation and Interpretation of Biological Statistics of Fish Populations. *Bul. Fish. Res. Board Can.* **1975**, *191*, 1–382.
48. Francis, R.I.C.C. Back-Calculation of Fish Length: A Critical Review. *J. Fish Biol.* **1990**, *36*, 883–902. [CrossRef]
49. Campana, S.E. How Reliable Are Growth Back-Calculations Based on Otoliths? *Can. J. Fish. Aquat. Sci.* **1990**, *47*, 2219–2227. [CrossRef]
50. Von Bertalanffy, L. A Quantitative Theory of Organic Growth (Inquiries on Growth Laws. II). *Hum. Biol.* **1938**, *10*, 181–213.
51. Gompertz, B., XXIV. On the Nature of the Function Expressive of the Law of Human Mortality, and on a New Mode of Determining the Value of Life Contingencies. In a Letter to Francis Baily, Esq. F. R. S. &c. *Philos. Trans. R. Soc. Lond.* **1825**, *115*, 513–583. [CrossRef]
52. Verhulst, P.F. Notice on the Law That a Population Follows in Its Growth. *Corr. Math. Phys.* **1838**, *10*, 113–121.
53. Knight, W. Asymptotic Growth: An Example of Nonsense Disguised as Mathematics. *J. Fish. Res. Board Can.* **1968**, *25*, 1303–1307. [CrossRef]
54. Akaike, H. A New Look at the Statistical Model Identification. *IEEE Trans. Autom. Control* **1974**, *19*, 716–723. [CrossRef]
55. Sakamoto, Y.; Ishiguro, M.; Kitagawa, G. *Akaike Information Criterion Statistics*; D. Reidel: Dordrecht, The Netherlands, 1986; Volume 81, p. 26853.
56. Pauly, D. On the Interrelationships between Natural Mortality, Growth Parameters, and Mean Environmental Temperature in 175 Fish Stocks. *ICES J. Mar. Sci.* **1980**, *39*, 175–192. [CrossRef]
57. Pauly, D. Gill Size and Temperature as Governing Factors in Fish Growth: A Generalization of von Bertalanffy's Growth Formula. Ph.D. Thesis, University of Kiel, Kiel, Germany, Institut für Meereskunde, Hamburg, Germany, 1979.
58. Pauly, D.; Munro, J.L. Once More on the Comparison of Growth in Fish and Invertebrates. *Fishbyte* **1984**, *2*, 1–21.
59. Husson, F.; Josse, J.; Le, S.; Mazet, J.; Husson, M.F. Package 'FactoMineR.' Multivariate Exploratory Data Analysis and Data Mining. 2017. Available online: http://cran.r-project.org/web/packages/FactoMineR/FactoMineR.pdf (accessed on 10 August 2022).
60. ICES. Joint EIFAAC/ICES/GFCM Working Group on Eels (WGEEL). *ICES Sci. Rep.* **2020**, *2*, 223. [CrossRef]
61. Bark, A.; Williams, B.; Knights, B. Current Status and Temporal Trends in Stocks of European Eel in England and Wales. *ICES J. Mar. Sci.* **2007**, *64*, 1368–1378. [CrossRef]
62. Feunteun, E.; Acou, A.; Guillouët, J.; Laffaille, P.; Legault, A. Spatial Distribution of an Eel Population (*Anguilla anguilla* L.) in a Small Coastal Catchment of Northern Brittany (France). Consequences of Hydraulic Works. *Bull. Fr. Pêche Piscic.* **1998**, *349*, 129–139. [CrossRef]
63. Laffaille, P.; Feunteun, E.; Baisez, A.; Robinet, T.; Acou, A.; Legault, A.; Lek, S. Spatial Organisation of European Eel (*Anguilla anguilla* L.) in a Small Catchment. *Ecol. Freshw. Fish* **2003**, *12*, 254–264. [CrossRef]
64. Davey, A.J.H.; Jellyman, D.J. Sex Determination in Freshwater Eels and Management Options for Manipulation of Sex. *Rev. Fish Biol. Fish.* **2005**, *15*, 37–52. [CrossRef]
65. Geffroy, B.; Bardonnet, A. Sex Differentiation and Sex Determination in Eels: Consequences for Management. *Fish Fish.* **2016**, *17*, 375–398. [CrossRef]
66. Drouineau, H.; Durif, C.; Castonguay, M.; Mateo, M.; Rochard, E.; Verreault, G.; Yokouchi, K.; Lambert, P. Freshwater Eels: A Symbol of the Effects of Global Change. *Fish Fish.* **2018**, *19*, 903–930. [CrossRef]
67. Laffaille, P.; Acou, A.; Guillouët, J.; Mounaix, B.; Legault, A. Patterns of Silver Eel (*Anguilla anguilla* L.) Sex Ratio in a Catchment. *Ecol. Freshw. Fish* **2006**, *15*, 583–588. [CrossRef]
68. Bevacqua, D.; Melià, P.; Schiavina, M.; Crivelli, A.J.; De Leo, G.A.; Gatto, M. A Demographic Model for the Conservation and Management of the European Eel: An Application to a Mediterranean Coastal Lagoon. *ICES J. Mar. Sci.* **2019**, *76*, 2164–2178. [CrossRef]
69. Bevacqua, D.; Melià, P.; Gatto, M.; De Leo, G.A. A Global Viability Assessment of the European Eel. *Glob. Chang. Biol.* **2015**, *21*, 3323–3335. [CrossRef] [PubMed]
70. Han, Y.-S.; Tzeng, W.-N. Use of the Sex Ratio as a Means of Resource Assessment for the Japanese Eel *Anguilla japonica*: A Case Study in the Kaoping River, Taiwan. *Zool. Stud.* **2006**, *45*, 255–263.
71. Jones, V.R.P.; Sinha, W.J. On the Sex and Distribution of the Freshwater Eel (*Anguilla anguilla*). *J. Zool.* **1966**, *150*, 371–385. [CrossRef]
72. Boulenger, C.; Acou, A.; Trancart, T.; Crivelli, A.J.; Feunteun, E. Length-Weight Relationships of the Silver European Eel, *Anguilla anguilla* (Linnaeus, 1758), across Its Geographic Range. *J. Appl. Ichthyol.* **2015**, *31*, 427–430. [CrossRef]
73. Barcala, E.; Romero, D.; Bulto, C.; Boza, C.; Peñalver, J.; María-Dolores, E.; Muñoz, P. An Endangered Species Living in an Endangered Ecosystem: Population Structure and Growth of European Eel *Anguilla anguilla* in a Mediterranean Coastal Lagoon. *Reg. Stud. Mar. Sci.* **2022**, *50*, 102163. [CrossRef]

74. Piria, M.; Šprem, N.; Tomljanović, T.; Slišković, M.; Jelić Mrčelić, G.; Treer, T. Length Weight Relationships of the European Eel *Anguilla anguilla* (Linnaeus, 1758) from Six Karst Catchments of the Adriatic Basin, Croatia. *Croat. J. Fish.* **2014**, *72*, 32–35. [CrossRef]
75. Vøllestad, L.A. Geographic Variation in Age and Length at Metamorphosis of Maturing European Eel: Environmental Effects and Phenotypic Plasticity. *Br. Ecol. Soc.* **1992**, *61*, 9. [CrossRef]
76. Robinet, T.; Feunteun, E. Sublethal Effects of Exposure to Chemical Compounds: A Cause for the Decline in Atlantic Eels? *Ecotoxicology* **2002**, *11*, 265–277. [CrossRef]
77. Acou, A.; Robinet, T.; Lance, E.; Gerard, C.; Mounaix, B.; Brient, L.; Le Rouzic, B.; Feunteun, E. Evidence of Silver Eels Contamination by Microcystin-LR at the Onset of Their Seaward Migration: What Consequences for Breeding Potential? *J. Fish Biol.* **2008**, *72*, 753–762. [CrossRef]
78. Moriarty, C. Age Determination and Growth Rate of Eels, *Anguilla anguilla* (L). *J. Fish Biol.* **1983**, *23*, 257–264. [CrossRef]
79. Morais, R.A.; Bellwood, D.R. Global Drivers of Reef Fish Growth. *Fish Fish.* **2018**, *19*, 874–889. [CrossRef]
80. Munro, J.L.; Pauly, D. A Simple Method for Comparing the Growth of Fishes and Invertebrates. *Fishbyte* **1983**, *1*, 5–6.
81. Kullmann, B.; Pohlmann, J.-D.; Freese, M.; Keth, A.; Wichmann, L.; Neukamm, R.; Thiel, R. Age-Based Stock Assessment of the European Eel (*Anguilla anguilla*) Is Heavily Biased by Stocking of Unmarked Farmed Eels. *Fish. Res.* **2018**, *208*, 258–266. [CrossRef]
82. Gordo, L.S.; Jorge, M.I. Age and growth of the european eel, *Anguilla anguilla* (Linnaeus, 1758) in the Aveiro Lagoon, Portugal. *Sci. Mar.* **1991**, *55*, 389–395.
83. O'Connor, W. Biology and Management of European Eel (*Anguilla anguilla*, L) in the Shannon Estuary, Ireland. Ph.D. Thesis, National University of Ireland, Galway, Ireland, 2003.
84. Mann, R.H.K.; Blackburn, J.H. The Biology of the Eel *Anguilla anguilla* (L.) in an English Chalk Stream and Interactions with Juvenile Trout *Salmo Trutta* L. and Salmon *Salmo salar* L. *Hydrobiologia* **1991**, *218*, 65–76. [CrossRef]
85. Rasmussen, G.; Therkildsen, B. Food, Growth and Production of *Anguilla anguilla* in a Small Danish Stream. *Rapp. Procès-Verbaux Réun. Cons. Int. Pour Explor. Mer* **1979**, *174*, 32–40.
86. Vøllestad, L.A.; Jonsson, B. Life-History Characteristics of the European Eel *Anguilla anguilla* in the Imsa River, Norway. *Trans. Am. Fish. Soc.* **1986**, *115*, 864–871. [CrossRef]
87. Paulovits, G.; Biro, P. Age Determination and Growth of Eel, *Anguilla anguilla* (L.), in Lake FertS, Hungary. *Fish* **1986**, *4*, 101–110. [CrossRef]
88. Le Pape, O.; Bonhommeau, S. The Food Limitation Hypothesis for Juvenile Marine Fish. *Fish Fish.* **2015**, *16*, 373–398. [CrossRef]
89. Denis, J.; Rabhi, K.; Le Loc'h, F.; Ben Rais Lasram, F.; Boutin, K.; Kazour, M.; Diop, M.; Gruselle, M.-C.; Amara, R. Role of Estuarine Habitats for the Feeding Ecology of the European Eel (*Anguilla anguilla* L.). *PLoS ONE* **2022**, *17*, e0270348. [CrossRef] [PubMed]
90. Feunteun, E. Management and Restoration of European Eel Population (*Anguilla anguilla*): An Impossible Bargain. *Ecol. Eng.* **2002**, *18*, 575–591. [CrossRef]
91. Haenen, O.L.M.; Mladineo, I.; Konecny, R.; Yoshimizu, M.; Groman, D.; Muñoz, P.; Saraiva, A.; Bergmann, S.M.; van Beurden, S.J. Diseases of Eels in an International Perspective: Workshop on Eel Diseases at the 15th International Conference on Diseases of Fish and Shellfish, Split, Croatia, 2011. *Bull. Eur. Assoc. Fish Pathol.* **2012**, *32*, 109–115.
92. Muñoz, P.; Barcala, E.; Peñalver, J.; Romero, D. Can Inorganic Elements Affect Herpesvirus Infections in European Eels? *Environ. Sci. Pollut. Res.* **2019**, *26*, 35266–35269. [CrossRef]
93. Delrez, N.; Zhang, H.; Lieffrig, F.; Mélard, C.; Farnir, F.; Boutier, M.; Donohoe, O.; Vanderplasschen, A. European Eel Restocking Programs Based on Wild-Caught Glass Eels: Feasibility of Quarantine Stage Compatible with Implementation of Prophylactic Measures Prior to Scheduled Reintroduction to the Wild. *J. Nat. Conserv.* **2021**, *59*, 125933. [CrossRef]
94. Legrand, M.; Briand, C.; Buisson, L.; Artur, G.; Azam, D.; Baisez, A.; Barracou, D.; Bourré, N.; Carry, L.; Caudal, A.-L.; et al. Contrasting Trends between Species and Catchments in Diadromous Fish Counts over the Last 30 Years in France. *Knowl. Manag. Aquat. Ecosyst.* **2020**, *421*, 7. [CrossRef]
95. Waldman, J.; Wilson, K.A.; Mather, M.; Snyder, N.P. A Resilience Approach Can Improve Anadromous Fish Restoration. *Fisheries* **2016**, *41*, 116–126. [CrossRef]
96. Lin, H.-Y.; Brown, C.J.; Dwyer, R.G.; Harding, D.J.; Roberts, D.T.; Fuller, R.A.; Linke, S.; Possingham, H.P. Impacts of Fishing, River Flow and Connectivity Loss on the Conservation of a Migratory Fish Population. *Aquat. Conserv. Mar. Freshw. Ecosyst.* **2018**, *28*, 45–54. [CrossRef]
97. Baudoin, J.-M.; Burgun, V.; Chanseau, M.; Larinier, M.; Ovidio, M.; Sremski, W.; Steinbach, P.; Voegtle, B. *Évaluer le Franchissement des Obstacles par les Poissons: Principes et Méthodes: Informations sur la Continuité Ecologique, ICE*; ONEMA: Vincennes, France, 2014; ISBN 979-10-91047-29-6.
98. Itakura, H.; Kitagawa, T.; Miller, M.J.; Kimura, S. Declines in Catches of Japanese Eels in Rivers and Lakes across Japan: Have River and Lake Modifications Reduced Fishery Catches? *Landsc. Ecol. Eng.* **2015**, *11*, 147–160. [CrossRef]
99. Costa, J.L.; Assis, C.A.; Almeida, P.R.; Moreira, F.M.; Costa, M.J. On the Food of the European Eel, *Anguilla anguilla* (L.), in the Upper Zone of the Tagus Estuary, Portugal. *J. Fish Biol.* **1992**, *41*, 841–850. [CrossRef]

 fishes

MDPI

Communication

Age and Growth of Quillback Rockfish (*Sebastes maliger*) at High Latitude

Camron J. Christoffersen, Dennis K. Shiozawa, Andrew D. Suchomel and Mark C. Belk *

Department of Biology, Brigham Young University, Provo, UT 84602, USA;
camron.christoffersen@gmail.com (C.J.C.); shiozawa@byu.edu (D.K.S.); drew@suchomel.com (A.D.S.)
* Correspondence: mark_belk@byu.edu; Tel.: +1-801-422-4154

Abstract: Data on age and growth of fishes is critical for effective management; however, growth rates documented in one location may not be representative of other locations, especially for species that occur across wide geographic ranges. *Sebastes maliger*, quillback rockfish, occur across a broad latitudinal range, but their growth patterns have been quantified only in the southern part of their range. To provide information for *S. maliger* in the more northern part of its range, we report age and growth patterns derived from otolith analysis from a population collected in southeast Alaskan waters. In southeast Alaska mean annual growth increments for years 1 and 2 range from 60–80 mm, and for ages 6–9 annual growth increments average about 20 mm. From age 10 on average the annual growth increment is about 5 mm. These data can be used in conjunction with harvest data to manage stocks of *S. maliger* in Alaskan waters.

Keywords: sagittal otolith; somatic growth; long-lived fishes

Citation: Christoffersen, C.J.; Shiozawa, D.K.; Suchomel, A.D.; Belk, M.C. Age and Growth of Quillback Rockfish (*Sebastes maliger*) at High Latitude. *Fishes* **2022**, *7*, 38. https://doi.org/10.3390/fishes7010038

Academic Editors: Josipa Ferri and Filipe Martinho

Received: 22 November 2021
Accepted: 3 February 2022
Published: 5 February 2022

1. Introduction

Information about species growth rates is crucial for effectively managing fish stocks [1]. Age-based growth models are used to assess most west coast groundfish stocks [2]. However, growth rates of fishes can vary among locations based on variation in water temperature, salinity, and food resource availability [3–6]. Water temperature is known to affect fish growth rates, but the relationship between temperature and growth can be positive, negative, or indeterminate among different species and locations [1,7–13]. Growth rates documented in one location may not be representative of other locations, especially for species that occur across wide geographic, and especially, latitudinal ranges [14].

Sebastes maliger, quillback rockfish, range from the Anacapa Passage (off the southern coast of California) to the Gulf of Alaska (Kenai Peninsula). They are a long-lived, benthic-oriented species reaching a maximum age of 95 years and a maximum size of 61 cm (Figure 1). *Sebastes maliger* constitute an important segment of the commercial rockfish fishery and the recreational charter boat fishery [15]. However, published age and growth information for *S. maliger* is limited to one study conducted near Puget Sound, Washington, USA, in the southern part of the species' range [16]. No age and growth studies have been conducted in more northerly parts of *S. maliger* range, and we assume that temperature and other factors may influence growth of *S. maliger* at higher latitudes as it does in other species [9,10,12]. To provide information for *S. maliger* in the more northern part of its range, we report age and growth patterns from a population collected in southeast Alaska.

Figure 1. Photograph of adult male *Sebastes maliger* captured in Frederick Sound, Alaska in June 2016 (photograph by Riley Nelson).

2. Materials and Methods

We obtained 31 *S. maliger* (16 male and 15 female) via hook-and-line sampling in Frederick Sound near Admiralty Island, Alaska (57.1365° N, 134.1200° W) in mid-June 2016. We measured standard length and total length to the nearest mm and assigned a unique ID number to each fish. We determined sex and reproductive maturity by inspection of the gonads. We removed sagittal otoliths and stored them dry in small coin envelopes for further analysis.

We determined age at capture by counting presumptive annuli on sagittal otoliths. The term "presumptive annuli" was used because we did not conduct an age validation study; however, otolith annuli have been validated in several other species of rockfishes [8,17]. We prepared otoliths for counting using standard methods for long-lived fishes [18,19]. First, we created a transverse section by grinding down half of each otolith using 500, 800, and 1200 grit silicon-carbide grinding paper on a water-fed polishing wheel, then we burned the transverse section to provide greater contrast for identifying and counting presumptive annuli. We counted presumptive annuli under a dissecting microscope (Leica Wild M10 and Leica Wild M3C; Leica Microsystems Inc. Buffalo Grove, IL USA) at 6X magnification on a black background with reflected light. The burned surface of the otolith was covered with immersion oil to reduce refraction and reflection (Figure 2a). Presumptive annuli were counted independently by two observers. Of the 31 individual otoliths, 6 (19.3%) were estimated to be the same age, 20 (64.5%) were estimated at one year difference, and 5 (16.1%) were estimated at two years difference in age by the two observers (mean difference of counts between observers = 0.97, SE = 0.11). When counts differed between observers, we resolved disparities by mutual inspection.

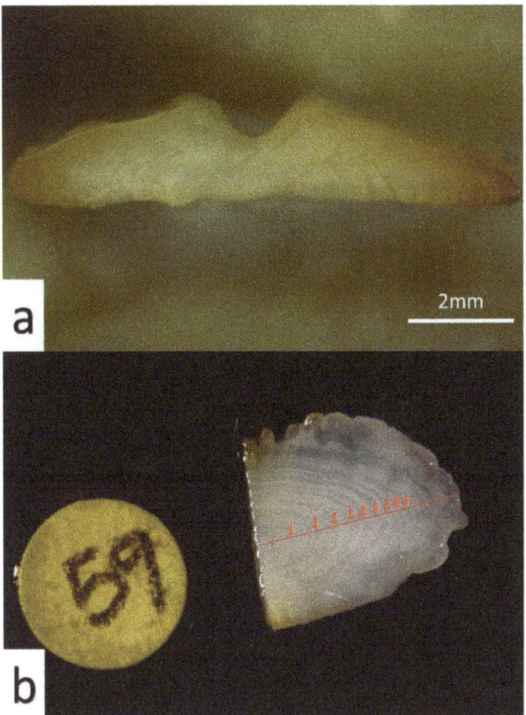

Figure 2. (**a**) Transverse cross section of otolith from *S. maliger*. This view was used to count presumptive annuli for an age estimate (photo by Andrew Suchomel). (**b**) Photograph of convex surface of the otolith (sagitta) used to measure the distance from the core of presumptive annuli for ages 1–9 (red arrows) along the longest axis (red line; photograph by Camron Christoffersen). Otoliths pictured in a and b do not belong to the same individual.

To estimate the growth curve of *S. maliger* we combined two methods. First, we plotted size-at-capture versus estimated age to characterize the growth curve for ages 10+. The youngest fish in the sample was 10 years of age, so to estimate the growth curve for ages 1-9 we back-calculated individual size at age for the first nine years for all individuals in the sample from measurements of otolith presumptive annuli. Age estimates were derived from counts of presumptive annuli on the transverse section, whereas annual growth increments were measured on the convex surface of the otolith from photographs of the otolith taken under a Zeiss Axioskop microscope with a Nikon D810 camera. We measured the total radius of each otolith and the radius from the core to each of the first nine presumptive annuli along the longest axis on the convex surface of the otolith (Digimizer, https://www.digimizer.com/download.php). Measurements were scaled (in mm) by reference to a known-size object placed in each of the photographs (Figure 2b). We were not able to consistently resolve growth increments on the otolith from one individual, so the sample size for size at ages 1–9 was 30. Size of otoliths increased with size of fish [20], and the correlation between radius of otolith and total length was significant ($t_{29} = 5.07$, $p = 0.00002$, R-squared = 0.48). We back-calculated lengths at ages 1-9 using a modified Fraser-Lee formula [21]:

$$L_x = L_0 + ((L_c - L_0)(R_x - R_0))/(R_c - R_0) \quad (1)$$

where: L_x = total length at age x; L_0 = total length at birth (4.5 mm; [16]); L_c = total length at capture; R_x = radius of otolith at age x; R_c = radius of otolith at capture; R_0 = radius of otolith at birth (estimated at 0.08 mm).

Lengths-at-age from each fish were averaged for ages 1–9 to produce the early part of the growth curve. We then combined this early trajectory with the best fit line (ordinary least squares regression) from the size-at-age captured plot to create the entire growth curve. To compare differences between sexes in size and age at capture we used a two-sample *t*-test. In addition, to facilitate comparison of the growth curve generated for *S. maliger* in southeastern Alaska waters, we fit the data to a von Bertalanffy growth equation [22].

3. Results

In our sample, estimated ages ranged from 10 to 38, and total length at capture ranged from 316 to 474 mm. Growth rates of *S. maliger* were highest during their first two years of life and decreased as age progressed. Mean annual growth increments for ages 1 and 2 ranged from 60 to 80 mm, and declined to about 20 mm per year at ages 6–9. At ages greater than 10 years the average annual growth increment was about 5 mm (Figure 3). There was no difference between sexes in size at capture (*t*-test, $p = 0.375$), or age at capture (*t*-test, $p = 0.1$). Parameter values for this sample from the von Bertalanffy model are: $k = 0.137$, $t_0 = -0.78$, and $L_{max} = 448$ mm (Figure 3).

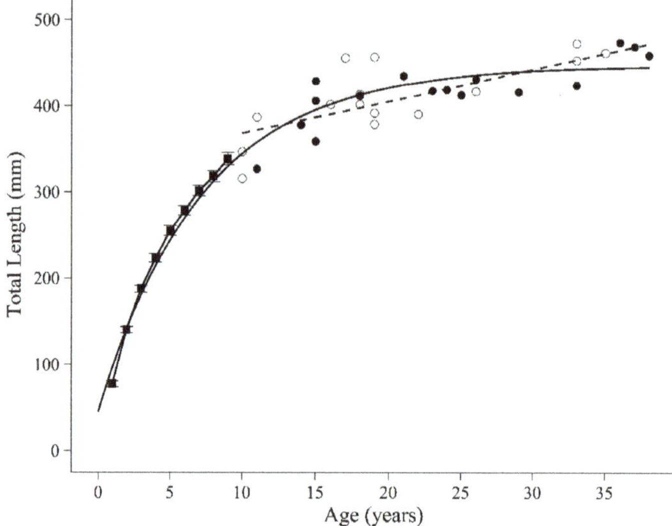

Figure 3. Two-part representation of growth curve for *S. maliger* captured in southeast Alaska USA, in 2016. The first nine years of growth are represented by mean total length (±1 standard error) for ages 1–9 (solid squares) obtained through back-calculations of growth increments of otolith annuli. Means are connected by a solid line representing the early growth trajectory. The second part of the growth curve represents growth from age 10+ as indicated by size at age captured. Females are represented by open circles, males are represented by closed circles, and the dotted line represents the best fit regression line. The von Bertalanffy growth function is indicated by the solid line that extends across all ages.

4. Discussion

Growth of *S. maliger* in southeast Alaskan waters is similar to patterns and rates identified in the southern part of the species range [16]. Surprisingly, growth of *S. maliger* does not vary substantially from populations at lower latitudes even though the physical environment varies considerably across this range. This pattern of similar growth curves across wide geographic ranges is consistent with patterns of two other species of rockfish compared across latitude [23], where even populations from distant locations exhibit a high degree of phenotypic similarity in growth rates and patterns [24]. Heterogeneity in growth rates across latitudes is likely, in part, a response to temperature, but may also be due to differences in upwellings, productivity, population density and competition in the local environment [8,25].

Knowing age and size structure specifically for this location allows managers to assess causes and effects of growth rate variation more accurately [25]. This information also allows researchers to detect changes in growth rate or age structure in the *S. maliger* population caused by climate change or other factors as those changes occur [26,27]. Our age-growth curve for *S. maliger* provides a baseline for comparison of future studies. Thus, these data are a valuable addition to the management of an economically important species with a latitudinally large and ecologically variable range.

While 31 individuals might be considered a sub-optimal number, we suggest that our analysis for these data is simple. We estimated means for ages 1-9, and we used an ordinary least squares (OLS) regression of size at capture on estimated age to characterize growth at ages 10+. Sample sizes of 30 are more than adequate to estimate means (as evidenced by the narrow 95% confidence intervals), and to estimate a best fit line using OLS regression [28].

5. Conclusions

We have not formally tested hypotheses. Our goal was to only estimate means and best-fit lines and to illustrate the relevance of such information. A formal testing of hypotheses requires the determination of precision generated by the sample size. But often one must weigh the cost of gathering data against the value of greater precision [28]. As such, we encourage researchers the world over to publish growth estimates derived from otolith analysis, even if sample sizes are modest (i.e., 30 or so, individuals). This is especially critical for poorly known species and for isolated locations. Such a collection of published growth data would be of great value for testing synthetic and long-standing hypotheses [24,29], as well as for management of populations [2,3].

Author Contributions: Conceptualization, M.C.B., C.J.C., D.K.S.; methodology, M.C.B., C.J.C.; formal analysis, M.C.B., C.J.C.; investigation, M.C.B., C.J.C., D.K.S.; resources, M.C.B., D.K.S.; data curation, A.D.S.; writing—original draft preparation, M.C.B., C.J.C.; writing—review and editing, A.D.S., M.C.B., C.J.C., D.K.S.; visualization, A.D.S., M.C.B.; supervision, M.C.B., D.K.S.; project administration, M.C.B.; funding acquisition, M.C.B. All authors have read and agreed to the published version of the manuscript.

Funding: This research received no external funding.

Institutional Review Board Statement: Collection of fish and the work performed on them was reviewed and supervised by the Brigham Young University Institutional Animal Care and Use Committee under BYU IACUC protocol 15-0602. We only used specimens caught by recreational fisherman that were already dead.

Data Availability Statement: The otoliths and data sets are accessioned in the fish range at the Life Science Museum at Brigham Young University. The data are available upon request.

Acknowledgments: We thank the Roger and Victoria Sant Foundation, Scott and Jody Jorgenson at Pybus Point Lodge, the College of Life Sciences and the Department of Biology at Brigham Young University for providing funding and support for this study. We also thank Mikaela Nielson, Yuka Yanagita, Michelle Nishiguchi, Aaron Esplin, Michael Sorenson, Peter Searle, Samantha Tilden, and Aaron Brooksby for their help with gathering data.

Conflicts of Interest: The authors declare no conflict of interest.

References

1. Laidig, T.E.; Pearson, D.E.; Sinclair, L.L. Age and Growth of Blue Rockfish (*Sebastes mystinus*) from Central and Northern California. *Fish. Bull.* **2003**, *101*, 800–809.
2. Pacific Fishery Management Council. *Pacific Coast Groundfish Fishery Management Plan, for the California, Oregon, and Washington Groundfish Fishery*; NA15NMF441016; Pacific Fishery Management Council: Portland, OR, USA, 2020.
3. Black, B.A.; Boehlert, G.W.; Yoklavich, M.M. Establishing Climate–Growth Relationships for Yelloweye Rockfish (*Sebastes ruberrimus*) in the Northeast Pacific Using a Dendrochronological Approach. *Fish. Oceanogr.* **2008**, *17*, 368–379. [CrossRef]
4. Kinne, O. Growth, Food Intake, and Food Conversion in a Euryplastic Fish Exposed to Different Temperatures and Salinities. *Physiol. Zool.* **1960**, *33*, 288–317. [CrossRef]
5. Sogard, S. Variability in Growth Rates of Juvenile Fishes in Different Estuarine Habitats. *Mar. Ecol. Prog. Ser.* **1992**, *85*, 35–53. [CrossRef]
6. Imsland, A.K.; Sunde, L.M.; Folkvord, A.; Stefansson, S.O. The Interaction of Temperature and Fish Size on Growth of Juvenile Turbot. *J. Fish Biol.* **1996**, *49*, 926–940. [CrossRef]
7. Pörtner, H.O. Climate Variations and the Physiological Basis of Temperature Dependent Biogeography: Systemic to Molecular Hierarchy of Thermal Tolerance in Animals. *Comp. Biochem. Physiol. Part A Mol. Integr. Physiol.* **2002**, *132*, 739–761. [CrossRef]
8. Thompson, J.E.; Hannah, R.W. Using Cross-Dating Techniques to Validate Ages of Aurora Rockfish (*Sebastes aurora*): Estimates of Age, Growth and Female Maturity. *Env. Biol. Fish.* **2010**, *88*, 377–388. [CrossRef]
9. Stocks, J.R.; Gray, C.A.; Taylor, M.D. Synchrony and Variation across Latitudinal Gradients: The Role of Climate and Oceanographic Processes in the Growth of a Herbivorous Fish. *J. Sea Res.* **2014**, *90*, 23–32. [CrossRef]
10. Md Mizanur, R.; Yun, H.; Moniruzzaman, M.; Ferreira, F.; Kim, K.; Bai, S.C. Effects of Feeding Rate and Water Temperature on Growth and Body Composition of Juvenile Korean Rockfish, *Sebastes schlegeli* (Hilgendorf 1880). *Asian-Australas J. Anim. Sci.* **2014**, *27*, 690–699. [CrossRef]
11. Gertseva, V.; Matson, S.E.; Cope, J. Spatial Growth Variability in Marine Fish: Example from Northeast Pacific Groundfish. *ICES J. Mar. Sci.* **2017**, *74*, 1602–1613. [CrossRef]
12. Ong, J.J.L.; Rountrey, A.N.; Black, B.A.; Nguyen, H.M.; Coulson, P.G.; Newman, S.J.; Wakefield, C.B.; Meeuwig, J.J.; Meekan, M.G. A Boundary Current Drives Synchronous Growth of Marine Fishes across Tropical and Temperate Latitudes. *Glob. Chang. Biol.* **2018**, *24*, 1894–1903. [CrossRef] [PubMed]
13. Matta, M.E.; Helser, T.E.; Black, B.A. Intrinsic and Environmental Drivers of Growth in an Alaskan Rockfish: An Otolith Biochronology Approach. *Env. Biol. Fish.* **2018**, *101*, 1571–1587. [CrossRef]
14. Conover, D.O.; Present, T.M.C. Countergradient Variation in Growth Rate: Compensation for Length of the Growing Season among Atlantic Silversides from Different Latitudes. *Oecologia* **1990**, *83*, 316–324. [CrossRef]
15. Love, M.S.; Yoklavich, M.; Thorsteinson, L.K. *The Rockfishes of the Northeast Pacific*; University of California Press: Berkeley, CA, USA, 2002; ISBN 978-0-520-23438-3.
16. West, J.E.; Helser, T.E.; O'Neill, S.M. Variation in Quillback Rockfish (*Sebastes maliger*) Growth Patterns from Oceanic to Inland Waters of the Salish Sea. *Bull. Mar. Sci.* **2014**, *90*, 747–761. [CrossRef]
17. Kimura, D.K.; Mandapat, R.R.; Oxford, S.L. Method, Validity, and Variability in the Age Determination of Yellowtail Rockfish (*Sebastes flavidus*), Using Otoliths. *J. Fish. Res. Bd. Can.* **1979**, *36*, 377–383. [CrossRef]
18. Christensen, J.M. Burning of Otoliths, a Technique for Age Determination of Soles and Other Fish. *ICES J. Mar. Sci.* **1964**, *29*, 73–81. [CrossRef]
19. Chilton, D.; Beamish, R.J. Age Determination for Fishes Studied by the Groundfish Program at the Pacific Biological Station. *Can. Spec. Publ. Fish. Aquat. Sci.* **1982**, *60*, 102.
20. Munk, K.M.; Smikrud, K.M. *Relationships of Otolith Size to Fish Size and Otolith Ages for Yelloweye Sebastes Ruberrimus and Quillback S. maliger Rockfishes, Report 5J02-05*; Alaska Department of Fish and Game, Regional Information Report: Juneau, AK, USA, 2002; pp. 1–50.
21. Campana, S.E. How Reliable Are Growth Back-Calculations Based on Otoliths? *Can. J. Fish. Aquat. Sci.* **1990**, *47*, 2219–2227. [CrossRef]
22. von Bertalanffy, L. Quantitative Laws in Metabolism and Growth. *Q. Rev. Biol.* **1957**, *32*, 217–231. [CrossRef]
23. Boehlert, G.; Kappenman, R. Variation of Growth with Latitude in Two Species of Rockfish (*Sebastes pinniger* and *S. diploproa*) from the Northeast Pacific Ocean. *Mar. Ecol. Prog. Ser.* **1980**, *3*, 1–10. [CrossRef]
24. Conover, D.O.; Schultz, E.T. Phenotypic Similarity and the Evolutionary Significance of Countergradient Variation. *Trends Ecol. Evol.* **1995**, *10*, 248–252. [CrossRef]
25. Fennie, H.W.; Sponaugle, S.; Daly, E.A.; Brodeur, R.D. Prey tell: What quillback rockfish early life history traits reveal about their survival in encounters with juvenile coho salmon. *Mar. Ecol. Prog. Ser.* **2020**, *650*, 7–18. [CrossRef]
26. McGreer, M.; Frid, A. Declining size and age of rockfishes (*Sebastes* spp.) inherent to indigenous cultures of Pacific Canada. *Ocean & Coastal Manag.* **2017**, *145*, 14–20. [CrossRef]
27. Markel, R.W.; Shurin, J.B. Contrasting effects of coastal upwelling on growth and recruitment of nearshore Pacific rockfish (genus Sebastes). *Can. J. Fish. Aquat. Sci.* **2020**, *77*, 950–962. [CrossRef]
28. Lenth, R.V. Some Practical Guidelines for Effective Sample Size Determination. *Am. Stat.* **2001**, *55*, 187–193. [CrossRef]
29. Belk, M.C.; Houston, D.D. Bergmann's rule in ectotherms: A test using freshwater fishes. *Am. Nat.* **2002**, *160*, 803–808. [CrossRef]

MDPI

Article

Age, Growth, and Utility of Otolith Morphometrics as Predictors of Age in the European Barracuda, *Sphyraena sphyraena* (Sphyraenidae): Preliminary Data

Josipa Ferri * and Anđela Brzica

Department of Marine Studies, University of Split, 21000 Split, Croatia; andela.brzica59@gmail.com
* Correspondence: josipa.ferri@unist.hr; Tel.: +385-21-558-195

Abstract: Age and growth of the European barracuda, *Sphyraena sphyraena*, were determined by examining sagittal otoliths belonging to fish sampled in the eastern Adriatic Sea, as the northernmost region of the Mediterranean. A total of 113 specimens (59 females, 53 males, and one individual of indeterminate sex), ranging from 23.4 to 42.5 cm in total length, were analyzed. The maximum observed age was 5 years for both females and males, and this barracuda population was dominated by 3 year old fish. Growth was described by the von Bertalanffy growth curve (L_∞ = 55.58 cm, K = 0.12 year^{-1}, t_0 = −4.29 year, R^2 = 0.580), and the growth performance index (Φ′) was 2.57. Otolith length, width, and mass were measured, and the utility of these morphometrics as predictors of age in *S. sphyraena* was evaluated. The results showed that counting otolith annuli produced a better estimation of age than proposed linear models based on relationships between observed fish age and otolith morphometrics. In comparison with age and growth data available in the literature for *S. sphyraena*, our results from the Adriatic Sea provide more insights into the life-history traits of this species and can be used in the future effective management and conservation.

Keywords: age; growth; otoliths; morphometry; *Sphyraena sphyraena*

Citation: Ferri, J.; Brzica, A. Age, Growth, and Utility of Otolith Morphometrics as Predictors of Age in the European Barracuda, *Sphyraena sphyraena* (Sphyraenidae): Preliminary Data. *Fishes* **2022**, *7*, 68. https://doi.org/10.3390/fishes7020068

Academic Editors: Pedro Morais and Yongjun Tian

Received: 27 January 2022
Accepted: 15 March 2022
Published: 17 March 2022

1. Introduction

The family Sphyraenidae (barracudas), which is typically represented in shelf waters of tropical and temperate seas in many regions of the world, comprises only one genus, *Sphyraena*, and 28 species [1–3]. They are all active predatory and voracious fishes that can adapt to diverse ecological conditions [4,5]. Large adult barracudas are often found to be solitary, whereas juveniles and small adults are gregarious and aggregate to form schools [6]. In the Mediterranean, six Sphyraenidae species have been recorded so far: three Lessepsian (*S. chrysotaenia* Klunzinger, 1884; *S. flavicauda* Rüppell, 1838; *S. obtusata* Cuvier, 1829), two Atlanto-Mediterranean (*S. sphyraena* (Linnaeus, 1758); *S. viridensis* Cuvier, 1829), and one new species identified from the Gulf of Taranto (*S. intermedia* Pastore, 2009) [7,8].

The European barracuda, *S. sphyraena*, is a widespread species found in the eastern Atlantic (Bay of Biscay to Angola), including the Mediterranean and Black Sea, Canary Islands, and Azores [9]. It is a pelagic species living close to the surface [10], although individuals can be observed in a midwater, down to depths of 100 m [11]. In the Mediterranean, *S. sphyraena* is a specialized piscivore fish reaching 165 cm of the total length, while its size is usually between 30 and 50 cm [12]. This barracuda is a temperate species and generally shares habitats and resources with the thermophilic *S. viridensis* [5,12]. In comparison to other Sphyraenidae, only a few investigations have been carried out on the life-history traits of *S. sphyraena*. The growth [4,13], otolith shape analysis [8], reproductive biology [14], and feeding ecology [12] have been investigated in the southeastern Mediterranean. Otolith shape analysis has been also undertaken in the southwestern Mediterranean [7], and some patterns of reproductive biology have been reported in the northwestern part of the basin [5]. However, information regarding this barracuda species in the northernmost region of the Mediterranean (Adriatic Sea) has not yet been described in detail.

This study is the first attempt to investigate the age and growth of the eastern Adriatic population of the European barracuda; therefore, the aim is to provide better insight into the life-history of this species. Fish age was determined by counting annuli in sagittal otoliths; however, because this is one of the most time-consuming aging methods, we also evaluated the utility of otolith morphometrics (length, width, and mass) as an alternative.

2. Materials and Methods

2.1. Sampling

Samples of the European barracuda were obtained monthly from the local fish market between November 2020 and February 2021 and belonged to catches taken by purse and boat seiners in the central eastern Adriatic Sea during that period. Each fish was measured to the nearest 0.1 cm total length (TL) and weighed (W) to the nearest 0.1 g. Sex was determined by macroscopic analysis of the gonads, and sagittal otolith pairs were removed, cleaned, and stored dry in labeled Eppendorf tubes for later analyses. Sex ratio was tested for equality (female to male = 1:1) using the chi-square goodness-of-fit test at the 0.05 significance level. Length frequency distributions of females and males were compared using the Kolmogorov–Smirnov two-sample test.

2.2. Otolith Morphometrics

Prior to polishing, otoliths were photographed using an Olympus DP-25 digital camera (Olympus Corporation, Hamburg, Germany) attached to an Olympus SZX10 stereo microscope. Otolith length (OL) and width (OW) were measured to the nearest 0.01 mm using Olympus Cell^A Imaging Software, version 2.6 (Build 1210). Otolith length was defined as the longest axis between anterior and posterior otolith edge, and width was defined as the distance from dorsal to ventral edge taken perpendicular to the length throughout the otolith focus. Otolith mass (OM) was weighed to the nearest 0.0001 g. Differences between left and right otoliths were tested by paired *t*-test, while ANCOVA was used to test for differences in otolith measures between females and males, considering fish length as a covariate. The statistical assumptions of these tests were tested using the Kolmogorov–Smirnov test of normality, and, when necessary, data were log-transformed. Relationships between fish total length and otolith morphometrics (length, width, and mass) were constructed using the linear model.

2.3. Age and Growth Estimation

The medial face of a single, randomly selected otolith from each pair was polished using a wet abrasive paper and photographed under reflected light against a dark background, using a stereo microscope coupled with an Olympus DP-25 digital camera. Growth rings were visible as alternating opaque and translucent zones, and ages were assigned to fish specimens on the basis of their counts. Therefore, one opaque zone combined with one subsequent translucent zone was interpreted as one annulus, as already observed in *Sphyraena* species [2,15]. Each otolith was read by one reader, without auxiliary data on the fish size. A second reading was carried out 1 week later, and, when readings differed by one or more years, a third reading was made. We used images for the age estimation because their quality and ability to adjust contrast and brightness made it easier to interpret zoning patterns than direct observations under the stereo microscope.

The von Bertalanffy growth model was fitted to the estimated age–length dataset using a nonlinear least-square procedure of a Gauss–Newton algorithm: $TL = L_\infty (1 - \exp(-K(t - t_0)))$, where TL is the total length of fish at age t, L_∞ is the estimated asymptotic length, K is a constant that determines the rate at which TL approaches L_∞, and t_0 is the hypothetical age at zero length. To compare the fish within the genus *Sphyraena*, the index Φ' was calculated as follows: $\Phi' = 2 \log L_\infty + \log K$, where Φ' is the growth performance of fish, L_∞ is the asymptotic length derived from the von Bertalanffy growth model, and K is the growth constant also derived from the von Bertalanffy growth model [16]. Once the age was estimated by counting annual growth rings, relationships between observed fish age and otolith morphometrics (length, width, and mass) were constructed using the linear model.

3. Results

Of the 113 individuals of the European barracuda sampled, 59 (52.2%) were females, 53 (46.9%) were males, and one (0.9%) individual was of indeterminate sex. The sex ratio was not significantly skewed toward females or males ($p = 0.57$). The total length of females and males ranged from 28.5 to 42.5 cm and from 23.4 to 38.6 cm, respectively (Table 1). A higher proportion of males were smaller than females. Males dominated in the ≤31.0 cm length classes, whereas females were more abundant in the ≥36.0 cm length classes (Figure 1). The Kolmogorov–Smirnov two-sample test showed a significant difference between length frequency distributions of females and males ($p < 0.01$). The mean length of females (34.6 ± 3.75 cm) was significantly greater than that of males (32.0 ± 3.56 cm) (t-test, df = 1, $p < 0.01$).

Table 1. Range of total length and weight and mean total length and weight with standard deviation (SD) of the European barracuda, *Sphyraena sphyraena*, in the eastern Adriatic.

	Total Length (cm)		Weight (g)	
	Range	**Mean ± SD**	**Range**	**Mean ± SD**
Females	28.5–42.5	34.6 ± 3.75	91.5–270.0	159.9 ± 47.09
Males	23.4–38.6	32.0 ± 3.56	47.3–224.2	131.7 ± 37.27
All specimens	23.4–42.5	33.4 ± 3.89	47.3–270.0	146.1 ± 44.95

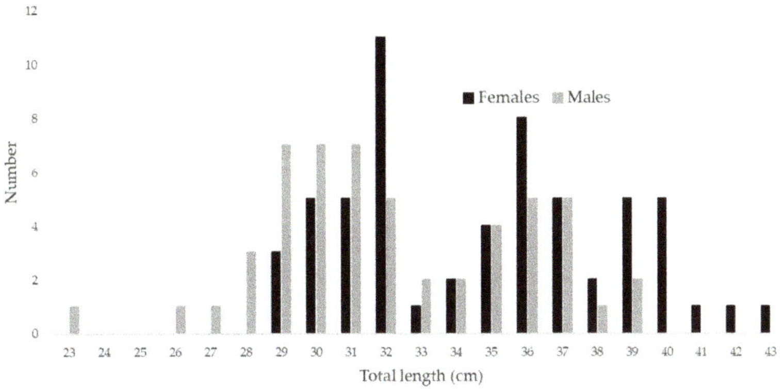

Figure 1. Length frequency distributions of the European barracuda, *Sphyraena sphyraena*, females and males in the eastern Adriatic.

Otoliths of *S. sphyraena* are spindle-shaped, with a short rostrum and poorly defined antirostrum. Their anterior region is peaked, while the posterior is oblique. No significant differences in morphometric measures (length, width, and mass) were found between left and right otoliths (paired t-test, $p > 0.05$ for all measures) and between females and males (ANCOVA, $p > 0.05$ for all measures); thus, data were pooled for both sexes, and mean values for each otolith pair were used in further analyses (Table 2). A linear model described relationships between fish total length and otolith length, width, and mass, with high coefficients of determination (≥0.805) (Table 3).

Table 2. Range of otolith measures (OL = otolith length, OW = otolith width, OM = otolith mass) and their mean values with standard deviation (SD) of the European barracuda, *Sphyraena sphyraena*, in the eastern Adriatic.

Otolith Measures	Range	Mean ± SD
OL (mm)	7.56–12.96	10.20 ± 1.10
OW (mm)	2.43–4.09	3.29 ± 0.34
OM (g)	0.0157–0.0769	0.0390 ± 0.01

Table 3. Parameters of the linear regression between fish total length (TL) and otolith morphometrics (OL = otolith length, OW = otolith width, OM = otolith mass) of the European barracuda, *Sphyraena sphyraena*, in the eastern Adriatic (a = intercept value; b = regression slope; R^2 = coefficient of determination).

Relationships	a	b	R^2
TL vs. OL	1.683	0.256	0.865
TL vs. OW	0.690	0.078	0.805
TL vs. OM	−0.073	0.003	0.851

Sagittal otoliths displayed alternating opaque and translucent zones under reflected light (Figure 2). The maximum age of the sampled European barracuda was 5 years for both females and males (Figure 3). Age distribution was similar for both sexes and was skewed to the right. The population was dominated by 3 year old fish, which represented almost 45% of the fish sampled. Fish younger than 3 years were poorly represented (we found only one 1 year old individual). Mean lengths (±standard deviation) of 1, 2, 3, 4, and 5 year old fish were 23.4 cm, 29.0 ± 0.41 cm, 32.4 ± 3.27 cm, 33.9 ± 3.29 cm, and 37.5 ± 3.15 cm, respectively. Females were larger than males at all ages. The von Bertalanffy growth curve was fitted to the age–length dataset estimated for all specimens (Figure 4). Estimated parameters were L_∞ = 55.58 cm, K = 0.12 year^{-1}, and t_0 = −4.29 year (R^2 = 0.580). The growth performance index (Φ′) was 2.57 for the sampled fish. Relationships between observed fish age and otolith morphometrics are presented in Table 4. The linear model explained between 23.6% and 27.7% of the variation in age.

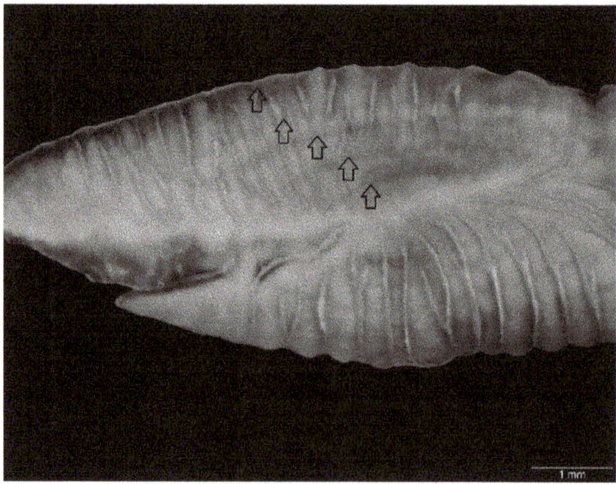

Figure 2. Sagittal otolith of the 5 year old European barracuda, *Sphyraena sphyraena*, in the eastern Adriatic.

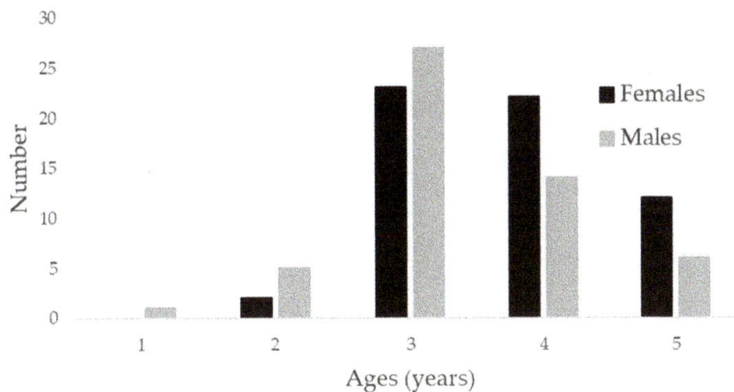

Figure 3. Age distributions of the European barracuda, *Sphyraena sphyraena*, females and males in the eastern Adriatic.

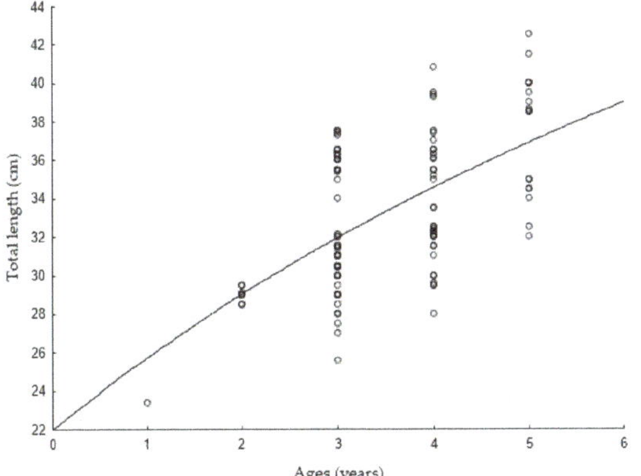

Figure 4. Fitted von Bertalanffy growth curve of the European barracuda, *Sphyraena sphyraena*, in the eastern Adriatic.

Table 4. Relationships between observed fish age and otolith morphometrics (OL = otolith length, OW = otolith width, OM = otolith mass) of the European barracuda, *Sphyraena sphyraena*, in the eastern Adriatic, described using the linear model (age = *a* (otolith morphometric) + *b*); R^2 = coefficient of determination.

Linear Model	R^2
Age = 0.425 OL − 0.761	0.277
Age = 1.294 OW − 0.694	0.255
Age = 28.912 OM + 2.445	0.236

4. Discussion

Despite its wide distribution, studies concerning the age of *S. sphyraena* are currently available only for the southeastern Mediterranean populations (Egyptian waters) [4,13]. The maximum age estimated in our study using whole otolith readings (5 years) was lower than the maximum age demonstrated by Allam et al. [13] (8 years) using the same method and a similar length range. However, only 4% of the specimens collected in Egyptian waters were older than 5 years [13]. On the other hand, our age readings were higher than those reported in another available study (3 years) [4]. The most striking difference in the life span of *S. sphyraena* seems to occur among populations from the same area. As far as growth parameters are concerned, there was good agreement, and parameters estimated in our study were very similar to the only available growth data on this species for Egyptian waters: L_∞ = 55.27 cm and K = 0.123 year^{-1} [13]. This similarity may be attributable to the dominance of individuals in the same age classes in both populations, in conjunction with the already mentioned small proportion of the oldest individuals (>5 years) in Egyptian waters. In addition, the growth increments for both populations were greatest in the first 2 years. All observed similarities were also reflected in almost the same values for Φ′: 2.57 and 2.58 [13] for the northern and southeastern Mediterranean populations, respectively. The low estimate of t_0 can be explained by the few fish ≤28.0 cm and a complete absence of the fish <23.0 cm in our samples. In fact, if smaller fish were sampled in both areas, t_0 would have been higher and closer to 0. Consequently, the population age structure shows a lack of fish in the first two age classes, principally due to accessibility to the resources on the fish market. Although females and males showed differences in growth, as females were larger than males at all ages sampled, we acknowledge that our study is not conclusive about possible differences in growth between sexes. More data, particularly from smaller and younger fish, would improve estimates and ensure modeling for *S. sphyraena* females and males separately. However, although the sample size was relatively small for young individuals, the sample covered a wide length and age range; thus, the results of this study are still reliable and valuable. By producing such preliminary growth estimates for *S. sphyraena* in the studied area, we take an important step toward facilitating future studies and management strategies.

The index of growth performance is a useful tool for comparing the growth curves of different populations of the same species and of different species belonging to the same genus or family [17]. Calculated from published data of L_∞ and K in Mediterranean *Sphyraena* species, the values of this index were the highest in *S. viridensis* (2.95 in Egyptian waters [13] and 3.33 in Algerian waters [15]) and the lowest in *S. chrysotaenia* (2.46 [13] and 2.62 [2] in Egyptian waters). In *S. flavicauda* from Egyptian waters, recorded values were 2.77 [13] and 2.82 [2]. These variations could be attributed to environmental conditions affecting distribution, population structure, and growth patterns [18]. The obtained Φ′ value in this study was within the range of values for different *Sphyraena* species in the Mediterranean (2.46 < Φ′ < 3.33). In general, growth studies providing the von Bertalanffy growth parameters for other *Sphyraena* species are scarce, and the main point of reference is represented by data provided for the southeastern Mediterranean populations.

Interpreting annual growth rings in otoliths, as in this study, is the most frequently used method for estimating individual fish age [19]. On the other hand, it has several disadvantages; it is subjective, laborious, time-consuming, and dependent on the readers' experience [20]. In that context, several studies have analyzed and demonstrated a close relationship between the size of the otoliths and fish age [20–25]. A linear model described all relationships between otolith measures and fish age, and the age of the investigated barracuda could be best predicted from the otolith length. However, we must emphasize that models did not provide precise estimates, and this could be explained by the overlap in lengths between fish of different age (especially in the age classes 3 and 4) and a lack of the fish in the first two age classes. Despite that, the observed otolith morphology of *S. sphyraena* generally corresponded to descriptions for this species from the eastern Atlantic and Mediterranean [7,8,26], and it can be used in the identification of species (e.g., in feeding studies or species separation within *Sphyraena* genus).

5. Conclusions

This study provides the first analysis of European barracuda age and growth in the northernmost region of the Mediterranean. We showed that otoliths can be used for aging, and our estimates of growth seems reasonable for a species with a moderate life span. Counting otolith annuli produced better estimation of age than proposed linear models based on relationships between observed fish age and otolith morphometrics. However, these preliminary results should be used with caution, and further studies with larger sample sizes collected throughout the year are required to refine the age and growth information for *S. sphyraena* in the Adriatic. Similar studies are, therefore, encouraged within the Mediterranean to achieve a more comprehensive understanding of the life-history traits of this temperate fish species.

Author Contributions: Conceptualization, J.F.; methodology, J.F.; software, J.F.; validation, J.F. and A.B.; formal analysis, J.F. and A.B.; investigation, J.F. and A.B.; resources, J.F. and A.B.; data curation, J.F.; writing—original draft preparation, J.F.; writing—review and editing, J.F.; visualization, J.F.; supervision, J.F. All authors read and agreed to the published version of the manuscript.

Funding: This research received no external funding.

Institutional Review Board Statement: The study was conducted according to the guidelines of the Code of Ethics of the University Department of Marine Studies; we only used specimens obtained from the fish market that were already dead.

Data Availability Statement: Data from this study are available from the corresponding author upon request (J.F.: josipa.ferri@unist.hr).

Conflicts of Interest: The authors declare no conflict of interest.

References

1. Nelson, J.S.; Grande, T.C.; Wilson, M.V.H. *Fishes of the World*, 5th ed.; Wiley: Hoboken, NJ, USA, 2016; pp. 468–472.
2. ElGanainy, A.; Amin, A.; Ali, A.; Osman, H. Age and growth of two barracuda species *Sphyraena chrysotaenia* and *S. flavicauda* (Family: Sphyraenidae) from the Gulf of Suez, Egypt. *Egypt. J. Aquat. Res.* **2017**, *43*, 75–81. [CrossRef]
3. Ballen, G.A. Nomenclature of the Sphyraenidae (Teleostei: Carangaria): A synthesis of fossil- and extant-based classification systems. *Zootaxa* **2019**, *4686*, 397–408. [CrossRef] [PubMed]
4. Wadie, W.F.; Rizkalla, S.I. Fisheries for the genus *Sphyraena* (Perciformes, Sphyraenidae) in the south-eastern part of the Mediterranean Sea. *Pak. J. Mar. Sci.* **2001**, *10*, 21–34.
5. Villegas-Hernández, H.; Muñoz, M.; Lloret, J. Life-history traits of temperate and thermophilic barracudas (Teleostei: Sphyraenidae) in the context of sea warming in the Mediterranean Sea. *J. Fish Biol.* **2014**, *84*, 1940–1957. [CrossRef] [PubMed]
6. Pastore, A. *Sphyraena intermedia* sp. nov. (Pisces: Sphyraenidae): A potential new species of barracuda identified from the central Mediterranean Sea. *J. Mar. Biol. Assoc.* **2009**, *89*, 1299–1303. [CrossRef]
7. Bourehail, N.; Morat, F.; Lecomte-Finiger, R.; Kara, M.H. Using otolith shape analysis to distinguish barracudas *Sphyraena sphyraena* and *Sphyraena viridensis* from the Algerian coast. *Cybium* **2015**, *39*, 271–278.
8. Yedier, S. Otolith shape analysis and relationships between total length and otolith dimensions of European barracuda, *Sphyraena sphyraena* in the Mediterranean Sea. *Iran. J. Fish. Sci.* **2021**, *20*, 1080–1096.
9. Froese, R.; Pauly, D. FishBase. World Wide Web Electronic Publication. Version (08/2021). Available online: www.fishbase.org (accessed on 10 January 2022).
10. Golani, D.; Öztürk, B.; Başusta, N. *The Fishes of Eastern Mediterranean*; Turkish Marine Research Foundation: Istanbul, Turkey, 2006; pp. 179–181.
11. Ben-Tuvia, A. Sphyraenidae. In *Fishes of the North-Eastern Atlantic and the Mediterranean*; Whitehead, P.J.P., Bauchot, M.L., Hureau, J.C., Nielsen, J., Tortonese, E., Eds.; UNESCO: Paris, France, 1986; pp. 1194–1196.
12. Kalogirou, S.; Mittermayer, F.; Pihl, L.; Wennhage, H. Feeding ecology of indigenous and non-indigenous fish species within the family Sphyraenidae. *J. Fish Biol.* **2012**, *80*, 2528–2548. [CrossRef] [PubMed]
13. Allam, S.M.; Faltas, S.N.; Ragheb, E. Age and growth of barracudas in the Egyptian Mediterranean waters. *Egypt. J. Aquat. Res.* **2004**, *30*, 281–289.
14. Allam, S.M.; Faltas, S.N.; Ragheb, E. Reproductive biology of *Sphyraena* species in the Egyptian Mediterranean waters of Alexandria. *Egypt. J. Aquat. Res.* **2004**, *30*, 255–270.
15. Bourehail, N.; Kara, M.H. Age, growth and mortality of the yellowmouth barracuda *Sphyraena viridensis* (Sphyraenidae) from eastern coasts of Algeria. *J. Mar. Biol. Assoc.* **2021**, *101*, 599–608. [CrossRef]
16. Munro, J.L.; Pauly, D. A simple method for comparing the growth of fishes and invertebrates. *Fishbyte* **1983**, *1*, 5–6.
17. Sparre, P.; Ursin, E.; Venema, S.C. *Introduction to Tropical Fish Stock Assessment. Part 1: Manual*; FAO Fisheries Technical Paper; FAO: Rome, Italy, 1987; pp. 75–76.
18. Wootton, R.J. *Ecology of Teleost Fishes*; Chapman and Hall: England, UK, 2012; pp. 117–158.
19. Matić-Skoko, S.; Ferri, J.; Škeljo, F.; Bartulović, V.; Glavić, K.; Glamuzina, B. Age, growth and validation of otolith morphometrics as predictors of age in the forkbeard, *Phycis phycis* (Gadidae). *Fish. Res.* **2011**, *112*, 52–58. [CrossRef]
20. Cardinale, M.; Arrhenius, F. Using otolith weight to estimate the age of haddock (*Melanogrammus aeglefinus*): A tree model application. *J. Appl. Ichthyol.* **2004**, *20*, 470–475. [CrossRef]

21. Pilling, G.M.; Grandcourt, E.M.; Kirkwood, G.P. The utility of otolith weight as a predictor of age in the emperor *Lethrinus mahsena* and other tropical fish species. *Fish. Res.* **2003**, *60*, 493–506. [CrossRef]
22. Pino, A.C.; Cubillos, L.A.; Araya, M.; Sepúlveda, A. Otolith weight as an estimator of age in the Patagonian grenadier, *Macruronus magellanicus*, in central-south Chile. *Fish. Res.* **2004**, *66*, 145–156. [CrossRef]
23. Doering-Arjes, P.; Cardinale, M.; Mosegaard, H. Estimating population age structure using otolith morphometrics: A test with known-age Atlantic cod (*Gadus morhua*) individuals. *Can. J. Fish. Aquat. Sci.* **2008**, *65*, 2341–2350. [CrossRef]
24. Steward, C.A.; DeMaria, K.D.; Shenker, J.M. Using otolith morphometrics to quickly and inexpensively predict age in the gray angelfish (*Pomacanthus arcuatus*). *Fish. Res.* **2009**, *99*, 123–129. [CrossRef]
25. Škeljo, F.; Ferri, J.; Brčić, J.; Petrić, M.; Jardas, I. Age, growth and utility of otolith morphometrics as a predictor of age in the wrasse *Coris julis* (Labridae) from the eastern Adriatic Sea. *Sci. Mar.* **2012**, *76*, 587–595. [CrossRef]
26. Tuset, V.M.; Lombarte, A.; Assis, C.A. Otolith atlas for the western Mediterranean, north and central eastern Atlantic. *Sci. Mar.* **2008**, *72*, 7–198. [CrossRef]

MDPI

Article

Automatic Fish Age Determination across Different Otolith Image Labs Using Domain Adaptation

Alba Ordoñez [1,*], Line Eikvil [1], Arnt-Børre Salberg [1], Alf Harbitz [2] and Bjarki Þór Elvarsson [3]

1 Norwegian Computing Center, 0373 Oslo, Norway; eikvil@nr.no (L.E.); salberg@nr.no (A.-B.S.)
2 Institute of Marine Research, 9294 Tromsø, Norway; alf.harbitz@hi.no
3 Marine and Freshwater Research Institute, 220 Hafnarfjordur, Iceland; bjarki.elvarsson@hafogvatn.is
* Correspondence: albao@nr.no

Abstract: The age determination of fish is fundamental to marine resource management. This task is commonly done by analysis of otoliths performed manually by human experts. Otolith images from Greenland halibut acquired by the Institute of Marine Research (Norway) were recently used to train a convolutional neural network (CNN) for automatically predicting fish age, opening the way for requiring less human effort and availability of expertise by means of deep learning (DL). In this study, we demonstrate that applying a CNN model trained on images from one lab (in Norway) does not lead to a suitable performance when predicting fish ages from otolith images from another lab (in Iceland) for the same species. This is due to a problem known as *dataset shift*, where the *source data*, i.e., the dataset the model was trained on have different characteristics from the dataset at test stage, here denoted as *target data*. We further demonstrate that we can handle this problem by using domain adaptation, such that an existing model trained in the source domain is adapted to perform well in the target domain, without requiring extra annotation effort. We investigate four different approaches: (i) simple adaptation via image standardization, (ii) adversarial generative adaptation, (iii) adversarial discriminative adaptation and (iv) self-supervised adaptation. The results show that the performance varies substantially between the methods, with adversarial discriminative and self-supervised adaptations being the best approaches. Without using a domain adaptation approach, the root mean squared error (RMSE) and coefficient of variation (CV) on the Icelandic dataset are as high as 5.12 years and 28.6%, respectively, whereas by using the self-supervised domain adaptation, the RMSE and CV are reduced to 1.94 years and 11.1%. We conclude that careful consideration must be given before DL-based predictors are applied to perform large scale inference. Despite that, domain adaptation is a promising solution for handling problems of dataset shift across image labs.

Keywords: fish age determination; Greenland halibut; deep learning; dataset shift; domain adaptation

Citation: Ordoñez, A.; Eikvil, L.; Salberg, A.-B.; Harbitz, A.; Elvarsson, B.Þ. Automatic Fish Age Determination across Different Otolith Image Labs Using Domain Adaptation. *Fishes* **2022**, *7*, 71. https://doi.org/10.3390/fishes7020071

Academic Editor: Josipa Ferri

Received: 14 February 2022
Accepted: 17 March 2022
Published: 18 March 2022

1. Introduction

Otoliths are small calcium carbonate structures that form part of the balance organ in the inner ear of fish [1]. Due to their rich content of various information, they could be referred to as the "tachograph" of the fish. Fish otolith sciences have evolved in a range of research fields, from paleontology to stock discrimination and fish age determination. For example, the isotopic fingerprints of fish otoliths have proven to give insight into the water bodies which have been previously occupied by fish [2]. In this area, fish otoliths as old as 172 million years have been used [3]. Otoliths have also been used to distinguish different species. This has been, for instance, useful for studying the diet of sea mammals [4] or seabirds [5] by examining otoliths from feces samples. Other interesting studies have been carried out for long-lived fish in the Pacific [6], where radio-carbon data obtained from otoliths have provided valuable information on carbon flux in the oceans and have also been used to validate fish age.

When it comes to fish age determination, otolith images from captured fish are manually read, with in the order of a million images read every year [7]. Age-readers count age zones analogous to counting age rings in a tree, typically using a microscope or high-resolution images. However, this process is challenging and time-consuming [8]. During the last decades, automatic techniques have been proposed to automate this tedious activity (e.g., [9]). More recently, the use of deep learning (DL) has received more and more attention with promising results shown for Greenland halibut [10,11], snapper and hoki [12], red mullet [13] and Atlantic salmon [14].

However, the construction of DL systems can be challenged and hindered by several factors. The developed system trained at one otolith image lab where it works well, may not necessarily work well when applied to otolith images from another lab for the same species. In such cases, the system may have trouble generalizing to unseen data from another lab, due to data characteristics that are different from those at the training stage.

The discrepancy between data coming from different labs may have a range of explanations. From personal communication with reader experts the following was noted: different countries manage stocks that might have otoliths with varying readabilities due to different fish environments and catch seasons. Ages may also be read using thin sectioning of otoliths in some labs instead of reading the whole otolith (e.g., for Greenland halibut). The conservation and preparation of the otoliths might influence the age readability as well, one option being the storage of the otolith in an envelope and another being to freeze the otolith in water [15]. Camera quality, lighting conditions and other imaging setup conditions may also affect image characteristics, in addition to the magnifying glass equipment.

Acquiring new annotations such that it is possible to train the DL system with a more diverse dataset can also be challenging, both in terms of effort and availability of expertise. This process requires trained age-reading experts. Multiple methods are employed to minimize error, such as age-reading workshops where age determined by different readers from various labs are compared [16]. Reference otoliths are also used in training and are periodically reread by readers to ensure consistency [17]. Lack of highly trained age-reading experts can nevertheless lead to backlogs of otoliths that have not been annotated.

To address those challenges and increase the willingness to automate the age-reading using DL, it could be of interest to show that an already trained system could generalize to novel otolith images coming from other labs, while not requiring extra annotation effort.

The challenging situation where we observe different characteristics between the data the model was trained on and the data used at test stage is often referred to as dataset shift [18,19]. Training a classifier which performs well in this scenario can be addressed using domain adaptation (DA) [20], i.e., adapting models trained on data from a source domain to a different domain, known as the target domain. DA is typically carried out in a setting where the training data from the source domain is labelled, but little or no labels are available from the target domain. The case where we do not have labels at all in the target domain is commonly described as unsupervised domain adaptation (UDA). As we could more frequently face the scenario where we would like to test trained models on unlabeled otolith images, in this paper, we will consider solving a UDA task for automating fish age determination across different image labs, using DL.

There has been an effort made to develop deep neural networks that could handle UDA. In a recent review study, Zhao et al. [21] divided UDA approaches into four categories: (i) discrepancy-based, (ii) adversarial discriminative methods, (iii) adversarial generative methods and (iv) self-supervised methods. The approaches from the two adversarial categories were said to score best on performance and have been very popular. The discrepancy-based approaches were said to have lower performance and be less applicable to complex datasets, while the self-supervision-based methods represented a class of newer approaches that were shown to be robust and applicable to complex datasets.

Based on this, the aim of this work has been to investigate different UDA approaches from categories (ii)–(iv). We intended to see how they performed for the problem of generalizing a model trained for fish age prediction on otolith images from a lab in one country, to do prediction on images from a different lab in a different country, without new manual annotation.

The experiments were carried out based on otolith images from Greenland halibut acquired by the Institute of Marine Research (Norway) and the Marine and Freshwater Research Institute (Iceland). We adapted a trained network using otolith images and labels from the Norwegian lab (source domain) to otolith images coming from the Icelandic lab (target domain).

2. Materials and Methods

2.1. Data

The source and target datasets were collected by the Institute of Marine Research (Norway) and by the Marine and Freshwater Research Institute (Iceland), respectively.

The source dataset was a subset of the one described in Moen et al. [10], it consisted of 4109 images of paired right and left otoliths (collected between 2006 and 2017) having a resolution of 2596 × 1944 pixels. Each otolith pair was separated leading to 8218 single right and left otolith images, with labeled ages ranging from 1 to 26 years. Only ages read by two experienced readers from the same lab were used and only one reader for each otolith. Co-readings between the two readers revealed a negligible between-reader bias, independent of age.

The target dataset was composed of 3501 right otolith images, that were obtained from images of paired right and left otoliths having a resolution of 2048 × 1536 pixels. The otoliths were collected between 2015 and 2020 from the Icelandic autumn ground fish survey [22] and labeled ages ranged from 1 to 20 years. The ages were determined by a single reader.

Examples of image variation across the Norwegian and the Icelandic lab are shown in Figure 1, where one can observe some differences in terms of background, lighting conditions and magnification. The age distribution estimated by readers for the Icelandic dataset was such that the age label space was inside the range of the label space estimated by the Norwegian lab (Figure 2).

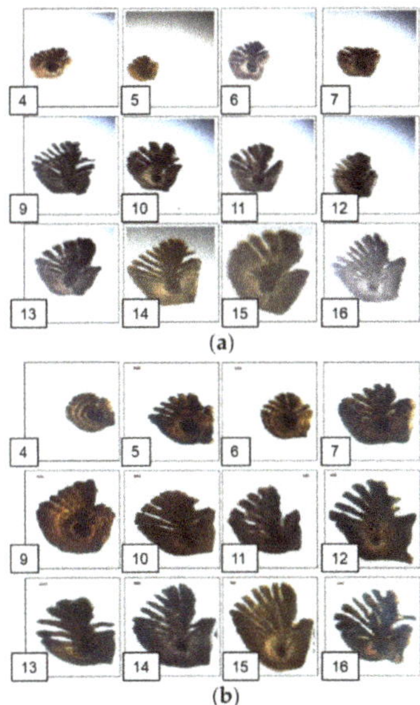

Figure 1. Examples of otolith images (resized to 224 × 224 pixels) corresponding to different ages (4–16 years) predicted by human readers. (**a**) Images acquired and annotated by the Norwegian lab; (**b**) Images acquired and annotated by the Icelandic lab.

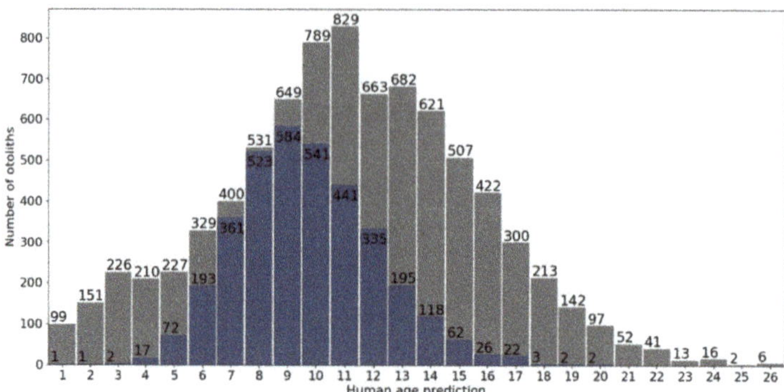

Figure 2. Age frequency distribution of predicted ages by human readers in the Norwegian dataset (gray) and the Icelandic dataset (blue).

2.2. UDA for Age Classification

We considered three different UDA approaches (Figure 3) to adapt an existing age classification system for otolith images to a new target domain, such that we could still have a suitable classification performance. For that, we exploited images from a source domain D_S and a target domain D_T as well as labels only from D_S. Common for the three approaches was a weight sharing constraint between a feature extractor for the D_S images and the D_T images, allowing us to learn a domain-invariant feature space. Another common module was the classifier learned from the D_S data. Since the output of the feature extractor was supposed to be domain-invariant after being optimized during the training process, we expected that using that module together with the classifier could produce meaningful age predictions on images from D_T.

2.2.1. Adversarial Generative Adaptation

The first method we used was a generative adversarial adaptation (Figure 3a), based on the coupled generative adversarial networks (CoGAN) proposed by Liu and Tuzel [23]. It consists of a pair of GANs and the idea of using more than one GAN was compelling to us by its originality. Moreover, the method showed promising results on UDA tasks when it was introduced [23].

Our CoGAN had generators that synthesized images by taking as input a 100-dimensional noise vector. We used the same networks as proposed in the implementation taken from [24].

Feature extractors were used to output feature representations for the discriminators. The architecture of the feature extractors was based on the commonly used convolutional neural network (CNN) ResNet [25] and took images with the default input size of 224×224 pixels. Those networks were initialized using the weights from the source-CNN (excluding the last classification layer), i.e., the CNN model trained to classify otolith ages using images and labels from D_S. The discriminator models were simply estimating the probabilities that the generated images were real or synthesized for each of the domains.

Following [23], we tied the weights of the first few layers of the generators as well as the weights from the feature extractors. This weight sharing allowed us to learn a domain-invariant feature space, without requiring the existence of any pairs of corresponding images in the two domains. For classifying otolith ages, the classifier was added to the feature extractor and corresponded to the last layer of the ResNet model.

We trained the whole architecture by jointly solving the CoGAN learning problem (CoGAN related loss [23]), involving images from D_S and D_T and the age classification problem in D_S (cross-entropy classification loss).

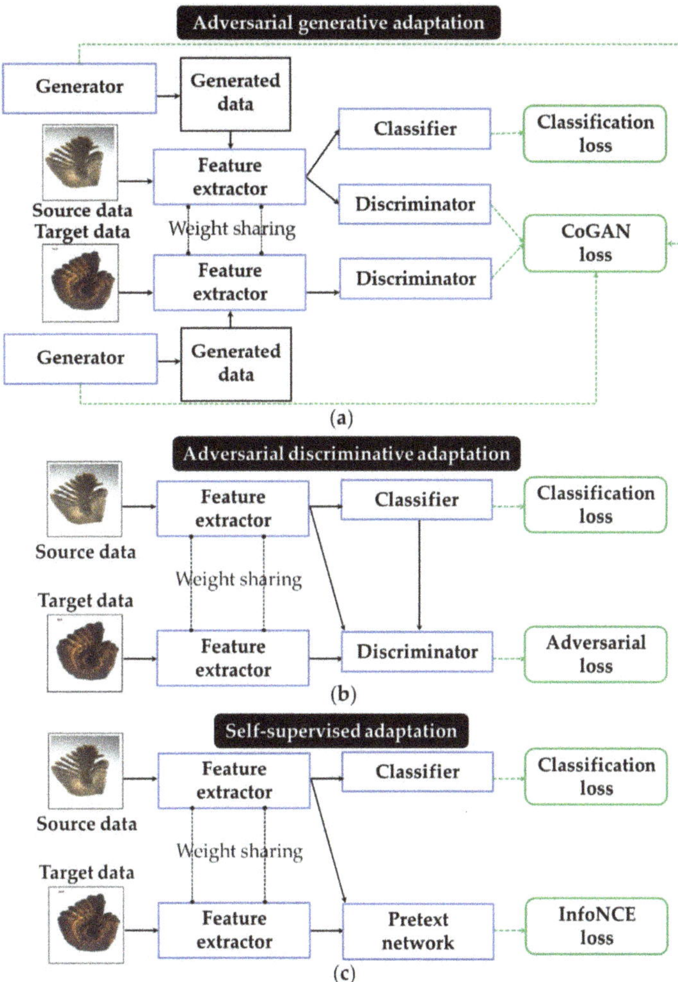

Figure 3. Illustration of the different UDA architectures utilized for automatic age determination of otoliths across Norwegian (source) and Icelandic (target) labs. (**a**) Adversarial generative adaptation (CoGAN); (**b**) Adversarial discriminative adaptation (CDAN); (**c**) Self-supervised adaptation (SimCLR).

2.2.2. Adversarial Discriminative Adaptation

The second method was an adversarial discriminative adaptation (Figure 3b) and differed from the previous approach in that it did not require generators. The implementation followed the state-of-the-art method conditional adversarial domain adaptation (CDAN) proposed by Long et al. [26]. As in the previous approach, the feature extractor was initialized with the weights of the source-CNN model (taking input images of size 224 × 224) and therefore shared the same architecture (excluding the last classification layer). The classifier predictor corresponded to the last layer of the ResNet model. The discriminator was a multilayer perceptron with two hidden layers having ReLU as a non-linear activation function. It received an input x defined as a joint variable of the extracted feature representations and the classifier predictions. This network architecture ($x \rightarrow 1024 \rightarrow 1024 \rightarrow 2$) was used to classify whether the input was coming from D_S or D_T. The size of the hidden layers was chosen to match the default implementation of CDAN available from [27].

This architecture was trained to solve a minimax optimization problem. The term to be minimized included the cross-entropy classification loss (using data and labels from D_S), combined with the adversarial loss derived from the discriminator which tried to match the distribution of the source and target domains involving images from D_S and D_T (we followed [26] and fixed the combination parameter to 1). This last term relating to the discriminator was the one that was also maximized in the minimax optimization problem (see [26] for further details).

2.2.3. Self-Supervised Adaptation

The last method we chose to test for UDA used self-supervision (Figure 3c) and differed from the previous approach in that the discriminator component was replaced by a pretext network. In self-supervised learning, a supervised task is created out of the unlabeled original data. Typically, the original images are transformed and the pretext network needs to learn how to predict certain aspects related to the transformations. Predicting image rotations [28], solving a jigsaw puzzle [29], retrieving colors from grayscale images [30] are examples of pretext tasks. The idea of including the pretext network was to be able to learn again a domain invariant feature representation using unlabeled otolith images from the source and target domains.

We based our implementation on the generic method proposed by Xu et al. [31] and available from [32]. As carried out in [31], we initially experimented with image rotation prediction as a pretext task. However, the pretext network was rapidly finding the solution during training and the learned feature representations from the otolith images did not help for DA. We decided instead to select a more challenging pretext task, using the simple framework for contrastive learning of visual representation (SimCLR) proposed by Chen et al. [33]. This method was chosen for its simplicity of implementation and its good performance that achieved significant advances in state-of-the-art self-supervised learning. In SimCLR, the pretext network is trained to recognize positive samples, i.e., different augmented views of the same image and distinguish them from the negatives, i.e., augmented views of other images from the dataset. We followed the recommendations from [33] and used random cropping, resizing, color distortions and Gaussian blur as data augmentations. The pretext network was a multilayer perceptron with one hidden layer having a ReLU as non-linearity. The network received as input extracted features x from the ResNet model feature extractor and outputted a 26-dimensional feature representation ($x \rightarrow 512 \rightarrow 512 \rightarrow 26$). The size of the hidden layer and the output were chosen to match the default implementation of SimCLR available from [34]. As for the adversarial approaches, the feature extractor was initialized with the weights of the source-CNN model.

The architecture of the self-supervised adaptation was trained to learn jointly the pretext task (using unlabeled images from D_S and D_T) and the age classification task, where the classifier predictor received images and labels from D_S. The losses relating to each of the tasks were added and minimized during training. The pretext task loss was the InfoNCE loss [35], which is a popular choice of loss function for contrastive learning aiming to pull feature representations that are close and push away representations that are different. As pointed out by [33], learning such a task benefits from having large batch sizes such that a large number of negative samples is obtained for comparison with every single positive sample. To be able to use contrastive learning to our advantage for DA, the feature extractor took as input images of size 96 × 96 pixels (instead of 224 × 224), ensuring a sufficiently large batch size.

2.2.4. Implementation Details

We implemented the above methods using the PyTorch framework on a single GTX 1080 Ti GPU with 11 GB memory. For a fair comparison, we used the same ResNet architecture for the feature extractors. Given our hardware resources, we chose a ResNet18 to be able to choose a large batch size for the self-supervised adaptation. For the layers that were not part of the feature extractors (initialized using the Source-CNN), the initialization was done using random values scaled according to the method proposed in [36]. For each of the models, we ran five trials where different random number generators were used for the initialization. This helped in building a certain confidence in the performance of the different algorithms when comparing them, by checking whether or not the results were obtained by chance (consistency of the model).

For all the approaches, we followed Moen et al. [10] and used a constant learning rate of 0.0004 and the Adam optimizer [37]. For adversarial DA approaches, we checked the default parameters from the CoGAN implementation taken from [38] and chose a batch size of 64. As self-supervised learning benefits from large batch sizes [33], we chose a batch size of 512 (according to our computing resources). For all the methods, a number of 200 epochs was selected.

2.3. Other Considered Classifiers

To better understand how the three UDA approaches performed, we considered comparing them with other classifiers, all defined by a ResNet18 architecture that was trained either on image data acquired from the Norwegian lab or the Icelandic lab.

First, we considered ResNet18 networks trained on otolith images acquired and annotated by the Norwegian lab. The models classified ages into one of the 26 categories (from 1 to 26 years as defined by the labels from the source dataset) and the classification was performed on two versions of resized images: 224×224 and 96×96. We directly deployed these models on otolith images from the Icelandic lab, without any adaptation or finetuning. These experiments were referred to as the *lower performance bound*, as we expected them not to perform that well on the target data given the observed image variation across the Norwegian and the Icelandic lab (Figure 1). We also applied these models on the Norwegian source data and we denoted the experiments as the *Norwegian bound*.

Next, we considered ResNet18 networks trained on the Icelandic data, where during training, we used the age labels provided by Iceland. This provided us with a *higher performance bound*, as we expected to have better results than not using labels from Iceland at all. These models classified ages into one of the 20 categories (from 1 to 20 years as defined by the labels from the target dataset).

Finally, we considered a model where we performed a simpler domain adaptation via a standardization preprocessing step. This was a semi-automatic approach derived by practitioners with the aim of reducing the variability in background and resolution between labs. The approach was based on a semi-automatic thresholding to remove the background. Then, images were resized so that the vertical extension of the otolith covered 90% of the vertical extension of the image. After standardizing the otolith images, we trained the ResNet18 models on the Norwegian dataset (images of size 224×224 and 96×96) and deployed them on the preprocessed images from Iceland. This provided us with another domain adaptation approach denoted *standardization*. It was simpler than the three other considered UDA methods (Figure 3) that required both images from the Norwegian and the Icelandic labs when training. Examples of standardized images are shown in Figure 4.

For all the above models, a small validation set was used to choose the batch size (128) and to control when to terminate the training process, setting a maximum number of epochs of 150, as done in Moen et al. [10]. For the experiments denoted as lower performance bound, Norwegian bound and the standardization approach, the validation set was extracted from the source dataset, whereas for the higher performance bound we used a validation set extracted from the target data. We chose the same learning rate/optimizer as for the UDA approaches and we ran five trials for each of the models. For each trial, different random number generators were used for the initialization of the networks that also followed the method from [36].

A summary of the different experiments carried out in this study, with their corresponding characteristics is reported in Table 1.

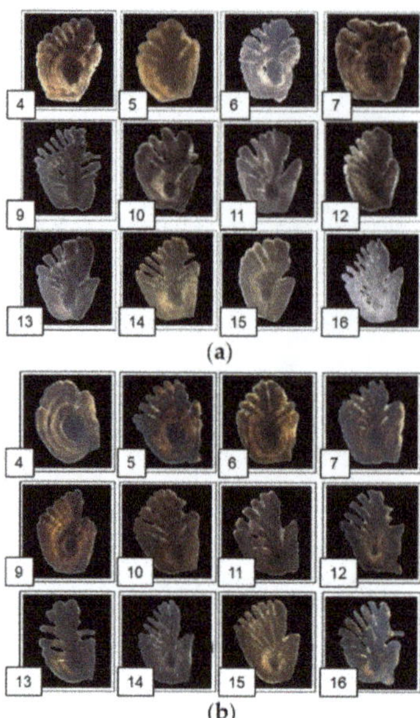

Figure 4. Examples of otolith images where the standardization preprocessing step was applied. (**a**) Standardized version of images from Figure 1a acquired and annotated by the Norwegian lab; (**b**) Standardized version of images from Figure 1b acquired and annotated by the Icelandic lab.

Table 1. Summary of the different experiments with their associated characteristics.

Experiment	Training Labels	Training Images	Test Images	Resolution	Image Pre-processing	Considered as DA
Norwegian bound 224	Nor.	Nor.	Nor.	224	No	No
Norwegian bound 96	Nor.	Nor.	Nor.	96	No	No
Lower performance 224	Nor.	Nor.	Ice.	224	No	No
Lower performance 96	Nor.	Nor.	Ice.	96	No	No
Higher performance 224	Ice.	Ice.	Ice.	224	No	No
Higher performance 96	Ice.	Ice.	Ice.	96	No	No
Standardization 224	Nor.	Nor.	Ice.	224	Yes	Yes
Standardization 96	Nor.	Nor.	Ice.	96	Yes	Yes
Adv. generative (CoGAN)	Nor.	Nor. and Ice.	Ice.	224	No	Yes
Adv. discriminative (CDAN)	Nor.	Nor. and Ice.	Ice.	224	No	Yes
Self-supervised (SimCLR)	Nor.	Nor. and Ice.	Ice.	96	No	Yes

2.4. *Performance Measurement*

For assessing the performance of the different classifiers, we considered as carried out in [11] the root mean squared error (RMSE) between age prediction and read age, as well as the mean coefficient of variation (CV) of independent estimators (human and DL system) calculated for each given otolith. For both metrics, the lower the values the better.

3. Results

The results from our experiments are summarized in Tables 2 and 3 and also illustrated in Figures 5–7. Rows 1 and 2 of Table 2 report the performance of age prediction by experimenting with the Norwegian bound, where models were trained and tested on otolith images acquired and annotated by the Norwegian lab. The classification was performed on the two versions of resized images: 224 × 224 and 96 × 96. Reasonably good RMSE/CV results were obtained considering five different trials (with relatively low variation over these splits ≤ 0.10 for the RMSE) and were comparable to the earlier study of Moen et al. [10] (RMSE = 1.65 years. and CV = 9%), although they tested on a smaller dataset. Resizing the images to different resolutions had a very limited effect on the performance.

Table 2. Summary of the performances achieved on the experiments that are not considered as domain adaptation, i.e., Norwegian bound and lower/higher performance bounds. For each experiment, the ResNet18 model was trained 5 times (using 5 different random number generators) and we reported the averaged RMSE/CV together with standard deviation over the 5 trials (quantity after ± sign).

Experiment	RMSE (Years)	CV (%)
Norwegian bound 224	2.08 ± 0.05	10.09 ± 0.63
Norwegian bound 96	2.18 ± 0.10	10.4 ± 0.35
Lower performance 224	5.12 ± 0.58	28.6 ± 2.3
Lower performance 96	5.95 ± 1.3	31.3 ± 4.9
Higher performance 224	1.50 ± 0.036	8.14 ± 0.12
Higher performance 96	1.48 ± 0.037	7.99 ± 0.28

Table 3. Summary of the performances achieved on the experiments that are considered as domain adaptation, i.e., simple standardization approach and the UDA methods. For each experiment, the model was trained 5 times (using 5 different random number generators) and we reported the averaged RMSE/CV and standard deviation over the 5 trials.

Experiment	RMSE (Years)	CV (%)
Standardization 224	3.57 ± 0.26	19.6 ± 1.2
Standardization 96	3.18 ± 0.28	17.1 ± 1.4
Adv. generative (CoGAN)	3.57 ± 0.72	21 ± 6.0
Adv. discriminative (CDAN)	2.18 ± 0.08	12.7 ± 0.54
Self-supervised (SimCLR)	1.94 ± 0.11	11.1 ± 0.62

Rows 3 and 4 of Table 2 report the results on the lower performance bound experiments, showing that without any DA involved, the performance on the Icelandic data was considerably lower than for the Norwegian data. In addition, a larger variation in performance was observed over the different trials (≥0.5 for the RMSE). The last two rows of Table 2 correspond to the higher performance bound. They demonstrated the potential for using a DL system to predict ages on the target data from the Icelandic lab, where good levels of RMSE/CV together with low variations over the five trials (≤0.04 for the RMSE) were achieved when training on these data. These observations were supported by the age-bias plots of Figure 5 displaying the age predictions (median over the five trials) of the lower and higher performance bounds (using images of size 224 × 224) against the human annotated age for the Icelandic data. We could observe that the lower performance bound (Figure 5a) completely overestimated the ages. The higher performance bound results were satisfactory (Figure 5b), although we noticed an underestimation for the older age groups, similar to that which was previously observed for the Norwegian data [10,11].

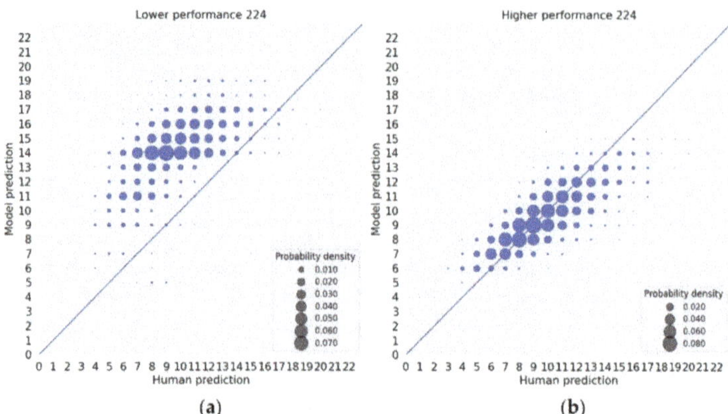

Figure 5. Model age predictions vs. human annotated age obtained on the Icelandic data for the lower and higher performance experiments (using images of size 224 × 224). For each model, the age predictions correspond to the calculated median from the predicted ages obtained over the 5 different trials. The scatters have an area proportional to the probability density of data. (**a**) Age predictions for the lower performance bound; (**b**) age predictions for the higher performance bound.

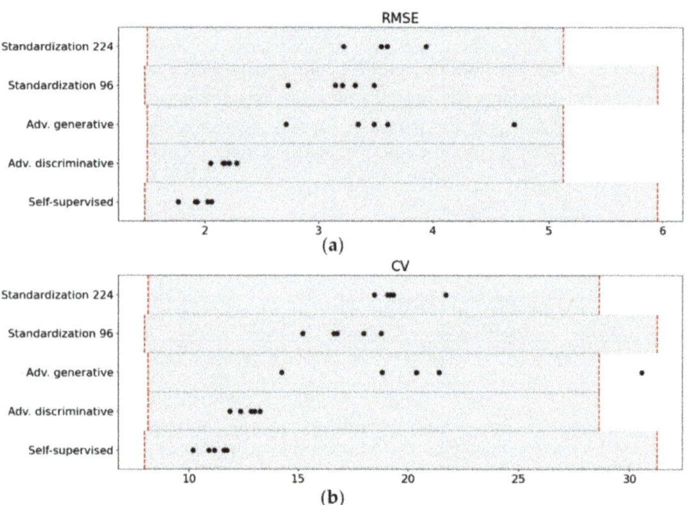

Figure 6. Visualization of the spread of the performance results obtained on the Icelandic data for the methods that are considered as domain adaptation. Each dot corresponds to one of the 5 trials. The red dashed lines delimiting the gray bars correspond to the higher (left) and lower (right) averaged performance bounds obtained with designated resolution (depending on the experiment). (**a**) Performance measured in terms of RMSE (years); (**b**) performance measured in terms of CV (%).

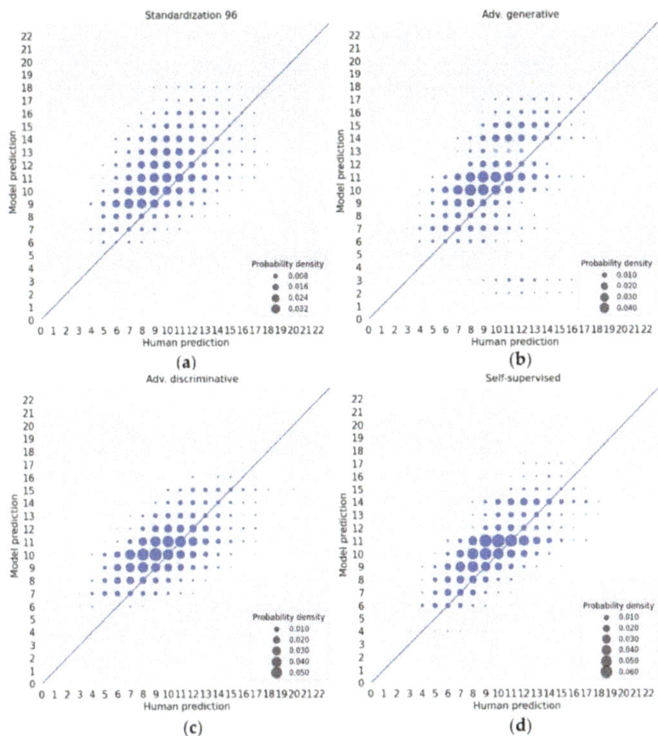

Figure 7. Age predictions of different adaptation models vs. human annotated age obtained from the Icelandic data. For each model, the age predictions correspond to the calculated median from the predicted ages obtained over 5 different trials. The scatters have an area proportional to the probability density of data. (**a**) Age predictions for the simple standardization approach (image resolution 96×96 pixels); (**b**) age predictions for the adversarial generative adaptation; (**c**) age predictions for the adversarial discriminative adaptation; (**d**) age predictions for the self-supervised adaptation.

In Table 3, we report the performance results obtained for the DA task, i.e., involving models trained on otolith images and labels from the Norwegian lab adapted to images from the Icelandic lab. For the different methods, we also visualize in Figure 6 the spread of the results for each of the five trials, represented as black dots. To be able to visualize the gap between domain adaptation methods and the higher and lower performance bounds that were tested on Icelandic images, each experiment was represented inside a gray bar. This was delimited to the left by the averaged RMSE/CV from the higher performance bound and to the right by the averaged results of the lower performance bound. In addition to this, a closer inspection of the age predictions for the different methods is presented in the age-bias plots of Figure 7.

From both Table 3 and Figure 6, we observed that the simple standardization approach improved on the lower performance bound and the results were comparable to those from the more complex adversarial generative approach. Similar patterns of age predictions were noticed when comparing the two approaches (Figure 7a,b). However, we found that the variation in performance was much lower for the standardization approach. For this case, we also noted that the performance for the low-resolution images was slightly better than for the high-resolution images.

The method based on the adversarial discriminative approach showed better performances than the generative approach both in terms of average RMSE/CV and associated variations. The best results were nonetheless obtained with self-supervised adaptation. It achieved the lowest average RMSE/CV and got closest to the higher performance bound, while the variations were on a level with that of the adversarial discriminative approach (Figure 6). When comparing the predicted ages from the two approaches (Figure 7c,d), less underestimation was observed with the self-supervised adaptation. This, despite the fact we used a resolution of only 96 × 96 pixels compared to the 224 × 224 used by adversarial approaches.

4. Discussion

Since the emergence of DL, several studies have been demonstrated to perform well on automatically predicting fish age from otolith images [10–14]. An important topic of this work was to assess whether a suitable performance from a DL system could be maintained on data from the same species but acquired in a different lab.

Using otolith images from Greenland halibut, we showed evidence in this paper that a DL system trained on data from a lab in Norway had difficulties, at test stage (Table 2, Figure 5a), for generalizing to novel otolith images from another lab in Iceland (lower performance bound model). The reason was a dataset shift, i.e., the images across the labs had different characteristics, although with a simple visual inspection (Figure 1) one could not have expected the bad performance obtained at test stage. By directly training a new system with data and labels from Iceland (higher performance bound model), we obtained a suitable classification performance (Table 2, Figure 5b). However, this required asking for manual annotations, which was not the optimal solution as the process had a cost in terms of effort and expertise.

For both lower and higher performance bounds, reducing the input image size from 224 × 224 to 96 × 96 pixels did not affect the performance. This finding was consistent with a previous study [11] showing that the internal structure of the otolith was not an important attribute for DL systems to determine accurately most of the ages. Thus, using lower-resolution images in those models was not problematic.

To address the performance challenges due to dataset shift, we investigated three different strategies of UDA to adapt the existing classifier trained on the Norwegian data to the Icelandic data, without requiring extra labeling effort from Iceland when training. The performance of the adversarial generative approach (CoGAN, Figure 7b) varied quite substantially compared to the adversarial discriminative approach (CDAN, Figure 7c) and the self-supervised approach (SimCLR, Figure 7d). The results from CoGAN were the worst and exhibited a higher variation across different runs of the model (Table 3, Figure 6), indicating the instability of this GAN-based approach. This contrasted with the review study of Zhao et al. [21] where generative approaches scored well in terms of performance. A possible explanation could be that in this study, the method was challenged when trying to generate images resembling the ones from the source and target domains, where perhaps high focus was dedicated to compensating for background differences observed across labs. More recent adversarial generative methods do not use the concept of training two GANs anymore and trying a more recent state-of-the art approach could be considered in future work. The other adversarial approach, CDAN, came closer to the higher performance bound and little variation across the different runs was observed (Table 3, Figure 6). SimCLR also provided stable results and led to the best performance on the target domain compared to the other two methods. In Zhao et al. [21], self-supervised approaches did not have the best performance score, but in our case it seemed that the pretext task of recognizing positive samples from negative samples was sufficiently good to make the model learn invariant representations across source and target domains. As suggested in [39], with the chosen pretext task, the attention was probably focused on low- and mid-level network representations that might have captured brightness and contrast characteristics. Those characteristics could be found with low-resolution images, that is why operating on images of size 96x96 pixels was not a problem.

Applying a standardization preprocessing to the images from the Norwegian lab, training a model with these data and deploying it on preprocessed data from the Icelandic lab performed worse than the CDAN and the SimCLR approaches (Table 3, Figure 7a). Moreover, slightly higher variations across the runs were observed (Figure 6). This reflected that finding a solution solely based on data preprocessing was not enough to handle the dataset shift. There was a need for more elaborate adjustments in the deep neural network architecture and UDA using self-supervised learning (SimCLR) seemed the most promising alternative.

Finding an approach for automatic age determination across labs without requiring additional human expertise when training could have implications in two major respects. First, since acquiring otolith image data is less cost demanding than the human interpretation, the present study could raise the possibility of carrying out age determination of backlogs of otoliths that have not yet been annotated in other labs. Second, having a tool for adapting age-reading from one lab to another could help reduce the observed between-lab differences. This may streamline comparisons conducted in age-reading workshops such as [15,16]. In this context, it could also be interesting to combine age readings from different methods and estimate the relative merits of each. By using the approach proposed in [40], one could combine estimated ages while accounting for the biases and imprecisions of the different used methods. In this case, the results would be properly weighted before using them for stock assessment population models. Those could possibly be improved by combining the predictions of independent estimators, which is the case of manual predictions, based on analysis of otolith age zones and DL predictions, which are triggered by other aspects [11].

The results presented in this paper must be seen nonetheless in light of some limitations. We considered adapting a classifier that was trained on images from the right and left otoliths to images only belonging to the right otolith. The reason was that in the earlier phase of the study, we decided to use the same paired data as in Moen et al. [10] (including right and left otoliths) but when we prepared the training data for Iceland, we limited ourselves to the right-separated pair that we resized to 224×224 pixels and dismissed the rest of the image. This decision was not reviewed at the later stage of the results evaluation, but we plan in the future to analyze what would happen when including the left otolith images as well. Furthermore, the correspondence with sex and length could also be examined. The age distribution for Greenland halibut females and males is known to differ, with females tending to live longer and with a growth that exceeds that of males [41]. Incorporating information about the sex and length in the DL network as proposed in [42] may also be helpful.

To compare the performance of the different DL approaches, we used the CV, the RMSE and the age-bias plots. However, in the context of stock assessment, these quantities might be too limiting measures. Analyzing age distributions and examining the final stock assessment results would be important tasks for the future. It might happen when comparing methods (including different manual age readings) that the stock assessment results appear virtually the same, despite possible large differences in CV, RMSE, age-bias plots or vice-versa.

Finally, we considered adapting a classifier from a source to a target domain where the target age categories were included in the source label space (Figure 2). In the proposed UDA methods, we tried to match the whole source and target domains as if they fully shared the label space, which was not the case. This could have led to a suboptimal transferability of source examples, where in our study some age categories could have been over-represented (e.g., ages 10–12 as seen in Figure 7b–d). Looking for solutions to handle this, as proposed for instance in [43,44], will be an important task in the future. Another realistic scenario to be examined would be when the target label space has age categories that are not present in the source label space (e.g., transferring the age classifier trained on Icelandic data to Norwegian data). It could then happen that the techniques investigated in this paper would not perform that well. In that case, alternative solutions as proposed in [45] should be taken into consideration.

5. Conclusions

In this work, the dataset shift across otolith images acquired from labs in Norway and Iceland was handled using domain adaptation strategies. We were able to adjust a DL model to provide satisfactory predictions on the Icelandic data. However, the performance depended strongly on the selected strategy. We observed that the CDAN and SimCLR approaches resulted in better performances, compared to the CoGAN or the simple adaptation method via standardization.

Even though common practice consists of validating a DL model on a holdout dataset during training, this step is not sufficient to guarantee that the model will have a well-defined behavior for unseen data. Hence, before DL-based predictors are considered to perform large scale inference on otolith images, a proper handling of dataset shift across different image labs is needed. The insights from this study pointed out that domain adaptation was a promising direction to consider, although analyzing model performance based on information shared between the source and target label spaces deserves further exploration. The hope is that such findings will contribute to the further development of DL techniques that could aid in reducing effort and availability of expertise in the otolith age-reading process.

Author Contributions: Conceptualization, A.O., L.E. and A.-B.S.; data curation, A.H. and B.Þ.E.; formal analysis, A.O.; funding acquisition, A.-B.S. and L.E.; investigation, A.O.; methodology, A.O., L.E. and A.-B.S.; project administration, A.-B.S.; resources, A.-B.S.; supervision, A.-B.S. and L.E; validation, A.H. and B.Þ.E.; writing—original draft, A.O., L.E., A.-B.S., A.H. and B.Þ.E.; writing—review and editing, A.O. and A.H. All authors have read and agreed to the published version of the manuscript.

Funding: This research was funded by the Research Council of Norway as part of the COGMAR project, grant number 270966.

Institutional Review Board Statement: Ethical review and approval were waived for this study as the otolith images were obtained from dead individuals that were not endangered or protected.

Data Availability Statement: The Norwegian otolith data presented in this study are openly available through the Norwegian Marine Data Center under the DOI https://doi.org/10.21335/NMDC1949633559. The Icelandic otolith data can be made available upon request to the Marine and Freshwater Research Institute (Iceland).

Acknowledgments: We thank the Researach Council of Norway for funding, but also Auður Súsanna Bjarnadóttir and Sigurlína Gunnarsdóttir from the Marine and Freshwater Research Institute for their involvement in preparing the otolith dataset from Iceland (photography and age-reading) that was used in this study.

Conflicts of Interest: The authors declare no conflict of interest.

References

1. Schulz-Mirbach, T.; Ladich, F.; Plath, M.; Heß, M. Enigmatic Ear Stones: What We Know about the Functional Role and Evolution of Fish Otoliths. *Biol. Rev. Camb. Philos. Soc.* **2019**, *94*, 457–482. [CrossRef] [PubMed]
2. Patterson, W.P.; Smith, G.R.; Lohmann, K.C. Continental Paleothermometry and Seasonality Using the Isotopic Composition of Aragonitic Otoliths of Freshwater Fishes. *Wash. DC Am. Geophys. Union Geophys. Monogr. Ser.* **1993**, *78*, 191–202. [CrossRef]
3. Patterson, W.P. Oldest Isotopically Characterized Fish Otoliths Provide Insight to Jurassic Continental Climate of Europe. *Geology* **1999**, *27*, 199–202. [CrossRef]
4. Enoksen, S.; Haug, T.; Lindstrøm, U.; Nilssen, K. Recent Summer Diet of Hooded Cystophora Cristata and Harp Pagophilus Groenlandicus Seals in the Drift Ice of the Greenland Sea. *Polar Biol.* **2016**, *40*, 931–937. [CrossRef]
5. Polito, M.J.; Trivelpiece, W.Z.; Karnovsky, N.J.; Ng, E.; Patterson, W.P.; Emslie, S.D. Integrating Stomach Content and Stable Isotope Analyses to Quantify the Diets of Pygoscelid Penguins. *PLoS ONE* **2011**, *6*, e26642. [CrossRef]
6. Kalish, J.M. Pre- and Post-Bomb Radiocarbon in Fish Otoliths. *Earth Planet. Sci. Lett.* **1993**, *114*, 549–554. [CrossRef]
7. Campana, S.; Thorrold, S. Otoliths, Increments, and Elements: Keys to a Comprehensive Understanding of Fish Populations? *Can. J. Fish. Aquat. Sci.* **2001**, *58*, 30–38. [CrossRef]
8. Morison, A.; Burnett, J.; McCurdy, W.; Moksness, E. Quality Issues in the Use of Otoliths for Fish Age Estimation. *Mar. Freshw. Res.* **2005**, *56*, 773–782. [CrossRef]

9. Fablet, R.; Josse, N. Automated Fish Age Estimation from Otolith Images Using Statistical Learning. *Fish. Res.* **2005**, *72*, 279–290. [CrossRef]
10. Moen, E.; Handegard, N.O.; Allken, V.; Albert, O.T.; Harbitz, A.; Malde, K. Automatic Interpretation of Otoliths Using Deep Learning. *PLoS ONE* **2018**, *13*, e0204713. [CrossRef]
11. Ordoñez, A.; Eikvil, L.; Salberg, A.-B.; Harbitz, A.; Murray, S.M.; Kampffmeyer, M.C. Explaining Decisions of Deep Neural Networks Used for Fish Age Prediction. *PLoS ONE* **2020**, *15*, e0235013. [CrossRef]
12. Moore, B.; Maclaren, J.; Peat, C.; Anjomrouz, M.; Horn, P.L.; Hoyle, S.D. *Feasibility of Automating Otolith Ageing Using CT Scanning and Machine Learning*; New Zealand Fisheries Assessment Report 2019/58; New Zealand Fisheries Assessment: Wellington, New Zealand, 2019; ISBN 978-1-990008-66-5.
13. Politikos, D.V.; Petasis, G.; Chatzispyrou, A.; Mytilineou, C.; Anastasopoulou, A. Automating Fish Age Estimation Combining Otolith Images and Deep Learning: The Role of Multitask Learning. *Fish. Res.* **2021**, *242*, 106033. [CrossRef]
14. Vabø, R.; Moen, E.; Smoliński, S.; Husebø, Å.; Handegard, N.O.; Malde, K. Automatic Interpretation of Salmon Scales Using Deep Learning. *Ecol. Inform.* **2021**, *63*, 101322. [CrossRef]
15. ICES. *Report of the Workshop on Age Reading of Greenland Halibut (WKARGH)*; ICES CM 2011/ACOM:41; International Council for Exploration of the Seas: Vigo, Spain, 2011.
16. ICES. *Report of the Workshop on Age Reading of Greenland Halibut 2 (WKARGH2)*; ICES CM 2016/SSGIEOM:16; International Council for Exploration of the Seas: Reykjavik, Iceland, 2016.
17. Morison, A.K.; Robertson, S.G.; Smith, D.C. An Integrated System for Production Fish Aging: Image Analysis and Quality Assurance. *N. Am. J. Fish. Manag.* **1998**, *18*, 587–598. [CrossRef]
18. Quiñonero-Candela, J. Dataset Shift in Machine Learning. In *Neural Information Processing Series*; MIT Press: Cambridge, MA, USA, 2009; ISBN 978-0-262-17005-5.
19. Torralba, A.; Efros, A.A. Unbiased Look at Dataset Bias. In *CVPR 2011*; IEEE: Colorado Springs, CO, USA, 2011; pp. 1521–1528.
20. Ben-David, S.; Blitzer, J.; Crammer, K.; Kulesza, A.; Pereira, F.; Vaughan, J.W. A Theory of Learning from Different Domains. *Mach. Learn.* **2010**, *79*, 151–175. [CrossRef]
21. Zhao, S.; Yue, X.; Zhang, S.; Li, B.; Zhao, H.; Wu, B.; Krishna, R.; Gonzalez, J.E.; Sangiovanni-Vincentelli, A.L.; Seshia, S.A.; et al. A Review of Single-Source Deep Unsupervised Visual Domain Adaptation. *IEEE Trans. Neural Netw. Learn. Syst.* **2020**, *33*, 473–493. [CrossRef]
22. Sólmundsson, J.; Kristinsson, K.; Steinarsson, B.; Jonsson, E.; Karlsson, H.; Björnsson, H.; Palsson, J.; Bogason, V.; Sigurdsson, T.; Hjörleifsson, E. *Manuals for the Icelandic Bottom Trawl Surveys in Spring and Autumn*; Hafrannsóknir nr. 156: Reykjavík, Iceland, 2010.
23. Liu, M.-Y.; Tuzel, O. Coupled Generative Adversarial Networks. In *Advances in Neural Information Processing Systems*; Curran Associates, Inc.: Barcelona, Spain, 2016.
24. Linder-Norén, E. PyTorch CoGAN. Available online: https://github.com/eriklindernoren/PyTorch-GAN (accessed on 2 March 2022).
25. He, K.; Zhang, X.; Ren, S.; Sun, J. Deep Residual Learning for Image Recognition. In Proceedings of the IEEE Conference on Computer Vision and Pattern Recognition (CVPR), Las Vegas, NV, USA, 27–30 June 2016; IEEE: Las Vegas, NV, USA, 2016; pp. 770–778.
26. Long, M.; CAO, Z.; Wang, J.; Jordan, M.I. Conditional Adversarial Domain Adaptation. In *Advances in Neural Information Processing Systems*; Curran Associates, Inc.: Montréal, QC, Canada, 2018; pp. 1647–1657.
27. Jiang, J.; Chen, B.; Fu, B.; Long, M. Transfer-Learning-Library. Available online: https://github.com/thuml/Transfer-Learning-Library (accessed on 31 January 2022).
28. Gidaris, S.; Singh, P.; Komodakis, N. Unsupervised Representation Learning by Predicting Image Rotations. *arXiv* **2018**, arXiv:1803.07728.
29. Noroozi, M.; Favaro, P. Unsupervised Learning of Visual Representations by Solving Jigsaw Puzzles. In *European Conference on Computer Vision*; Springer International Publishing: Amsterdam, The Netherlands, 2016.
30. Zhang, R.; Isola, P.; Efros, A.A. Colorful Image Colorization. In *European Conference on Computer Vision*; Springer International Publishing: Amsterdam, The Netherlands, 2016; pp. 649–666.
31. Xu, J.; Xiao, L.; Lopez, A.M. Self-Supervised Domain Adaptation for Computer Vision Tasks. *IEEE Access* **2019**, *7*, 156694–156706. [CrossRef]
32. Jiaolong, X. Self-Supervised Domain Adaptation. Available online: https://github.com/Jiaolong/self-supervised-da (accessed on 1 February 2022).
33. Chen, T.; Kornblith, S.; Norouzi, M.; Hinton, G. A Simple Framework for Contrastive Learning of Visual Representations. In Proceedings of the 37th International Conference on Machine Learning, Virtual Event, 12–18 July 2020; pp. 1597–1607. Available online: http://proceedings.mlr.press/v119/chen20j.html (accessed on 13 March 2022).
34. Silva, T. PyTorch SimCLR: A Simple Framework for Contrastive Learning of Visual Representations. Available online: https://github.com/sthalles/SimCLR/blob/1848fc934ad844ae630e6c452300433fe99acfd9/models/resnet_simclr.py (accessed on 1 February 2022).
35. Van den Oord, A.; Li, Y.; Vinyals, O. Representation Learning with Contrastive Predictive Coding. *arXiv* **2019**, arXiv:Abs/1807.03748.

36. He, K.; Zhang, X.; Ren, S.; Sun, J. Delving Deep into Rectifiers: Surpassing Human-Level Performance on ImageNet Classification. In Proceedings of the IEEE International Conference on Computer Vision (ICCV), Santiago, Chile, 7–13 December 2015; IEEE: Santiago, Chile, 2015; pp. 1026–1034.
37. Kingma, D.P.; Ba, J. Adam: A Method for Stochastic Optimization. *arXiv* **2015**, arXiv:1412.6980.
38. Liu, M. CoGAN. Available online: https://github.com/mingyuliutw/CoGAN (accessed on 31 January 2022).
39. Zhao, N.; Wu, Z.; Lau, R.W.H.; Lin, S. What Makes Instance Discrimination Good for Transfer Learning. *arXiv* **2021**, arXiv:Abs/2006.06606.
40. Punt, A.E.P.E.; Smith, D.C.S.C.; KrusicGolub, K.K.; Robertson, S.R. Quantifying Age-Reading Error for Use in Fisheries Stock Assessments, with Application to Species in Australia's Southern and Eastern Scalefish and Shark Fishery. *Can. J. Fish. Aquat. Sci.* **2008**, *65*, 1991–2005. [CrossRef]
41. Nedreaas, K.; Soldal, A.V.; Bjordal, Å. Performance and Biological Implications of a Multi-Gear Fishery for Greenland Halibut (Reinhardtius Hippoglossoides). *J. Northwest Atl. Fish. Sci.* **1996**, *19*, 59–72. [CrossRef]
42. Martinsen, I. Deep Learning Applied to FIsh Otolith Images. Master's Thesis, The Arctic University of Norway, Tromsø, Norway, 2021.
43. Cao, Z.; Ma, L.; Long, M.; Wang, J. Partial Adversarial Domain Adaptation. In Proceedings of the European Conference on Computer Vision (ECCV), Munich, Germany, 8–14 September; Springer International Publishing: Munich, Germany, 2018.
44. Cao, Z.; You, K.; Long, M.; Wang, J.; Yang, Q. Learning to Transfer Examples for Partial Domain Adaptation. In Proceedings of the IEEE/CVF Conference on Computer Vision and Pattern Recognition (CVPR), Long Beach, CA, USA, 15–20 June 2019; IEEE: Long Beach, CA, USA, 2019; pp. 2980–2989.
45. Busto, P.P.; Gall, J. Open Set Domain Adaptation. In Proceedings of the IEEE International Conference on Computer Vision (ICCV), Venice, Italy, 22–29 October 2017; IEEE: Venice, Italy, 2017; pp. 754–763.

fishes

MDPI

Article

DeepOtolith v1.0: An Open-Source AI Platform for Automating Fish Age Reading from Otolith or Scale Images

Dimitris V. Politikos [1,*], Nikolaos Sykiniotis [2], Georgios Petasis [3], Pavlos Dedousis [1], Alba Ordoñez [4], Rune Vabø [5], Aikaterini Anastasopoulou [1], Endre Moen [5], Chryssi Mytilineou [1], Arnt-Børre Salberg [4], Archontia Chatzispyrou [1] and Ketil Malde [5,6]

1 Institute of Marine Biological Resources and Inland Waters, Hellenic Centre for Marine Research, 16452 Argyroupoli, Greece; pdedou@hcmr.gr (P.D.); kanast@hcmr.gr (A.A.); chryssi@hcmr.gr (C.M.); a.chatzispyrou@hcmr.gr (A.C.)
2 Hellenic Navy, General Staff, 11525 Athens, Greece; n.sikiniotis@gmail.com
3 Institute of Informatics and Telecommunications, National Centre for Scientific Research "Demokritos", Agia Paraskevi, 60228 Athens, Greece; petasis@iit.demokritos.gr
4 Department of Statistical Analysis, Machine Learning and Image Analysis, Norwegian Computing Center, 0373 Oslo, Norway; albao@nr.no (A.O.); salberg@nr.no (A.-B.S.)
5 Institute of Marine Research, 5005 Bergen, Norway; rune.vaboe@hi.no (R.V.); endre.moen@hi.no (E.M.); ketil.malde@hi.no (K.M.)
6 Department of Informatics, University of Bergen, 5008 Bergen, Norway
* Correspondence: dimpolit@hcmr.gr

Citation: Politikos, D.V.; Sykiniotis, N.; Petasis, G.; Dedousis, P.; Ordoñez, A.; Vabø, R.; Anastasopoulou, A.; Moen, E.; Mytilineou, C.; Salberg, A.-B.; et al. DeepOtolith v1.0: An Open-Source AI Platform for Automating Fish Age Reading from Otolith or Scale Images. *Fishes* **2022**, *7*, 121. https://doi.org/10.3390/fishes7030121

Academic Editor: Josipa Ferri

Received: 29 April 2022
Accepted: 27 May 2022
Published: 29 May 2022

Abstract: Every year, marine scientists around the world read thousands of otolith or scale images to determine the age structure of commercial fish stocks. This knowledge is important for fisheries and conservation management. However, the age-reading procedure is time-consuming and costly to perform due to the specialized expertise and labor needed to identify annual growth zones in otoliths. Effective automated systems are needed to increase throughput and reduce cost. DeepOtolith is an open-source artificial intelligence (AI) platform that addresses this issue by providing a web system with a simple interface that automatically estimates fish age by combining otolith images with convolutional neural networks (CNNs), a class of deep neural networks that has been a dominant method in computer vision tasks. Users can upload otolith image data for selective fish species, and the platform returns age estimates. The estimates of multiple images can be exported to conduct conclusions or further age-related research. DeepOtolith currently contains classifiers/regressors for three fish species; however, more species will be included as related work on ageing will be tested and published soon. Herein, the architecture and functionality of the platform are presented. Current limitations and future directions are also discussed. Overall, DeepOtolith should be considered as the first step towards building a community of marine ecologists, machine learning experts, and stakeholders that will collaborate to support the conservation of fishery resources.

Keywords: fish otoliths; deep learning; CNN; age determination; web tool

1. Introduction

Every year, the ages of thousands of fish are determined from otoliths or scales in the framework of national data collection fishery programs, supporting several objectives, such as the estimation of parameters for the demographic and population dynamics of fish stocks and the fitting of stock assessment models and length-at-age growth curves [1,2]. Age information is manually extracted by expert readers who count daily or annual growth zones in otoliths using a microscope or high-resolution images [2,3]. However, this is a labor-intensive, time-consuming, and costly process to perform [4]. This limits the number of fish that can be age-analyzed, and monitoring programs must—to a larger extent—rely on growth-at-age models to determine the age composition of fish populations [1]. This underlines the need for automated tools that could be practically used by otolith scientists to facilitate a more streamlined analysis in their laboratories.

Methods to automatically read otoliths for fish age extraction have been proposed using diverse data, such as images, fish biological features (e.g., fish length, catch data, sex) and geometrical features (shape and the opaque and translucent zonation patterns) [4–6]. Although these methods have shown good predictability, they require additional biological and geometric information beyond otoliths and a complex preprocessing stage to extract certain features from otolith images (e.g., measurement of translucent and opaque rings) before being able to determine fish age. This tends to limit their applicability as automated solutions for fish ageing.

Deep learning is a subfield of artificial intelligence (AI) that has revolutionized automation in a wide range of real-world applications related to images, text, audio, and videos [7,8]. Convolutional neural networks (CNNs) are a dominant class of deep neural networks for processing images that are being widely used, among others, for image classification (note: Image classification is the task of identifying the class that an image represents, e.g., predicting gender from a face image. In our case, prediction of age as a class category from an otolith image is a classification task.) [9] and image regression (note: Image regression is the task of predicting a continuous variable from an image, e.g., predicting house price from a house image. In our case, prediction of age as a numeric value from an otolith image is a regression task.) [10,11]. CNNs receive images as input, while the whole learning process is carried out in the network; they learn sequentially from simple shapes (lines, edges, etc.) in the first layers, to more detailed patterns in the next layers, and finally, classes of objects or numeric features in the final layers. This is achieved through a stack of alternately arranged convolutional, pooling, and activation layers, followed by a fully connected layer that performs the image classification or regression task (Figure A1, Appendix A). The key advantage of CNNs lies in their efficiency in capturing the spatial interaction between adjacent pixels in an image and, hence, extracting meaningful features that are able to correctly resolve the computer vision task [12]. For more details about CNNs, the reader is referred to the recent review paper [9].

In recent years, CNNs have received increased attention for automating fish age estimation from otolith images, as shown for Greenland halibut (*Reinhardtius hippoglossoides*) [13,14], snapper (*Pagrus auratus*) and hoki (*Macruronus novaezelandiae*) [15], Atlantic salmon (*Salmo salar*) [16], and red mullet (*Mullus barbatus*) [17]. These studies provided evidence that deep learning could offer an automated methodology for the analysis of otolith images, although with varying levels of accuracy, and provide cost-efficient and effective support towards the sustainability and management of fishery resources.

Despite the aforementioned AI advancements in fishery science, otolith researchers may not have the computing expertise or budget to take advantage of AI tools. In addition, even when AI algorithms are employed, without a community, scientists may face obstacles in implementing their case studies, collaborating with each other, and, on the whole, contributing to the field. DeepOtolith (http://otoliths.ath.hcmr.gr/, accessed on 1 March 2022, Figure 1) brings together AI researchers, fish scientists, and software developers to bridge the gap between state-of-the-art computing techniques and otolith research, providing a simple web interface that automatically estimates fish age by combining otolith images with deep learning. The user can export age estimates from multiple images to conduct further age-related research. At present, the platform contains models for three fish species (Table 1, Figure 2). Additional species, however, can be incorporated as related works will be published in the future. This image analysis platform is a Python application (https://www.python.org, accessed in 2001) that uses Python Flask (https://flask.palletsprojects.com/en/2.1.x/, accessed in 2010) and ReactJS [18] to operate as a webserver at the front-end. The source code for DeepOtolith, as well as sample otolith/scale images for experimentation, are available at: https://github.com/dimpolitik/Deep-Otolith (accessed on 1 March 2022). Below, the platform and its structure are introduced; the three otolith case studies that are currently available on the platform are reviewed; and finally, the functionality of the platform is demonstrated.

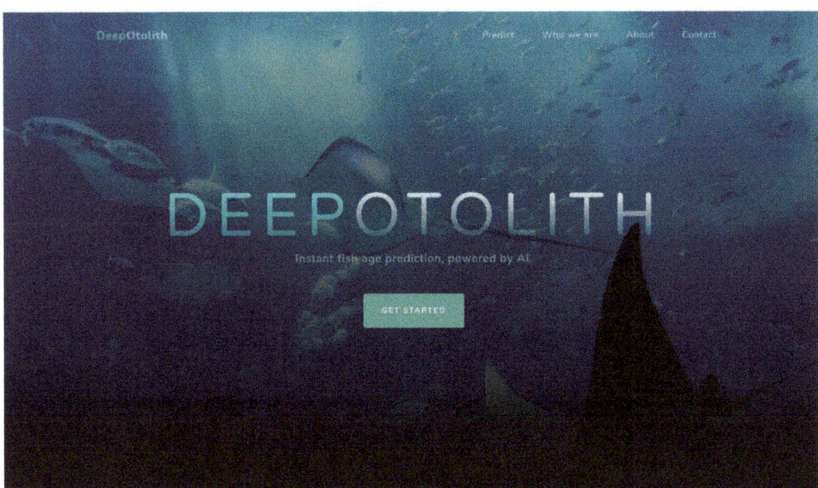

Figure 1. Landing page of the DeepOtolith platform (http://otoliths.ath.hcmr.gr/, accessed on 1 March 2022), which applies deep learning techniques to predict fish age from otolith images.

Table 1. Fish species that are currently available on the DeepOtolith platform.

Species	Age Groups	Region	References
Greenland halibut (*Reinhardtius hippoglossoides*)	1–26	Norway	[14]
Atlantic salmon (*Salmo salar*)	1–6 (river age) 1–9 (sea age)	Norway	[16]
Red mullet (*Mullus barbatus*)	0–5+	Greece	[17]

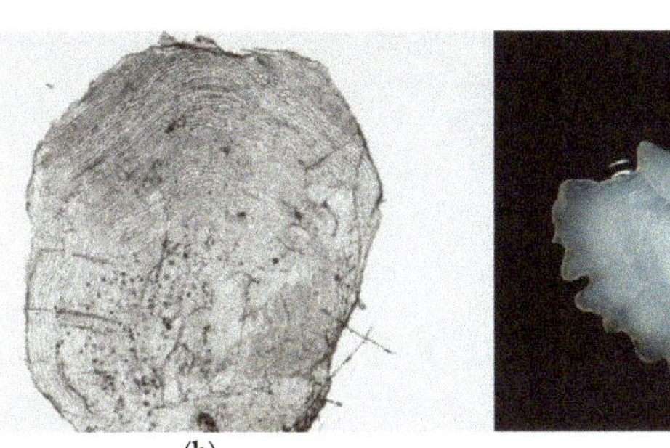

(a) (b) (c)

Figure 2. Fish species that are currently processed through the DeepOtolith platform: (**a**) Greenland halibut (*Reinhardtius hippoglossoides*) otolith, (**b**) Atlantic salmon (*Salmo salar*) scale, and (**c**) red mullet (*Mullus barbatus*) otolith.

2. Materials and Methods

2.1. Platform Architecture

The architecture of DeepOtolith is shown in Figure 3. The tool consists of two main components: the front-end, visible to the end user, and the back-end, where all processing takes place. On the front-end, the user can select one of the three currently available fish species and then consecutively upload otolith images (note: Images can be uploaded as separate files, not as a directory or a zip file. Images should be in .png or .jpg format.) of the selected species. The uploaded images are transferred to the back-end, where they are preprocessed according to the requirements of each respective species model; for example, images may be resized and loaded to the resolution expected by the corresponding trained CNN model. After preprocessing has been completed, the weights of the CNN for the selected fish species are loaded (Figure 3); the trained networks have been stored offline, and no new training is required. Then, the loaded model is applied to the preprocessed images to make the predictions, which are returned to the front-end. For each user-provided image, the probabilities of each predicted fish age group are displayed in a bar-plot. The age group with the highest probability defines the resulting fish age prediction. Beyond visualization, the user can download age predictions as a CSV file. Each row in the CSV file contains: the image name, the probability in % for each age group, and the resulting age prediction based on the age group with the highest probability. When the two highest probabilities in age prediction are below 10%, then the message "->Difference of two largest probabilities <10%: Additional validation by experts is recommended" is returned as a message. We note that a limit rate of 30 images per minute was imposed to prevent upload errors due to latency (enforced by the access control layer, as shown in Figure 3).

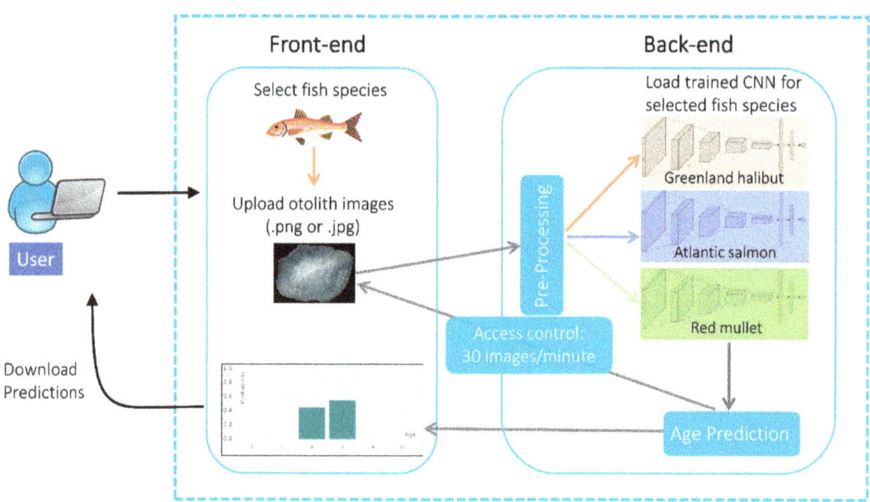

Figure 3. The architecture of the DeepOtolith platform.

2.2. Case Studies

2.2.1. Greenland Halibut (*Reinhardtius hippoglossoides*)

The automatic age determination of Greenland halibut otoliths was based on the work of [14], who focused on explaining the decisions of deep neural networks used for fish age prediction. The considered dataset was a subset of the one described in Moen et al. (2018), which consisted of pre-existing otolith images from the Institute of Marine Research (IMR, Bergen, Norway) collected between 2006 and 2017. For the acquisition of the images, the whole paired right and left otoliths were first collected and put into plastic trays for transportation, where salt or water was added to keep them moist until they could be frozen in the lab. This was done for preservation reasons prior to image capture. After the thawing and cleaning processes, the paired otoliths (or single otoliths if the corresponding pair was damaged or lost) were immersed in water and placed under a stereomicroscope on a white background with transmitted light such that the digital images could be taken. Those were finally imported into Photoshop and calibrated to a 10 mm scale. The resolution of the images was 2596 × 1944 pixels.

In the deep learning training process, ref. [14] adopted the VGG19 CNN architecture [19]. They used 8218 training sample images of right and left otoliths. The original images, were resized to fit a 224 × 224 square, as expected in the VGG19 model. The model was trained to classify ages into one of 26 categories (from 1 to 26 years), and age labels were provided by human experts based on current recommendations in the field. Using a test set of right otoliths composed of 165 samples, the trained model attained a root mean square error (RMSE) of 1.69 years between age prediction and age reading by experts. The predicted error was comparable to the earlier study of Moen et al. (2018) (RMSE = 1.65 years), which used a regression network based on the Inception v3 architecture [20] to automate the Greenland halibut ageing.

2.2.2. Atlantic Salmon (*Salmo salar*)

Vabø et al. [16] utilized an implementation of the EfficientNetB4 CNN network [21], with an input image resolution of 380 × 380, to automate the age estimation of Atlantic salmon (*Salmo salar*) scales. EfficientNetB4 was trained using transfer learning, with pre-trained weights from ImageNet [22]. The dataset used consisted of a total of 9056 high-resolution images of salmon scales sampled by the Institute of Marine Research in Bergen (IMR), Norway (from 2015 to 2018) and Rådgivende Biologer (from 2016 to 2017) in rivers along the coast of Norway. Salmon scale photos were taken using a Nikon SMZ25 stereomicroscope with a Nikon Digital Sight DS-Fi2 camera using an SHR Plan Apo 1× objective. The images were captured with a resolution of 2560 × 1920 pixels on a light gray background and postprocessed using NIS Elements D software. The images were annotated both for sea and river ages by expert readers. Two independent networks were trained for separately predicting river and sea age. each task. Age prediction was treated as a regression problem, returning a decimal number that was rounded to the nearest integer age and compared with the ground truth. From the total dataset, 8286 images were annotated with sea age, and 6238 were annotated with river age. Sea ages ranged from 1 to 9 years, with 2 years being the most frequent age (50.6%), followed by age 1 and age 3. River ages ranged from 1 to 6 years, with age 3 seen most frequently (56.5%), followed by age 2.

The prediction of sea age obtained an accuracy of 86.99%, while the predictive accuracy of river age was 63.2%. The study also included a test of the network's performance in comparison with six human readers on an additional dataset of 150 scales. This revealed that the ground truth estimates of river age by expert readers exhibited higher variance and lower levels of agreement compared to sea age, and this may indicate why this task appeared more difficult for the CNN to attain high accuracy [16]. Additionally, the CNN overpredicted the age of 1 year, whereas predictions were best for 2 and 3 years sea age and 3 years river age. This can be partially attributed to the imbalanced distribution of ages in the salmon imagery. Specifically, 90.2% of images had a river age of 2 or 3 years, when only 6% were 4-year-olds and 3% were 1-year-olds.

2.2.3. Greek Red Mullet (*Mullus barbatus*)

The automatic age estimation of the Greek red mullet (*Mullus barbatus*) was based on [17]. The dataset included 5027 otolith images, provided by the Hellenic Centre for Marine Research (HCMR) database, along with the age readings and fish length (body size in mm) of each individual fish. For the acquisition of the images, the whole otolith was used without any treatment; it was placed in a petri dish with the inner face looking upwards and immersed in water. The petri dish was placed under a stereoscope on a black background. Reflected cold LED light 50 W was used, provided by two photonic goosenecks, and adjusted to illuminate the whole surface of the otolith. Digital images were taken under a magnification of 16× with a resolution 768 × 576 pixels. Since different lighting conditions, zoom levels, or backgrounds of the tested images may impact the performance of the CNNs, the webpage users should consider, as possible, the above protocols to attain a more reliable age estimation. The age of red mullet in the dataset ranged from 0 to 11 years old. Due to the low number of specimens aged >5 years old (~6%), these were merged into the 5+ age group.

The Inception v3 CNN model [20] was trained using transfer learning with otolith images of resolution 400 × 400 as input, considering fish age estimation as a multi-class classification task with six age groups (Age-0, Age-1, Age-2, Age-3, Age-4, Age-5+). The potential benefit of multitask learning was also explored to improve the network's predictability, with the auxiliary task being the prediction of fish size. The enhanced neural network simultaneously received as input the otolith images and predicted fish age and length. The results showed that, without multitask learning, the ages of the red mullet were predicted correctly by 64.4%, performing better in the younger Age-0 and Age-1 classes (F1 score > 0.8) and moderately in the older age classes (F1 score between 0.50–0.54). Multitask learning increased the correct age prediction to 69.2%, with an additional 28.2% being within 1 year of error; this also proved a better approach to estimate older age groups, increasing accuracy between 3–23%. Additionally, the multitask network achieved a root mean square error (RMSE) of 0.56 years between predicted and human-based age predictions. The moderate accuracy in predicting older age groups can be attributed to the objective difficulty in distinguishing the growth zones in the otoliths of older fish, as well as to the low number of older-fish otoliths in the dataset. This was verified in age-reading workshops [23,24], where age estimations of older fish showed high variability amongst reader experts.

3. Results

To demonstrate the functionality of the platform, 30 images from each fish species were used, along with age estimates from human readers. For instance, the red mullet species was selected, and the images were uploaded on the platform (note: When the file size of the images is large (>2 MB), it is suggested to upload them consecutively through the "Add Files" button (Figure 4)). The platform estimates the fish age for each image separately (Figure 4). Then, age predictions can be extracted into a CSV file using the "Export to CSV" button (default name: "export.csv"; this name can be changed to a different filename) (Figure 4). The content form of the CSV file after being loaded into Excel can be found in the Supplementary Materials (Suppl_red_mullet.xls); the "Refresh page" button allows the user to restart age prediction after completing the limit of 30 images per minute or testing a new species.

Outside of the platform, the predicted age frequency of red mullet was compared with human age estimates (Figure 5a). Although the dataset is small enough to extract general conclusions, we noticed that model predictions tend to underestimate the age-5 class, leading to higher occurrences of ages 3 and 4. This can be partially explained by the fact that during the training of the CNN model for the red mullet, the age-5 class also included ages 5 to 9, due to their small representation in the dataset.

The corresponding results of human against AI age estimates and the exported CSV file for Greenland halibut can be found in Figure 5b and Supplementary Materials (Suppl_Greenland_halibut.xls); for Atlantic salmon (river age), the results can be found in Figure 5c and Supplementary Materials (Suppl_Atlantic_salmon_rive_age.xls); and for Atlantic salmon (river age), the results can be found in Figure 5d and Supplementary Materials (Suppl_Atlantic_salmon_sea_age.xls).

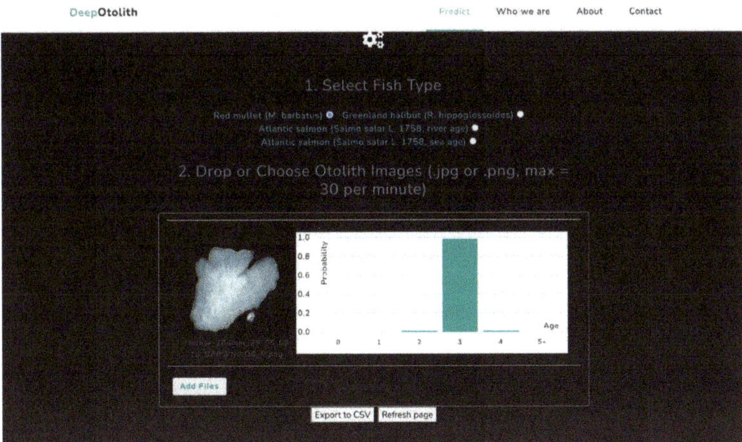

Figure 4. Snapshot of otolith images for the red mullet species, along with predicted probabilities for each age group. The highest probability implies the predicted age for the specific image. The user can upload at once or consecutively 30 images per minute through the "Add Files" button and extract age predictions in a CSV file (default name: "export.csv") using the "Export to CSV" button. The "Refresh page" button allows the user to restart the age prediction process after completing the limit of 30 images per minute or testing a new species.

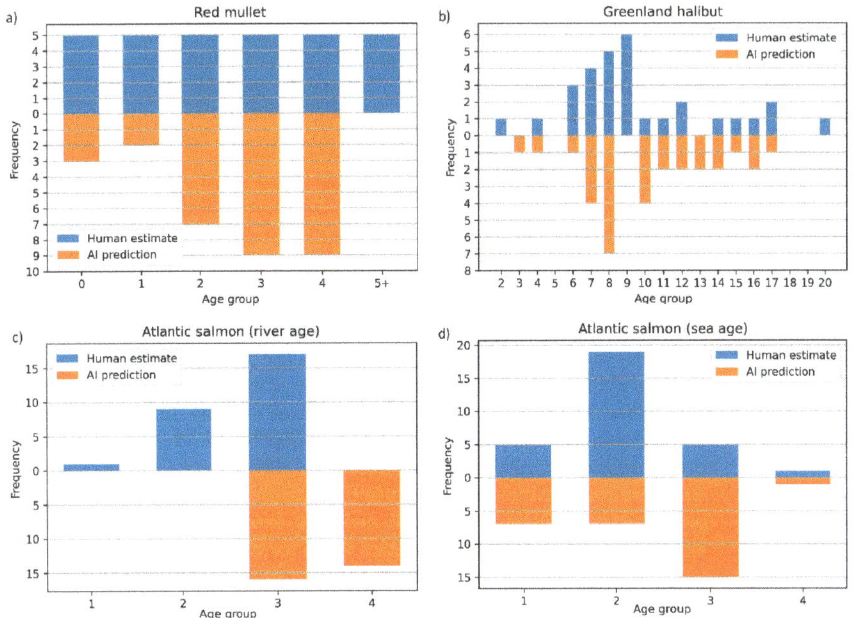

Figure 5. Age frequency distributions of human estimates against AI predictions for (**a**) red mullet (*Mullus barbatus*), (**b**) Greenland halibut (*Reinhardtius hippoglossoides*), (**c**) Atlantic salmon (*Salmo salar*; river age), and (**d**) Atlantic salmon (*Salmo salar*; sea age). For each species, we used a dataset of 30 images.

Institutional Review Board Statement: Ethical review and approval were waived for this study as the otolith images were obtained from dead individuals that were not endangered or protected. We also declare that the uploaded images were neither stored nor sent to third parties.

Data Availability Statement: The data applied to build the model for Greenland halibut are openly available through the Norwegian Marine Data Center under the DOI https://doi.org/10.21335/NMDC1949633559 (accessed on 1 March 2022). Sample otolith/scale images for experimenting with DeepOtolith are available at: https://github.com/dimpolitik/Deep-Otolith (accessed on 1 March 2022).

Acknowledgments: We thank HCMR-NOC for the technical support to release the website, and the Research Council of Norway for funding part of this study. We express our gratitude to the two reviewers for taking the time to review the present work and their useful comments. The authors have declared that no competing interests exist.

Conflicts of Interest: The authors declare no conflict of interest.

Appendix A

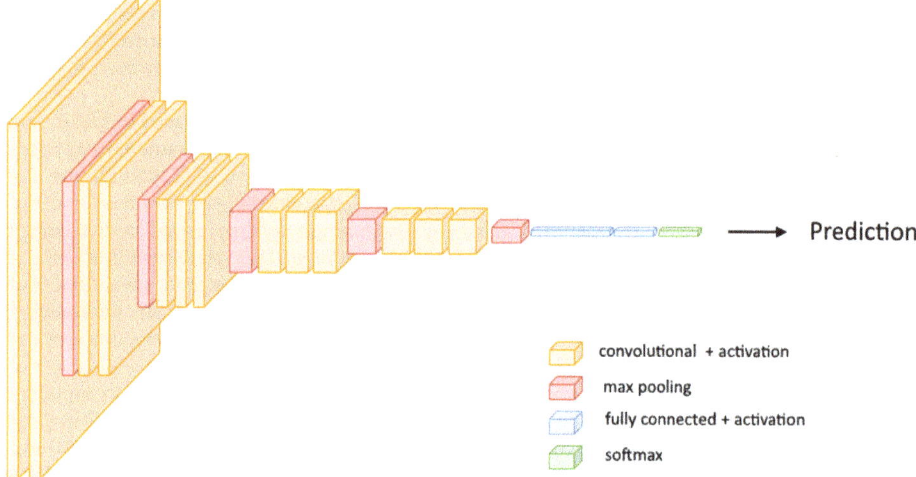

Figure A1. Schematic diagram of a convolutional neural network architecture, composed by a stack of alternately arranged convolutional, pooling, and activation layers, followed by fully connected layers and a softmax function.

References

1. Carbonara, P.; Follesa, M.C. *Handbook on Fish Age Determination: A Mediterranean Experience*; Studies and Reviews; FAO: Rome, Italy, 2019; p. 192.
2. Wang, C.-H.; Benjamin, D.; Walther, B.D.; Gillanders, B.M. Introduction to the 6th International Otolith symposium. *Mar. Freshw. Res.* **2019**, *70*, i. [CrossRef]
3. Williams, T.; Bedford, B.C. The use of otoliths for age determination. In *The Ageing of Fish. Proceedings of the International Symposium*; Bagenal, T.B., Ed.; Allen & Unwin: London, UK, 1974; pp. 114–123.
4. Fisher, M.; Hunter, E. Digital omaging techniques in otolith data capture, analysis and interpretation. *MEPS* **2018**, *598*, 213–231. [CrossRef]
5. Robertson, S.; Morison, A. *Development of an Artificial Neural Network for Automated Age Estimation*; Department of Natural Resources and Environment: Victoria, Australia, 2001; p. 299, ISBN 0-7311-5038-4.
6. Fablet, R.; Le Josse, N. Automated fish age estimation from otolith images using statistical learning. *Fish. Res.* **2005**, *72*, 279–290. [CrossRef]
7. LeCun, Y.; Bengio, Y.; Hinton, G. Deep learning. *Nature* **2015**, *521*, 436–444. [CrossRef] [PubMed]
8. Mohammed, M.; Khan, M.B.; Bashier, E.B.M. *Machine Learning: Algorithms and Applications*; CRC Press: Boca Raton, FL, USA, 2016.
9. Chen, L.; Li, S.; Bai, Q.; Yang, J.; Jiang, S.; Miao, Y. Review of image classification algorithms based on convolutional neural networks. *Remote Sens.* **2021**, *13*, 4712. [CrossRef]

10. Rothe, R.; Timofte, R.; Van Gool, L. Deep expectation of real and apparent age from a single image without facial landmarks. *Int. J. Comput. Vis.* **2018**, *126*, 144–157. [CrossRef]
11. Sabottke, C.F.; Breaux, M.A.; Spieler, B.M. Estimation of age in unidentified patients via chest radiography using convolutional neural network regression. *Emerg. Radiol.* **2020**, *27*, 463–468. [CrossRef] [PubMed]
12. Goodfellow, I.J.; Bengio, Y.; Courville, A. *Deep Learning*; MIT Press: Cambridge, MA, USA, 2015; p. 433.
13. Moen, E.; Handegard, N.O.; Allken, V.; Albert, O.T.; Harbitz, A.; Malde, K. Automatic interpretation of otoliths using deep learning. *PLoS ONE* **2018**, *13*, e0204713. [CrossRef] [PubMed]
14. Ordoñez, A.; Eikvil, L.; Salberg, A.B.; Harbitz, A.; Murray, S.M.; Kampffmeyer, M.C. Explaining decisions of deep neural networks used for fish age prediction. *PLoS ONE* **2020**, *15*, e0235013. [CrossRef] [PubMed]
15. Moore, B.R.; McLaren, J.; Peat, C.; Anjomrouz, M.; Horn, P.L.; Hoyle, S. Feasibility of Automating Otolith Ageing Using CT Scanning and Machine Learning. *New Zealand Fish. Assess. Rep.* **2019**, *58*, 23.
16. Vabø, R.; Moen, E.; Smolinksi, S.; Husebo, A.; Handegard, N.O.; Malde, K. Automatic interpretation of salmon scales using deep learning. *Ecol. Inform.* **2021**, *63*, 101322. [CrossRef]
17. Politikos, D.V.; Petasis, G.; Chatzispyrou, A.; Mytilineou, C.; Anastasopoulou, A. Automating fish age estimation combining otolith images and deep learning: The role of multitask learning. *Fish. Res.* **2021**, *242*, 106033. [CrossRef]
18. Banks, A.; Porcello, E. *Learning React*, 2nd ed.; O'Reilly Media, Inc.: Newton, MA, USA, 2020.
19. Simonyan, K.; Vedaldi, A.; Zisserman, A. Deep inside convolutional networks: Visualising image classification models and saliency maps. *arXiv* **2013**, arXiv:1312.6034.
20. Szegedy, C.; Vanhoucke, V.; Ioffe, S.; Shlens, J.; Wojna, Z. Rethinking the inception architecture for computer vision. In Proceedings of the IEEE Conference on Computer Vision and Pattern Recognition (CVPR), Las Vegas, NV, USA, 27–30 June 2016.
21. Tan, M.; Le, Q. Efficientnet: Rethinking model scaling for convolutional neural networks. In Proceedings of the 36th International Conference on Machine Learning, Beach, CA, USA, 9–15 June 2019.
22. Deng, J.; Dong, W.; Socher, R.; Li, L.-J.; Li, K.; Fei-Fei, L. ImageNet: A large-scale hierarchical image database (In CVPR09). In Proceedings of the 2009 IEEE Conference on Computer Vision and Pattern Recognition, Miami, FL, USA, 20–25 June 2009. [CrossRef]
23. *ICES CM 2012/ACOM:60*; Report of the Workshop on Age Reading of Red Mullet and Striped Red Mullet. ICES: Boulogne-sur-Mer, France, 2012; p. 52.
24. *ICES CM 2017/SSGIEOM:31*; ICES, 2017. Workshop on Ageing Validation Methodology of Mullus Species (WKVALMU). Conversano. ICES: Conversano, Italy, 2017; p. 74.
25. Vitale, F.; Clausen, L.W. *Handbook of Fish Age Estimation Protocols and Validation Methods*; ICES Cooperative Research Report No. 346. Available online: http://doi.org/10.17895/ices.pub.5221(accessed on 1 November 2021).
26. Ordoñez, A.; Eikvil, L.; Salberg, A.-B.; Harbitz, A.; Elvarsson, B.Þ. Automatic fish age determination across different otolith image labs using domain adaptation. *Fishes* **2022**, *7*, 71. [CrossRef]
27. Moore, B.R.; Ámar, Z.T.; Schimel, A.C.G.; Maolagáin, C.Ó.; Hoyle, S.D. *Development of Deep Learning Approaches for Automating Age Estimation of Hoki and Snapper*; New Zealand Fisheries Assessment Report 2021/69. 2021; p. 38. Available online: https://www.researchgate.net/publication/356601174_Development_of_deep_learning_approaches_for_automating_age_estimation_of_hoki_and_snapper_New_Zealand_Fisheries_Assessment_Report_202169 (accessed on 1 November 2021).
28. Salimi, N.; Loh, K.H.; Kaur Dhillon, S.; Chong, V.C. Fully-automated identification of fish species based on otolith contour: Using short-time Fourier transform and discriminant analysis (STFT-DA). *PeerJ* **2016**, *4*, e1664. [CrossRef] [PubMed]
29. Lombarte, A.; Òscar, C.; Parisi-Baradad, V.; Olivella, R.; Piera, J.; García-Ladona, E. A Web-based Environment for Shape Analysis of Fish Otoliths. The AFORO database. *Sci. Mar.* **2006**, *70*, 147–152. [CrossRef]
30. Van Gansbeke, W.; Vandenhende, S.; Georgoulis, S.; Proesmans, M.; Van Gool, L. SCAN: Learning to Classify Images without Labels. *arXiv* **2020**, arXiv:2005.12320.
31. Blount, D.; Gero, S.; Van Oast, J.; Parham, J.; Kingen, C.; Scheiner, B.; Stere, T.; Fisher, M.; Minton, G.; Khan, C.; et al. Flukebook: An open-source ai platform for cetacean photo identification. *Mamm. Biol.* **2021**, *5*. [CrossRef]

Article

The Lifetime Migratory History of Anadromous Brook Trout (*Salvelinus fontinalis*): Insights and Risks from Pesticide-Induced Fish Kills

Scott D. Roloson [1,2,*], Kyle M. Knysh [2], Sean J. Landsman [3], Travis L. James [4], Brendan J. Hicks [5] and Michael R. van den Heuvel [2]

1 Salmon and Diadromous Fish Section, Fisheries and Oceans Canada, 165 John Yeo Drive, Charlottetown, PE C1A 7M8, Canada
2 Department of Biology, Canadian Rivers Institute, University of Prince Edward Island, 550 University Avenue, Charlottetown, PE C1A 4P3, Canada; kknysh@upei.ca (K.M.K.); mheuvel@upei.ca (M.R.v.d.H.)
3 Institute of Environmental and Interdisciplinary Science, Carleton University, 1125 Colonel By Drive, Ottawa, ON K1S 5B6, Canada; landsman.sean@gmail.com
4 Parks Canada, Resource Conservation, Prince Edward Island National Park, 40 Dalvay Cresent, York, PE C0A 1P0, Canada; travis.jamespca@gmail.com
5 Division of Health, Engineering, Computing and Science, School of Science, Environmental Research Institute, The University of Waikato, Hamilton 3240, New Zealand; hicksbj@waikato.ac.nz
* Correspondence: scott.roloson@dfo-mpo.gc.ca

Abstract: Brook trout populations in Prince Edward Island, Canada, have experienced over 50 pesticide-related fish kills since the 1960s. Life history evaluation of large sea-run brook trout recovered following two fish kill events was compared with a reference river using strontium:calcium otolith microchemistry. This study examined the dual hypotheses that anadromous brook trout are more likely to arise from sea-run mothers, and that freshwater entry timing makes them vulnerable to pesticide-induced fish kills. A total 89% of the fish exhibited an anadromous life history, and 77% of these were offspring of anadromous mothers, suggesting that anadromy is dominant in progeny of sea-run mothers. This study adds to our understanding of the maternal inheritance of anadromy in sea-run brook trout populations. Additionally, freshwater entry precedes the majority of fish kill events, illustrating that the overlap between migration and pesticide runoff contributes to the cumulative population risks to sea-run brook trout.

Keywords: brook trout; fish kill; anadromy; otolith microchemistry

Citation: Roloson, S.D.; Knysh, K.M.; Landsman, S.J.; James, T.L.; Hicks, B.J.; van den Heuvel, M.R. The Lifetime Migratory History of Anadromous Brook Trout (*Salvelinus fontinalis*): Insights and Risks from Pesticide-Induced Fish Kills. *Fishes* **2022**, *7*, 109. https://doi.org/10.3390/fishes7030109

Academic Editor: Josipa Ferri

Received: 12 April 2022
Accepted: 10 May 2022
Published: 11 May 2022

1. Introduction

The health of many aquatic ecosystems has been compromised by the cumulative impacts of human activities. Biodiversity in freshwater is declining faster than in terrestrial or marine environments [1]. While estuaries, which link fresh and marine waters, experience reduced biodiversity in systems affected by human development [2]. Diadromous fish species move between fresh and marine waters and factors that compromise the timing or restrict the spatial extent of movements pose a threat to migration [3–5]. Due to the multiple and cumulative threats to the habitat of diadromous fishes, many populations have been subject to pronounced declines and many species have experienced population extirpations within their historic range [6,7]. As human activities, including climate change, continue to infringe upon these environments, coherent conservation strategies are needed to protect diadromous species and the ecosystems they occupy [8,9].

Brook trout (*Salvelinus fontinalis*) are native to eastern North America and coastal populations are known to adopt a mixture of resident and sea-run life histories (i.e., facultative anadromy) [3,10]. Due to the spatial and temporal complexity of the habitat of anadromous brook trout, populations have been subject to declines across much of their native range, particularly south of the Bay of Fundy [11,12]. The sea-run life history strategy is thought to have an adaptive basis as anadromous mothers produce more eggs with enhanced fitness-related traits such as a shorter developmental time, faster juvenile growth rates, and higher survival [13]. Moreover, the progeny of sea-run females are the dominant proportion of young-of-year fish in areas where anadromous and resident spawners overlap [14]. Linking the contribution of anadromous parents to subsequent generations has primarily been explored in young-of-year offspring via stable-isotope analysis [14,15]. However, the marine isotopic signatures from anadromous mothers begin to change when juveniles commence exogenous feeding, making it difficult to detect the contribution of anadromous parents beyond early life stages [14,15]. Genetic differentiation has been used as a means of evaluating anadromous contributions [13] and sympatric sea-run and resident brook trout have been found to be genetically divergent, which might occur if sea-run progeny are themselves more likely to exhibit this phenotype [16]. In one brook trout study, Theriault et al. [17] used genetic methods to infer the heritability of the anadromous tactic as sea-run parentage was positively correlated with ultimate size at age 1. However, there are still knowledge gaps on the overall contribution of anadromous parents to subsequent generations and the perpetuation of the migratory life history.

Prince Edward Island (PEI), in eastern Canada, is well known for its anadromous brook trout populations, but the province also has a history of pesticide-related fish kills, where brook trout are the dominant species affected [18]. Since the 1960s, there have been at least 50 fish kills related to pesticide runoff. One study examining changes in the fish community after a fish kill found that brook trout suffered higher mortality than non-native rainbow trout (*Oncorhynchus mykiss*) and that the effects were strongest on young-of-year fish, but all sizes were affected [19]. Atlantic salmon (*Salmo salar*) were formerly widespread in the province, but overharvest and habitat deterioration have reduced their populations, making them less common in pesticide-related fish kills [18]. Overall, there is a lack of understanding of how these events influence the distribution, abundance and population demographics in local species assemblages.

Otoliths, or ear stones, incorporate certain environmental constituents in proportion to ambient concentration, and chemical analysis can be used to reconstruct chronological life history patterns [20]. In salmonid fishes, and some other species, elevated strontium concentrations are associated with marine occupation and prior to spawning, maternal signatures can be passed to eggs, and by examining otolith primordia of offspring, the maternal life history of offspring can be elucidated [21,22]. To date, opportunities to study the anadromous life history of large, sexually mature brook trout through otolith chemistry have been rare because otolith sampling is lethal and sacrificing large numbers of sexually mature fish would be ethically questionable given the significant reproductive contribution of these individuals to subsequent generations [13,23].

The overall aim of this study was to explore why brook trout are predisposed to the risk of pesticide-related fish kills and to use otolith microchemistry to explore the demographics of anadromous life history in coastal brook trout populations. Firstly, we examined how the timing of freshwater entry of anadromous brook trout relates to the timing of pesticide fish kills by analyzing 11 years of counting-fence data from an anadromous population on Montague River. It was predicted that freshwater entry would precede the occurrence of pesticide runoff events. Additionally, otolith microchemistry was used to establish the anadromous life history patterns of brook trout collected following fish kills in comparison to the signatures of samples taken from the Montague River counting fence. The prediction was that anadromous individuals would be more likely to arise from anadromous mothers, and that anadromous individuals would conduct repeated migrations once the anadromous life history strategy was adopted.

2. Materials and Methods

2.1. Study Sites

PEI has an area of 5560 km^2 and is characterized by many short watersheds fed predominantly by groundwater. Temperatures in most rivers are suitable for salmonids, remaining below 18 °C throughout the year [24–26]. Following Holocene sea level rise, many estuaries are considered to be drowned river valleys, and these relatively large and productive estuaries are highly conducive for anadromous fishes [27,28]. Agricultural land use influences fish populations in the province as approximately 50% of the province is used for agricultural production [29,30]. As a consequence of the highly agricultural land-use sedimentation, nitrogen loading from fertilizers and pesticide runoff events are common phenomena in PEI waterways [31–33]. The present study took place in three PEI watersheds, Mill River (50.1 km^2), Montague River (65.4 km^2), and Trout River (51.9 km^2) (Figure 1).

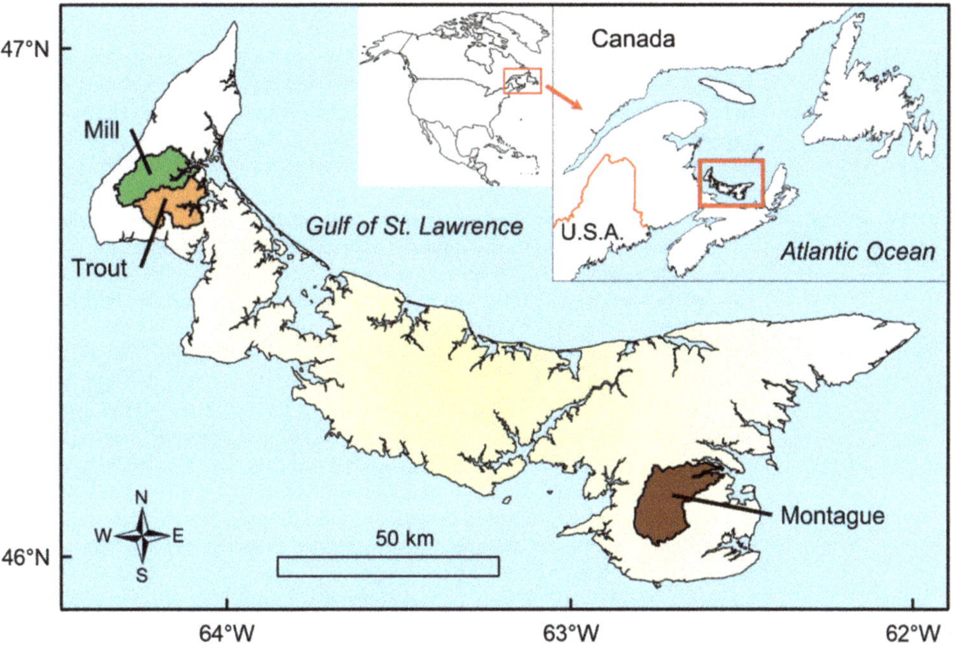

Figure 1. Map showing the location of Prince Edward Island, Canada, and the three study watersheds.

2.2. Fish Community

The freshwater fish community on PEI is generally dominated by salmonids [29]. Brook trout are ubiquitous and the primary sportfish in the province, especially anadromous brook trout which have strong cultural and socioeconomic value [29]. Current land-use patterns, and historic activities play a dominant role in structuring current salmonid communities in PEI watersheds [26]. Atlantic salmon were widespread prior to human colonization but are now lost from over 60% of the watersheds the species once occupied and now populations are stronger in forested watersheds with abundant cobble substrate [34,35]. Non-native rainbow trout are present in approximately 30 rivers across the province. Rainbow trout tend to thrive in higher sloped reaches and in watersheds with higher agricultural land use, while brook trout juveniles tend to be dominant in headwater reaches [25]. Other common species in PEI, which have been recovered following pesticide fish kills include American eel (*Anguilla rostrata*), white perch (*Morone americana*) and several stickleback species (Gasterosteus spp.) [18]. Other diadromous species that are seasonally present in PEI waterways during spawning migrations including rainbow smelt (*Osmerus mordax*), blueback herring (*Alosa aestivalis*), and alewife (*Alosa pseudoharengus*) [18].

2.3. Timing of Brook Trout Movements and Fish Kills

Given the paucity of data on run timing for anadromous brook trout, a historic dataset from a fish trap at Knox's Dam, Montague River was evaluated to establish the timing of freshwater return of anadromous brook trout. The fish trap was located in a technical concrete pool-and-weir-style fishway located at the head of tide. Fish moving upstream through the trap were considered to be moving from saltwater into freshwater. Historic data from the period 1982–1993 were provided by D. Cairns (Department of Fisheries and Oceans) and the counting fence was operated by the authors from 2011 to 2013. Fish were enumerated and individually measured (fork length, mm) and released upstream of the barrier. In total, 11 years of brook trout movement and abundance were evaluated including the total number of brook trout moving upstream each year, including 1982–1985, 1989–1990, 1991, 1993, and 2011–2013. Entry date was determined by finding the day of the year (Julian day) at which 50% of brook trout tend to be entering the river had arrived. Large female brook trout tend to be among the first to enter freshwater (3; S. Roloson, see results); therefore, the 50% cutoff includes the majority of the large breeding females. This date was used to compare the dates of fish kill events, in order to determine potential overlap between the timing of fish entry into freshwater and the incidence of pesticide runoff.

Data on fish kill timing were accessed from public records available online [18]. Of the 57 fish kills with information available from 1962 to 2019, 55 of these events list a specific date and two others list "summer" without a specific date. The Julian day of the fish kills with a specific date was used to enable comparisons of the 'day of the year' between subsequent years and with brook trout movement timing.

2.4. Fish Sampling and Otolith Microchemistry

Brook trout were opportunistically collected from two rivers following pesticide runoff events. The first fish kill sampled was on Trout River, which experienced a pesticide runoff event on 5 July 2012, and a subsample of 15 trout from 21.4 to 36.9 cm were collected on 6 July 2012 for otolith sampling. The Mill River fish kill occurred on 28 July 2013. Specimens were collected by local watershed officials who visually observed deceased trout on streambanks and on the streambed. Specimens were transported back to the lab for measuring fork length (mm) and weight (g) and extraction of saggital otoliths. Following removal, otoliths were rinsed with deionized water, left to dry for 24 h and stored in acid-washed 50 mL glass vials.

The third brook trout population was sampled from Montague River, where a counting fence captured fish as they ascended into freshwater from the estuary, so this river served as a reference as these fish were known to exhibit an anadromous life history. A subset of 31 fish from 15.6–38.1 cm were sampled for otolith microchemistry between 6 June and 29 June 2012. While fish at Montague River were sampled randomly during freshwater entry, samples at fish kills were opportunistically collected, and smaller fish could have been underrepresented due to the difficulty in finding and recovering smaller individuals.

Otoliths were processed at the Mass Spectroscopy Suite University of the Waikato in Hamilton, New Zealand. Prior to sampling, otoliths were stored in acid-washed glass vials and handled with non-metallic utensils to avoid trace metal contamination. Otoliths were mounted sulcus side up and sequentially polished with 1000, 2000, and 4000 grain wetted silicone carbide waterproof sandpaper until primordial nuclei were exposed. Decontamination procedures are outlined in Roloson et al. 2019 [36]. Once polished, otoliths were mounted on a single glass slide in batches [15–20] to allow for more efficient laser operation. Otoliths were analyzed with a Perkin Elmer Elan SCIEX DRCII laser ablation inductively coupled mass spectrometer (LAICP–MS) with a New Wave Research Nd:YAG

213 nm wavelength laser. To account for instrument drift, National Institute of Standards and Technology (NIST) 612 standard reference material was ablated prior, during and following sample ablation and for 1 min of every 15 min of laser operation time. The otolith core was identified using a transmitted light microscope and life history transects were oriented from the otolith primordia to the outer edge. The laser was set to a 20 μm beam diameter width at a frequency of 20 Hz and a travelling speed of 10 $m \cdot s^{-1}$. Brook trout otoliths were analyzed with the same protocols and run during the same sample runs as anadromous rainbow trout from PEI [36].

2.5. Statistical Analysis

Across all years, brook trout fresh water entry timing was summarized as the mean percent of total run entering freshwater each day. The mean daily proportion of entry was fitted using a Gaussian function: daily entry = $a \cdot \exp(-(\text{Julian Day} - b)^2/(2 \cdot c^2)$, where a represents the maximum daily entry proportion, b the day on which the maximum occurred, and c the variance around the day. Fish kill date was represented as a cumulative proportion of fish kills for each Julian day. The median fish kill day was determined by fitting a two parameter logistic regression to the data of the form: cumulative proportion of fish kills = $100/(1 + 10^{(\log 10(\text{Median Julian Day} - \text{Julian Day}) \cdot (\text{Hillslope})})$, where hillslope is the shape parameter and median Julian day the day when half of the fish kills occurred. All summary statistics and curve fits were performed in Statistica v. 13.5.

Assessment of life history signatures was based upon previously established methods [37–39]. To assess natal core signatures, we took the mean otolith core Sr:Ca values from the first 25 sequential data points located in first 200 μm of the transect [37,40]. Rather than using a threshold to designate freshwater or marine core signatures, we used the univariate *K*-means clustering in the "factoextra" package in R, which uses an algorithm to separate core signatures into two groups based on differences in the data [37,40]. Differences in anadromous migratory history are much more explicit upon visual examination of line scans as it is common for migratory individuals to have Sr:Ca peaks reaching 4–7 mmol/mol. Therefore, to designate migratory life history, we followed Austin et al [38], who studied the closely related bull trout (*Salvelinus confluentus*). In that study, individuals with a Sr:Ca ratio that remained below 1 mmol/mol Sr:Ca were considered freshwater residents, and those with a ratio over 1 mmol/mol Sr:Ca were considered anadromous.

3. Results

3.1. Timing of Brook Trout Movement into Fresh Water

The timing of brook trout movements was determined from 11 years of abundance counts at Knox's Dam (Montague River), which occurred intermittently between 1982 and 2013. Overall, the mean ± SD number of brook trout was 1268 ± 1020, with a high of 3069 brook trout counted in 1988 and a low of 193 in 1983. In the first time period, from 1982 to 1985, the mean annual brook trout count was 398 ± 234; between 1988 and 1993, the mean count was 2236 ± 515; and from 2011 to 2013, the mean count was 1138 ± 874. The timing of 50% entry spanned over a 43 day range from Julian Day 170 (18 June 1984) to Day 214 (2 August 1993). The median Julian Day of entry was 194 ± 13 SD. Peak freshwater entry is presented over the 11 year period, summarized in Figure 2, showing the daily run proportion entering freshwater each day. Figure 3 depicts the size distribution of brook trout entering freshwater when the authors operated the counting fence from 2011–2013 and it illustrates that larger individuals tend to be amongst the first to enter freshwater.

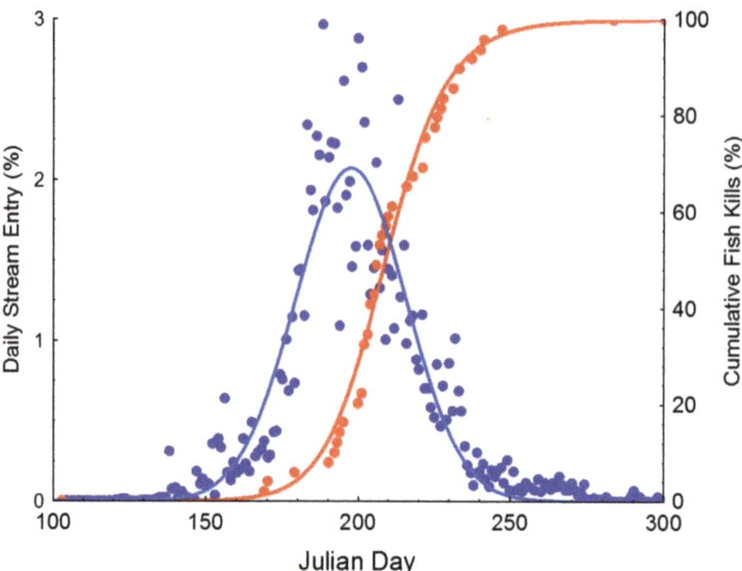

Figure 2. Timing of freshwater entry based upon 11 years of abundance data from Montague River (blue) and the date of occurrence of pesticide fish kill events (red).

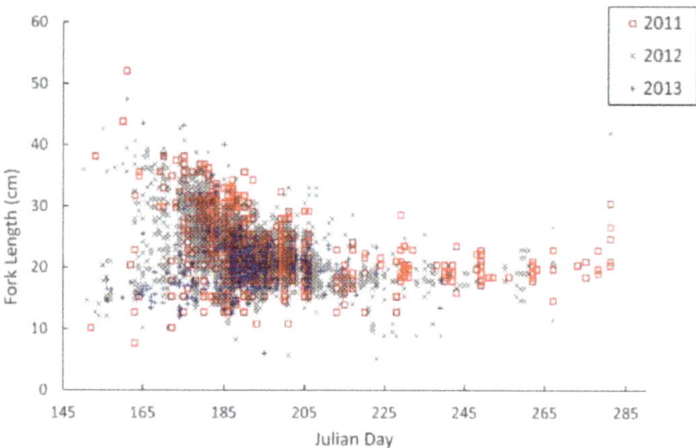

Figure 3. Size vs. timing of brook trout (n = 853) travelling upstream through the counting fence at Knox's Dam on Montague River in 2011.

3.2. Timing and Occurrence of Pesticide-Related Fish Kills

The timing of runoff events closely coincided with the timing of brook trout movements into freshwater. Mean ± SD Julian Day of all fish kill events was 212 ± 20. Fish kill dates ranged from Julian Day 169 (18 June 1977) to Day 283 (10 October 2017), with 29 occurring in July and 21 in August. The two rivers sampled following fish kills, Mill River and Trout River have numerous historic records of fish kills. Trout River has had six fish kills, which is the most of any watershed in the province, including fish kills in three consecutive years (2011, 2012, and 2013). In the 2011 event on Trout River (23 July 2011), 356 brook trout were collected with a mean ± SD size of 30.4 ± 9.5 cm. Following another runoff event on 5 July 2012, 1036 brook trout were collected with a mean size of 15 ± 6 cm SD. On 25 July 2013, 102 brook trout were collected with a mean ± SD size of

19.6 ± 8.6 cm. In the Mill River fish kill, which occurred on 27 July 2013, 71 fish were collected with a mean ± SD size of 33.8 ± 9.8 cm. Unfortunately, population surveys are not available following fish kills, which could help evaluate the percentage of the population that was affected by the runoff event. The extensive loss of brook trout following pesticide runoff events is reflected in publicly available summary reports for six recent fish kills, where brook trout were the dominant species found, constituting between 71% and 99% of fish collected following runoff events [18].

3.3. Otolith Chemistry and Anadromous Life History

From the microchemical evaluation of maternal phenotype a bimodal distribution of otolith cores was evident. Overall the *K*-means clustering algorithm estimated a cutoff value of 0.87 mmol/mol Sr:Ca, whereby individuals with core values above were designated as being from anadromous mothers, and those below were considered to be from freshwater mothers. Those identified to be descendants of anadromous mothers (n = 48), had a mean value of 1.26 mmol/mol Sr:Ca (±0.27 SD), and those from freshwater mothers (n = 16) had a mean Sr:Ca value of 0.47 mmol/mol (±0.23 SD), (Table 1).

Table 1. Summary statistics and phenotypes of 64 brook trout sampled for otolith microchemistry where individual lengths were measured in fork length (FL, cm). Phenotypes are designated from Sr:Ca ratios as anadromous (An) or freshwater (Fw) for maternal life history (<200 μm) and migratory life history (>500 μm).

Site	Mean FL (SE, n) (cm)	Maternal Origin/Adult Phenotype			
		An/Fw	An/An	Fw/Fw	Fw/An
Mill	43.6 (0.8, 18)	1	16	0	1
Montague	28.5 (1.1, 31)	2	17	3	9
Trout	29.0 (1.4, 15)	1	11	0	3

When lifetime migratory history patterns were considered, 89% of fish (57/64) exhibited an anadromous life history (Table 1). Of these, 77% (44/57) were from anadromous mothers and 23% (13/57) had freshwater mothers. Eleven percent (7/64) of individuals expressed a freshwater life history strategy, four of these arose from anadromous mothers and three from freshwater mothers.

The occurrence of repeated anadromous migrations was apparent at all sites as explicit annual Sr:Ca peaks were common (Figure 4). On Montague River, of the 31 individuals collected, 13 took one marine migration, 8 made two marine migrations and 4 made at least three. On Mill River, of the 18 individuals collected, 7 made two migrations, and another 7 showed evidence of three migrations; a single fish from the Mill river potentially made four marine migrations. On Trout River, of the 15 individuals sampled, 6 made a single migration, 5 made two migrations, and 3 made at least three migrations to the marine environment. At each site, there were individuals that exhibited non-typical Sr:Ca peaks suggestive of longer exposure to elevated salinity, or changing migratory patterns over the lifetime of the individual. On the Mill River, the representative non-typical migrant appeared to have multiple discontinuous migrations to higher-salinity environments, which may reflect different seasonal habitat use than typical individuals. Other atypical migrants showed increasing or decreasing Sr:Ca ratios over time, suggesting changes in migratory patterns throughout the individual's lifetime.

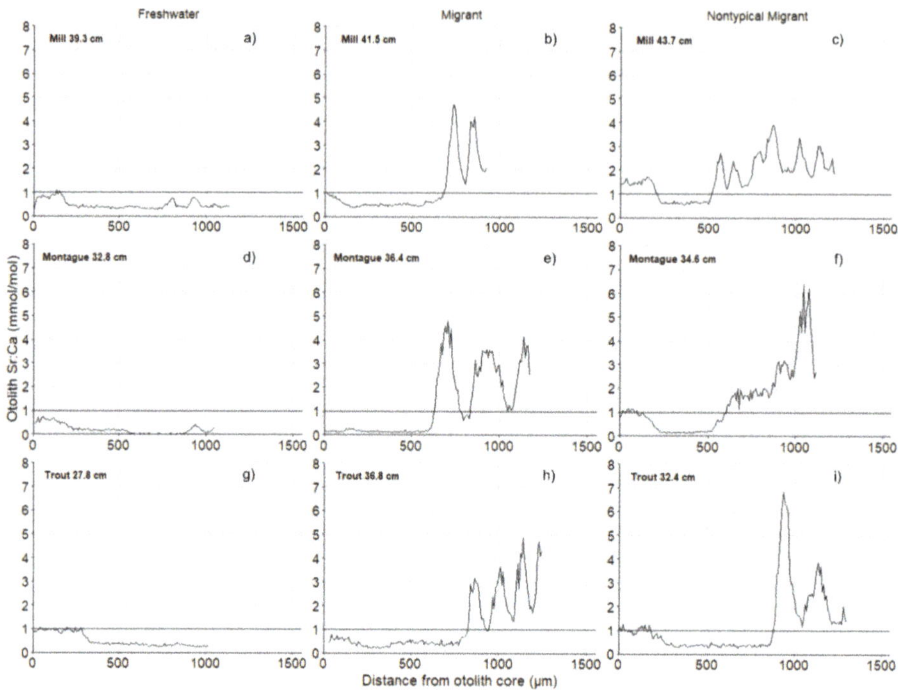

Figure 4. Representative otolith microchemistry line scans for freshwater (**a**,**d**,**g**), migrant (**b**,**e**,**h**) and nontypical migrant (**c**,**f**,**i**) life histories from Mill, Montague and Trout River. Horizontal line shows threshold for determination of anadromous migratory strategy. Size of fish reported in fork length (cm).

4. Discussion

This study illustrates how freshwater entry for anadromous brook trout precedes the occurrence of fish kills and may explain why brook trout represent the most numerous species recovered from fish kills on PEI. Opportunistic life history evaluation via otolith microchemistry revealed the prevalence of anadromy in PEI brook trout populations killed by pesticides. Prior studies have attempted to trace the heritability and fitness benefits of an anadromous life history, but the connections have been tentative [14,15]. To date, previous studies have not examined the entire life history and parental contribution in anadromous brook trout. Given that the majority of individuals had anadromous mothers and exhibited an anadromous life history, this study highlights the significance of anadromy in coastal brook trout populations.

Otolith microchemistry has not been previously used to examine the inheritance of anadromy or the lifetime anadromous history in brook trout. Early attempts to demonstrate the maternal influence using stable isotopes were tenuous because those isotopic signatures do not persist long beyond exogenous feeding [14,15]. Otolith microchemistry showed that sea-run migration is often repeated throughout the individual's lifespan and often arises from anadromous maternal origin, making otoliths a powerful tool for the study of anadromous life history [37–39]. These results using otolith microchemistry have the additional advantage of being able to show repeated patterns of anadromy over the life of the fish and many individuals undertook annual migrations to sea, once the strategy was adopted. Atypical migrant patterns could be explained by other studies that found anadromous brook trout leaving freshwater in the fall after spawning and overwintering under ice and in some cases returning to lower-salinity habitats during environmental events such as extreme cold [41,42]. Another similar study on PEI used otolith chemistry in conjunction with acoustic tracking to study non-native rainbow trout and showed that anadromous individuals regularly travelled between fresh and saltwater, possibly reflecting the presence of environmental constraints to habitat occupation [36].

The factors that influence the timing and demography of the freshwater return of anadromous brook trout are poorly understood despite decades of study. Previous studies, as well as the current study, illustrate a trend for large-bodied female brook trout to be the first individuals to return to freshwater; these females may dominate recruitment based upon the production of a greater number of eggs that tend to be larger and exhibit faster growth [17,43]. One reason that is speculated for freshwater return in brook trout, particularly those that return in early summer well in advance of fall spawning, is an obligate period of freshwater rearing for egg maturation [44,45]. Other drivers of entry into freshwater could be proximate environmental variables such as temperature, photoperiod, or dissolved oxygen that may differentially affect brook trout physiology based upon individual size. In their comprehensive review of anadromy in brook trout, Curry et al. [3] suggested that anadromous brook trout returned earlier in rivers where summer temperatures are high, and that these populations migrate to headwaters in order to find thermal refugia [46]. Thus, there may be river-specific attributes which affect the spatial distribution of sea-run brook trout that have returned to fresh water but it is unclear how this affects the potential susceptibility to pesticide runoff events.

While anadromy in brook trout can be seen to be maternally heritable, this does not necessarily imply clear genetic linkages. A study on brook trout microsatellite loci revealed that while there is differentiation between anadromous and resident populations, gene flow is likely mediated by freshwater resident males spawning with sea-run females, and so anadromy may not be a genetically mediated trait [17].That study found, sea-run parentage was correlated with larger juvenile size at age 1, suggesting that size, over genetics, could be a determining factor in the decision to become anadromous. Larger females will have larger and more numerous eggs, which are more likely to produce large juveniles. Despite the strong maternal–progeny relationship for anadromy, small numbers of brook trout with freshwaters mothers were observed to become anadromous (and vice versa) in all watersheds studied. Given that the species commonly persists as a mixture of resident and anadromous forms, multiple strategies are likely important to mitigate catastrophic disturbance events such as fish kills. The long-term effects of these events are unknown and further study is warranted.

While pesticide-related fish kills have obvious negative impacts on fish populations, other attributes of agriculture may lead to increases in density. Curry et al. [15] found that the density of juvenile brook trout on PEI was not related to the contribution of anadromous parents at all, possibly because nitrate enrichment is driving the abundance of juveniles more than the contribution of anadromous mothers. This is supported by Gormley et al. [19], who found that juvenile brook trout density was positively correlated with the amount of land in potato agriculture in the watershed area. Previous studies have found a positive association between experimentally elevated nutrient levels, namely nitrate and juvenile salmonid growth [47]. On PEI, elevated nitrate concentrations are associated with increased invertebrate abundance of certain taxa [48], as well as increases in brook trout density [19], suggesting that juvenile brook trout are capable of capitalizing on the added resources associated with nutrient enrichment. Enhanced growth could thereby contribute to the continuance of anadromy provided density does not increase substantially, which would limit growth. Ultimately, there is a poor understanding of how nutrient enrichment, sedimentation, and pesticide-induced fish kills influence populations of brook trout in PEI rivers.

Pesticide-induced fish kills are always preceded by rainfall and a recent pesticide application. There are a number of interventions that can reduce this risk including increasing buffers along streams and reducing or eliminating the use of particular compounds that pose a high risk. Watershed models demonstrate that increases in buffer width from the current 15 m can achieve a significant reduction in sediment load to PEI streams [49];

however, the impact on pesticide runoff during rain events is less clear. A study specifically examining pesticide reduction by buffer strips on PEI showed that pesticides could be reduced further if buffers were increased to 30 m [50].

Regulating specific high-risk compounds can also reduce risk to fishes. Between 1977 and 2002, the insecticide azinphos-methyl was responsible for just over half of documented fish kills on PEI [51]. While azinphos-methyl was not banned in Canada, it is in the USA and much of the EU. PEI introduced restrictions on its use near waterways, and there has not been a fish kill attributed to azinphos-methyl since 2002. Measured residues of this compound have since declined across the region [33]. However, from 2002 onward, the fungicide chlorothalonil is the suspected toxic substance in at least 70% of fish kills [51]. As of the last available pesticide sales data for PEI between 2014 and 2016, chlorothalonil (125–140 tonnes) was second only to the fungicide mancozeb (303–330 tonnes) in terms of the quantity of active ingredient sold [52]. Thus, reductions in the use of chlorothalonil, which ranked the highest in a release-weighted risk assessment [53], would likely have the highest probability of protecting brook trout populations. The Canadian Pest Management Regulatory Authority has recently re-evaluated chlorothalonil and changed its regulations to allow a maximum of three applications per year, though the implications of this on chlorothalonil usage are still unclear [54].

5. Conclusions

This study demonstrates the risk to brook trout populations associated with timing of freshwater entry and the temporal overlap with pesticide runoff events. Additional anthropogenic risk factors, primarily associated with agriculture, are linked to sedimentation in streams [31,55,56], which can affect critical life cycle stages such as egg incubation. The general population status of brook trout on PEI is presumed to be healthy, but angler capture of brook trout on PEI was observed to have declined by 33% between 1973 and 1994 [57]. However, relatively little monitoring of these populations makes their status uncertain. The threats to brook trout in this region are consistent with observations elsewhere in their range that land-use change is as significant factor in the decline of the species [58]. Given the manner in which sea-run brook trout life history and anthropogenic stress coincide, enhanced monitoring of sea-run brook trout populations and exploration of alternative pest management strategies in agriculture are warranted.

Author Contributions: Conceptualization, S.D.R., methodology, S.D.R., B.J.H. and M.R.v.d.H.; data collection T.L.J., S.D.R. and K.M.K.; formal analysis, S.D.R., K.M.K., S.J.L. and M.R.v.d.H.; resources, B.J.H. and M.R.v.d.H.; writing—original draft preparation, S.D.R., K.M.K. and M.R.v.d.H.; writing—review and editing, S.D.R., K.M.K., S.J.L., B.J.H. and M.R.v.d.H.; supervision, M.R.v.d.H. funding acquisition, S.D.R. and M.R.v.d.H. All authors have read and agreed to the published version of the manuscript.

Funding: This research was funded in part by an NSERC Industrial Postgraduate Scholarship to S.D.R., a University of Prince Edward Island Internal Research Grant and by the Prince Edward Island Wildlife Conservation Fund.

Institutional Review Board Statement: All fish were handled in accordance with approved University of Prince Edward Island animal care protocols. All fish collection activities complied with DFO Gulf Region License to Fish for Scientific Purposes.

Acknowledgments: We would like to thank all of those who assisted in the collection of field data especially M. Coffin and many other UPEI students as well as provincial government officials, R. MacFarlane and L. Jones. This manuscript is dedicated to the memory Daryl Guignion whose 40+ year career at UPEI had a profound influence on those involved in watershed ecology on PEI, including the authors of this study.

Conflicts of Interest: The authors declare no conflict of interest.

References

1. Su, G.; Logez, M.; Xu, J.; Tao, S.; Villéger, S.; Brosse, S. Human impacts on global freshwater fish biodiversity. *Science* **2021**, *371*, 835–838. [CrossRef] [PubMed]
2. Ahn, H.; Kume, M.; Terashima, Y.; Ye, F.; Kameyama, S.; Miya, M.; Yamashita, Y.; Kasai, A. Evaluation of fish biodiversity in estuaries using environmental DNA metabarcoding. *PLoS ONE* **2020**, *15*, e0231127. [CrossRef] [PubMed]
3. Curry, A.A.; Bernatchez, L.; Whoriskey, F.; Audet, C. The origins and persistence of anadromy in brook charr. *Rev. Fish Biol. Fish.* **2010**, *20*, 557–570. [CrossRef]
4. Goetz, F.A.; Beamer, E.; Connor, E.J.; Jeanes, E.; Kinsel, C.; Chamberlin, J.W.; Morello, C.; Quinn, T.P. The timing of anadromous bull trout migrations in estuarine and marine waters of Puget Sound, Washington. *Environ. Biol. Fishes* **2021**, *104*, 1073–1088. [CrossRef]
5. Roloson, S.D.; Coffin, M.R.S.; Knysh, K.M.; van den Heuvel, M.R. Movement of non-native rainbow trout in an estuary with periodic summer hypoxia. *Hydrobiologia* **2021**, *848*, 4001–4016. [CrossRef]
6. Limburg, K.E.; Waldman, J.R. Dramatic declines in north Atlantic diadromous fishes. *Bioscience* **2009**, *59*, 955–965. [CrossRef]
7. Merg, M.L.; Dézerald, O.; Kreutzenberger, K.; Demski, S.; Reyjol, Y.; Usseglio-Polatera, P.; Belliard, J. Modeling diadromous fish loss from historical data: Identification of anthropogenic drivers and testing of mitigation scenarios. *PLoS ONE* **2020**, *15*, e0236575. [CrossRef]
8. Charron, C.; St-Hilaire, A.; Ouarda, T.B.M.J.; van den Heuvel, M.R. Water temperature and hydrological modelling in the context of environmental flows and future climate change: Case study of the wilmot river (Canada). *Water* **2021**, *13*, 2101. [CrossRef]
9. Verhelst, P.; Reubens, J.; Buysse, D.; Goethals, P.; Van Wichelen, J.; Moens, T. Toward a roadmap for diadromous fish conservation: The Big Five considerations. *Front. Ecol. Environ.* **2021**, *19*, 396–403. [CrossRef]
10. MacCrimmon, H.; Campbell, S. World Distribution of Brook Trout, (*Salvelinus fontinalis*). *J. Fish. Res. Board Can.* **1969**, *26*, 1699–1725. [CrossRef]
11. Eastern Brook Trout Joint Venture. *Range-Wide Assessment of Brook Trout at the Catchment Scale: A Summary of Findings*; Eastern Brook Trout Joint Venture. 2016. Available online: https://www.easternbrooktrout.org (accessed on 10 December 2021).
12. Snook, E.L.; Letcher, B.H.; Dubreuil, T.L.; Zydlewski, J.; O'Donnell, M.J.; Whiteley, A.R.; Hurley, S.T.; Danylchuk, A.J. Movement patterns of Brook Trout in a restored coastal stream system in southern Massachusetts. *Ecol. Freshw. Fish* **2016**, *25*, 360–375. [CrossRef]
13. Perry, G.M.L.; Audet, C.; Bernatchez, L. Maternal genetic effects on adaptive divergence between anadromous and resident brook charr during early life history. *J. Evol. Biol.* **2005**, *18*, 1348–1361. [CrossRef] [PubMed]
14. Jardine, T.D.; Chernoff, E.; Curry, R.A. Maternal transfer of carbon and nitrogen to progeny of sea-run and resident brook trout (*Salvelinus fontinalis*). *Can. J. Fish. Aquat. Sci.* **2008**, *65*, 2201–2210. [CrossRef]
15. Curry, R.A. Assessing the reproductive contributions of sympatric anadromous and freshwater-resident brook trout. *J. Fish Biol.* **2005**, *66*, 741–757. [CrossRef]
16. Fraser, D.J.; Bernatchez, L. Allopatric origins of sympatric brook charr populations: Colonization history and admixture. *Mol. Ecol.* **1995**, *14*, 1497–1509. [CrossRef] [PubMed]
17. Thériault, V.; Garant, D.; Bernatchez, L.; Dodson, J.J. Heritability of life-history tactics and genetic correlation with body size in a natural population of brook charr (*Salvelinus fontinalis*). *J. Evol. Biol.* **2007**, *20*, 2266–2277. [CrossRef]
18. PEI Department of Environment, Energy and Climate Action. Fish Kill Information and Statistics. 2020. Available online: www.princeedwardisland.ca/en/information/environment-water-and-climate-change/fish-kill-information-and-statistics (accessed on 21 December 2021).
19. Gormley, K.L.; Teather, K.L.; Guignion, D.L. Changes in salmonid communities associated with pesticide runoff events. *Ecotoxicology* **2005**, *14*, 671–685. [CrossRef]
20. Walther, B.D.; Limburg, K.E. The use of otolith chemistry to characterize diadromous migrations. *J. Fish Biol.* **2012**, *81*, 796–825. [CrossRef]
21. Zimmerman, C.E.; Edwards, G.W.; Perry, K. Maternal Origin and Migratory History of Steelhead and Rainbow Trout Captured in Rivers of the Central Valley, California. *Trans. Am. Fish. Soc.* **2009**, *138*, 280–291. [CrossRef]
22. Engstedt, O.; Engkvist, R.; Larsson, P. Elemental fingerprinting in otoliths reveals natal homing of anadromous Baltic Sea pike (*Esox lucius* L.). *Ecol. Freshw. Fish.* **2014**, *23*, 313–321. [CrossRef]
23. Goodwin, J.C.A.; Andrew King, R.; Iwan Jones, J.; Ibbotson, A.; Stevens, J.R. A small number of anadromous females drive reproduction in a brown trout (*Salmo trutta*) population in an English chalk stream. *Freshw. Biol.* **2016**, *61*, 1075–1089. [CrossRef]
24. Jiang, Y.; Somers, G. Modeling effects of nitrate from non-point sources on groundwater quality in an agricultural watershed in Prince Edward Island, Canada. *Hydrogeol. J.* **2009**, *17*, 707–724. [CrossRef]
25. Knysh, K.M.; Giberson, D.J.; van den Heuvel, M.R. The influence of agricultural land-use on plant and macroinvertebrate communities in springs. *Limnol. Oceanogr.* **2016**, *61*, 518–530. [CrossRef]
26. Roloson, S.D.; Knysh, K.M.; Coffin, M.R.S.; Gormley, K.L.; Pater, C.C.; van den Heuvel, M.R. Rainbow trout (*Oncorhynchus mykiss*) habitat overlap with wild Atlantic salmon (*Salmo salar*) and brook trout (*Salvelinus fontinalis*) in natural streams: Do habitat and landscape factors override competitive interactions? *Can. J. Fish. Aquat. Sci.* **2018**, *75*, 1949–1959. [CrossRef]
27. van der Poll, H.W. *Geology of Prince Edward Island*; Report 83-1; Department of Energy and Forestry, Energy and Minerals Branch: Charlottetown, PE, Canada, 1983.

28. Shaw, J. Geomorphic evidence of postglacial terrestrial environments on Atlantic Canadian Continental Shelves. *Géographie Phys. Quat.* **2005**, *59*, 141–154. [CrossRef]
29. Guignion, D.; Dupuis, T.; Teather, K.; MacFarlane, R. Distribution and Abundance of Salmonids in Prince Edward Island Streams. *Northeast. Nat.* **2010**, *17*, 313–324. [CrossRef]
30. PEI Department of Environment, Energy and Forestry. *Resource Inventory. Corporate Land Use Inventory 2010*; PEI Department of Environment, Energy and Forestry: Charlottetown, PE, Canada, 2010.
31. Purcell, L.A.; Giberson, D.J. Effects of an azinphos-methyl runoff event on macroinvertebrates in the Wilmot River, Prince Edward Island, Canada. *Can. Entomol.* **2007**, *139*, 523–533. [CrossRef]
32. Alberto, A.; St-Hilaire, A.; Courtenay, S.C.; van den Heuvel, M.R. Monitoring stream sediment loads in response to agriculture in Prince Edward Island, Canada. Environ. *Monit. Assess.* **2016**, *188*, 415. [CrossRef]
33. Lalonde, B.; Garron, C. Temporal and Spatial Analysis of Surface Water Pesticide Occurrences in the Maritime Region of Canada. *Arch. Environ. Contam. Toxicol.* **2020**, *79*, 12–22. [CrossRef]
34. Cairns, D.K.; MacFarlane, R.E.; Guignion, D.L.; Dupuis, T. *The Status of Atlantic Salmon (Salmo salar) on Prince Edward Island (SFA 17) in 2011*; DFO Canadian Science Advisory Secretariat Research Document 2012/090; Canadian Science Advisory Secretariat: Ottawa, ON, Canada, 2012.
35. Cairns, D.K.; MacFarlane, R.E. *The Status of Atlantic Salmon (Salmo salar) on Prince Edward Island (SFA 17) in 2013*; DFO Canadian Science Advisory Secretariat Research Document 2015/019; Canadian Science Advisory Secretariat: Ottawa, ON, Canada, 2015.
36. Roloson, S.D.; Landsman, S.J.; Tana, R.; Hicks, B.J.; Carr, J.W.; Whoriskey, F.; Van Den Heuvel, M.R. Otolith microchemistry and acoustic telemetry reveal anadromy in non-native rainbow trout (*Oncorhynchus mykiss*) in prince Edward Island, Canada. *Can. J. Fish. Aquat. Sci.* **2020**, *77*, 1117–1130. [CrossRef]
37. Courter, I.I.; Child, D.B.; Hobbs, J.A.; Garrison, T.M.; Glessner, J.J.G.; Duery, S. Resident rainbow trout produce anadromous offspring in a large interior watershed. *Can. J. Fish. Aquat. Sci.* **2013**, *70*, 701–710. [CrossRef]
38. Austin, C.S.; Bond, M.H.; Smith, J.M.; Lowery, E.D.; and Quinn, T.P. Otolith microchemistry reveals partial migration and life history variation in a facultatively anadromous, iteroparous salmonid, bull trout (*Salvelinus confluentus*). *Environ. Biol. Fishes* **2019**, *102*, 95–104. [CrossRef]
39. Thibault, I.; Hedger, R.D.; Dodson, J.J.; Shiao, J.C.; Iizuka, Y.; and Tzeng, W.N. Anadromy and the dispersal of an invasive fish species (*Oncorhynchus mykiss*) in Eastern Quebec, as revealed by otolith microchemistry. *Ecol. Freshw. Fish.* **2010**, *19*, 348–360. [CrossRef]
40. Liberoff, A.L.; Quiroga, A.P.; Riva-Rossi, C.M.; Miller, J.A.; Pascual, M.A. Influence of maternal habitat choice, environment and spatial distribution of juveniles on their propensity for anadromy in a partially anadromous population of rainbow trout (*Oncorhynchus mykiss*). *Ecol. Freshw. Fish.* **2014**, 424–434. [CrossRef]
41. Smith, M.W.; Saunders, J.W. Movements of Brook Trout, *Salvelinus fontinalis* (Mitchill), Between and Within Fresh and Salt Water. *J. Fish. Res. Board Canada* **1958**, *15*, 1403–1449. [CrossRef]
42. Spares, A.D.; Dadswell, M.J.; MacMillan, J.; Madden, R.; O'Dor, R.K.; Stokesbury, M.J.W. To fast or feed: An alternative life history for anadromous brook trout *Salvelinus fontinalis* overwintering within a harbour. *J. Fish Biol.* **2014**, *85*, 621–644. [CrossRef]
43. Castonguay, M.; Fitzgerald, J. Life history and movements of anadromous brook charr, *Salvelinus fontinalis*, in the St-Jean River, Gaspe, Quebec. *Can. J. Zool.* **1982**, *60*, 3084–3091. [CrossRef]
44. Whoriskey, F.G.; Naiman, R.J.; Montgomery, W.L. Experimental sea ranching of brook trout Salvelinus fontinals. *J. Fish. Biol.* **1981**, *19*, 637–651. [CrossRef]
45. McCormick, S.D.; Naiman, R.J. Hypoosmoregulation in an anadromous teleost: Influence of sex and maturation. *J. Exp. Zool.* **1985**, *234*, 193–198. [CrossRef]
46. Curry, R.A.; Sparks, D.; van de Sande, J. Spatial and Temporal Movements of a Riverine Brook Trout Population. *Trans. Am. Fish. Soc.* **2002**, *131*, 551–560. [CrossRef]
47. Johnston, N.T.; Perrin, C.J.; Slaney, P.A.; Ward, B.R. Increased juvenile salmonid growth by whole-river fertilization. *Can. J. Fish. Aquat. Sci.* **1990**, *47*, 862–872. [CrossRef]
48. Purcell, L.A. The River Runs Through It: Evaluation of the Effects of Agricultural Land Use Practices on Macroinvertebrates in Prince Edward Island Streams Using Both New and Standard Methods. Master's Thesis, University of Prince Edward Island, Charlottetown, PE, Canada, 2003.
49. Sirabahenda, Z.; St-Hilaire, A.; Courtenay, S.C.; van den Heuvel, M.R. Assessment of the effective width of riparian buffer strips to reduce suspended sediment in an agricultural landscape using ANFIS and SWAT models. *CATENA* **2020**, *195*, 104762. [CrossRef]
50. Dunn, A.M.; Julien, G.; Ernst, W.R.; Cook, A.; Doe, K.G.; Jackman, P.M. Evaluation of buffer zone effectiveness in mitigating the risks associated with agricultural runoff in Prince Edward Island. *Sci. Total Environ.* **2011**, *409*, 868–882. [CrossRef] [PubMed]
51. PEI Department of Environment, Energy and Climate Change. Island Fish Kill from 1962–2019. 2019. Available online: https://www.princeedwardisland.ca/sites/default/files/publications/pei_fish_kills_summary_1962-2017.pdf (accessed on 20 December 2021).
52. PEI Department of Environment, Energy and Climate Change. Prince Edward Island 2015–2016 Retail Pesticide Sales Report. 2020; 7p. Available online: https://www.princeedwardisland.ca/sites/default/files/publications/2015-2016_pesticide_sales_report_final_rfw.pdf (accessed on 19 November 2021).

53. Dunn, A.M. *A Relative Risk Ranking of Pesticides Used in Prince Edward Island*; Environmental Protection Branch Report Series EPS-5-AR-04-03; Environment Canada: Dartmouth, NS, Canada, 2004; 48p.
54. Health Canada Pest Management Regulatory Agency. Re-Evaluation Decision RVD2018-11 Chlorothalonil and Its Associated End-use Products for Agricultural and Turf Uses Final Decision. 2018. Available online: https://ipmcouncilcanada.org/wp-content/uploads/2018/07/RVD2018-11-chlorothalonil.pdf (accessed on 19 November 2021).
55. Sirabahenda, Z.; St-Hilaire, A.; Courtenay, S.C.; Alberto, A.; van den Heuvel, M.R. A modelling approach for estimating suspended sediment concentrations for multiple rivers influenced by agriculture. *Water* **2017**, *62*, 2209–2221. [CrossRef]
56. Alberto, A.; Courtenay, S.C.; St-Hilaire, A.; van den Heuvel, M.R. Factors influencing brook trout (*Salvelinus fontinalis*) egg survival and development in streams influenced by agriculture. *J. Fish. Sci.* **2017**, *11*, 9–20.
57. *Effort, Harvest, and Expenditures of Trout and Salmon Anglers on Prince Edward Island in 1994, from a Mail-Out Survey*; Canadian Technical Report of Fisheries and Aquatic Sciences No. 2367; Cairns, D.K. (Ed.) Fisheries and Oceans Canada (DFO): Charlottetown, PE, Canada, 1996; 45p.
58. Stranko, S.A.; Hilderbrand, R.H.; Morgan, R.P.; Staley, M.W.; Becker, A.J.; Roseberry-Lincoln, A.; Perry, E.S.; Jacobson, P.T. Brook trout declines with land cover and temperature changes in Maryland. *N. Am. J. Fish. Manag.* **2008**, *28*, 1223–1232. [CrossRef]

Article

Fidelity to Natal Tributary Streams by Kokanee Following Introduction to a Large Oligotrophic Reservoir

J. Mark Shrimpton [1,*], Paige W. Breault [2] and Luc A. Turcotte [3]

[1] Department of Ecosystem Science & Management, University of Northern British Columbia, 3333 University Way, Prince George, BC V2N 4Z9, Canada
[2] Great Lakes Institute for Environmental Research, University of Windsor, 401 Sunset Avenue, Windsor, ON N9B 3P4, Canada; pbreault@uwindsor.ca
[3] Stock Assessment Fraser River & Interior Area, Fisheries and Oceans Canada, 985 McGill Place, Kamloops, BC V2C 6X6, Canada; luc.turcotte@dfo-mpo.gc.ca
* Correspondence: mark.shrimpton@unbc.ca; Tel.: +1-250-960-5991

Abstract: The WAC Bennett Dam was completed in 1968 and impounded the Upper Peace River to form the Williston Reservoir in north central British Columbia. In 1990, an enhancement project was initiated to stock Columbia River Kokanee (non-anadromous Sockeye Salmon; *Oncorhynchus nerka*) from southeastern British Columbia into tributary streams that drained into regions of the reservoir that were accessible by anglers. The current distribution of spawning Columbia-origin Kokanee in the Williston Reservoir watershed, however, does not reflect the locations where these fish were initially stocked and suggests extensive straying. Whether or not Kokanee will develop fidelity to specific spawning locations is not known, but it is important information to effectively manage these introduced fish. We used otolith microchemistry to estimate fidelity to natal streams by Columbia-origin Kokanee in the Williston Reservoir. Otolith elemental signatures for the region of the otolith that formed during the larval period and characterized the natal redd environment showed considerable variation among samples. Natal signatures tended to cluster for each river but not for all spawners, suggesting elemental signatures from other rivers. Homing to one of the four natal streams we examined was classified to be 73% based on linear discriminant analysis, although variation in the elemental signatures within each group suggests that homing by Kokanee to specific natal streams may be as low as 55%. Based on similarity of water elemental signatures for tributaries within large rivers, however, the proportion of fish that returned to their general region was likely higher and estimated to be approximately 83%. The result of regional homing could be reproductive isolation and adaptation to local conditions. It is unclear, however, if the current estimated level of straying will limit genetic differentiation and prevent local adaptation.

Keywords: otolith microchemistry; Kokanee; homing; stocking

Citation: Shrimpton, J.M.; Breault, P.W.; Turcotte, L.A. Fidelity to Natal Tributary Streams by Kokanee Following Introduction to a Large Oligotrophic Reservoir. *Fishes* **2022**, *7*, 123. https://doi.org/10.3390/fishes7030123

Academic Editor: Josipa Ferri

Received: 19 April 2022
Accepted: 25 May 2022
Published: 30 May 2022

1. Introduction

Homing to natal streams for spawning is an important life history characteristic for Pacific salmon within the genus *Oncorhynchus* to ensure success of their progeny [1]. A location that historically served for successful incubation of embryonic and larval fish is likely to provide the greatest potential for survival for the next generation. Consequently, salmonids often show fidelity to specific spawning locations within a watershed—even at the microhabitat scale [2,3]. Selection of a spawning site is crucial because the highest rates of mortality in salmonids generally occur during the incubation period [1], and this mortality is closely related to the features of the spawning and incubation site. Intragravel variables influence spawning site selection at the microhabitat scale [4,5]—particularly temperature, where intragravel water temperature is less variable than surface water [6]. Colonization of new habitat and range expansion, however, is also important as the area of spawning habitat is linked to effective population size—an indicator of population resilience to stochasticity [7].

A major stocking program of Kokanee (non-anadromous Sockeye Salmon; *Oncorhynchus nerka*) to the Williston Reservoir provided us with an opportunity to assess fidelity to spawning locations for fish introduced to a novel environment. The Williston Reservoir was formed in 1968 following the construction of the WAC Bennett Dam in the canyon near Hudson's Hope, British Columbia, Canada. The Dam impounded the headwaters of the Peace River and lower reaches of the Parsnip River, Finlay River, and many major tributary rivers, creating the largest lentic freshwater system in British Columbia. Kokanee were among the assemblage of species that naturally colonized the Williston Reservoir from headwater lakes when it formed and increased in numbers over time [8]. To increase angling opportunities and provide a prey source for large piscivorous fish species, a stocking program was initiated into five tributaries of the Williston Reservoir that were more accessible to anglers using Kokanee from the Columbia River [8–10]. An aerial enumeration study conducted from 2002 to 2006 found that the distribution and abundance of Kokanee in tributaries to the Williston Reservoir poorly reflected the stocking patterns from the 1990s. By 2006, Kokanee were reported to spawn in at least 68 rivers and streams from the Parsnip River watershed to the Finlay River watershed [10]. More recent aerial surveys indicate that these patterns have changed even further in the past decade [11,12]. The spawning Kokanee in tributary streams possessed bright red bodies and emerald green heads typical of Columbia origin fish rather than the rusty reddish brown coloration of mature native Williston Kokanee [13]. It is possible that the locations where Columbia origin Kokanee were introduced did not contain preferred habitat for the species, although small numbers of spawning Kokanee still return to some of the tributary streams where they were stocked [10–12]. Kokanee have shown a strong potential for straying following initial stocking into the Williston Reservoir watershed, but whether they have now selected specific habitat locations for spawning or continue to stray is not known.

Recently, we have shown that natal stream elemental signatures in otoliths of Coho Salmon (*O. kisutch*) were specific to spawning locations [3]. Otolith microchemistry, therefore, has strong potential to define natal location in adult Kokanee spawners and determine if fish are homing or straying. Kokanee migrate from lakes to tributary streams to spawn from July to November in British Columbia depending on latitude [14]. Fertilized eggs develop in redds until hatch and alevin remain in the gravel, and until the yolk is fully absorbed, usually emerging in May. Fry for most populations migrate downstream into the lake immediately after emergence to begin exogenous feeding [14]. Time within the natal streams is usually restricted to just the life history stage within the redd environment and the intragravel water defines the natal water elemental signature. Elemental signatures in streams are based on differences in the underlying bedrock, which varies with location [15]. Bedrock geology differs between the east and west sides of the Williston Reservoir, as well as between the north and south [16], and bedrock formations in the upper Peace River watershed were formed during different time periods and are composed of materials that have variable chemical composition [17]. The distinctive chemical signatures that vary among freshwater streams and rivers also correlate with elemental ratios in otoliths of freshwater species (Atlantic Salmon *Salmo salar* [15]; Cutthroat Trout *O. clarkii* [18]; Arctic Grayling *Thymallus arcticus* [19]; Coho Salmon [20]). Consequently, elemental signatures in otoliths of fish have been used to assess habitat shifts, population structure, and location of origin, but also have been used to trace the extent and pattern of movement of individual fish. Earlier work on differences in elemental signature for otoliths from Arctic Grayling [19] and Slimy Sculpin *Cottus cognatus* [21] captured among tributaries to the Williston Reservoir indicated large differences in elemental chemistry within the watershed. We used otolith microchemistry, therefore, to identify whether an introduced population of Kokanee exhibit homing and return to natal streams to spawn or stray from their natal locations.

2. Materials and Methods

2.1. Water Chemistry

To assess spatial variation in water chemistry signatures, we collected water samples from major rivers in the Williston Reservoir watershed where Columbia origin Kokanee have been observed to spawn: Omineca River, Ingenika River, Finlay River, and Ospika River (Figure 1). Chemical signatures of intragravel water where embryos and alevin develop (natal redd signatures) did not differ significantly from surface waters at multiple locations where Coho Salmon spawned in the Coldwater River, BC [3]. Consequently, we used surface water elemental signatures to assess differences among rivers and spawning locations within the Williston Reservoir watershed. We also collected water from nearby tributaries to assess spatial variation within each river system and from locations in the reservoir. Water samples were collected using the methods outlined by [22] as modified by [19]. Bottles and syringes were rinsed with ultrapure water, then soaked for 5 days in a 2% solution of high-purity nitric acid. Water samples were collected in duplicate during the period of summer river low flows, from 7 to 11 August 2020 and from 21 to 22 August 2021, when groundwater represents a large proportion of river water. Water samples collected when flows are low better represent the elemental signatures present as there is minimal dilution from overland surface water runoff [23].

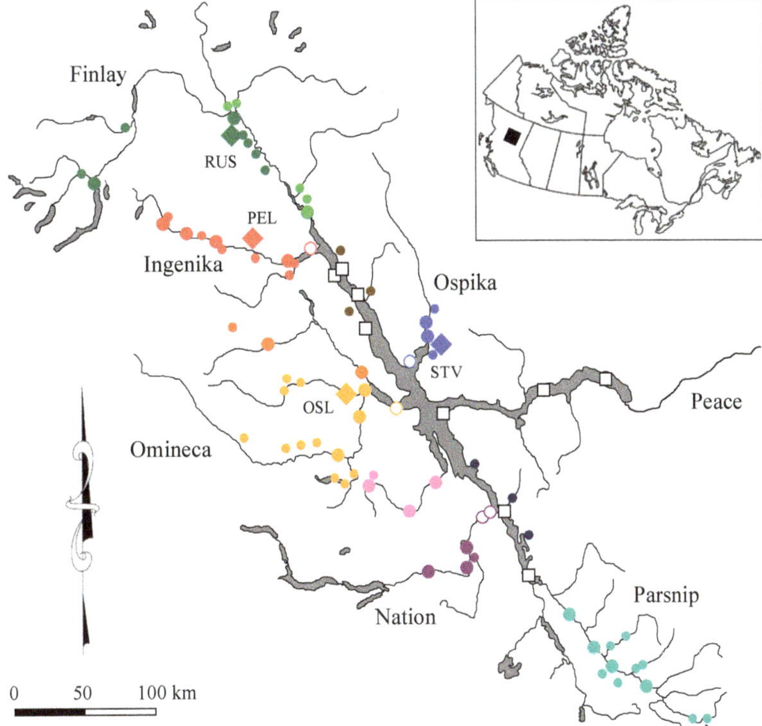

Figure 1. Sampling sites in the Williston Reservoir watershed where water samples and mature Kokanee spawners were collected. Water sampling sites are indicated as large, closed circles from mainstem rivers, small, closed circles from tributary streams, open large circles from embayments with color defining the regional groups, and open squares for samples collected from different locations in the reservoir. Locations where Kokanee were collected are shown as solid diamonds: Russel Creek (RUS), Pelly Creek (PEL), Osilinka River (OSL), and Stevenson Creek (STV).

Water samples were filtered and preserved in the field with 600 μL of HNO_3 added to a 30 mL water sample. An inductively coupled plasma-optical emission spectrometer (ICP-OES) (model 5100 Agilent Technologies, Santa Clara, CA, USA) was used to analyze water samples for calcium (Ca), manganese (Mn), strontium (Sr), and barium (Ba). Elemental values are ratios to Ca but expressed as the individual element.

Concentrations of elements vary seasonally with discharge, but the ratio of each element to Ca shows greater consistency seasonally and among years [24,25]. We incorporated published and unpublished data from earlier work, therefore, to assess temporal stability in water chemistry signatures. Samples collected in 2002 [25], 2003 and 2004 [19], 2011 (AD Clarke, unpublished data), and 2017 (MD Stamford, unpublished data) were also included in the analysis.

Discriminant function analysis (DFA) using jack-knife resampling was used to assess geographic separation of water chemistry data for Sr and Ba to Ca elemental ratios from watersheds where Columbia origin Kokanee are known to spawn or where native populations of Kokanee exist (SYSTAT version 7). Mn was not used as it was below detection for some locations. Water chemistries were obtained from both the tributaries and the mainstem river of each watershed, and over multiple years for some locations. Jack-knife re-sampling was used to validate the robustness of the discriminant functions.

2.2. Otolith Analysis

Spawning locations for Kokanee used to assess homing to natal streams were chosen based on the difference in stream elemental signatures from the signature of the reservoir and availability of archived otoliths. Mature Kokanee spawners sampled from Russel Creek in 2006 and 2018, Pelly Creek in 2006 and 2018, Osilinka River in 2016, and Stevenson Creek in 2016 and 2017 (Figure 1) were frozen prior to otolith extraction. All fish used in our study were genotyped and identified to be Columbia origin Kokanee [26]. Otoliths were removed from Kokanee carcasses after they were partially thawed.

Otoliths were mounted in epoxy, then polished with lapping papers with 320, 600, and 1200 grit sizes. After final sanding, samples were rinsed with clean distilled water prior to ablation. Elemental analysis on otoliths was conducted at two labs: School of Earth & Ocean Sciences (University of Victoria) and TrichAnalytics (Saanichton, BC, Canada). Material was extracted from the otoliths with a VG Elemental PQ II S + high sensitivity ICP-MS (Thermo Electron Corporation, Waltham, MA, USA) coupled to a Merchantek UV laser ablation system (New Wave Research, Fremont, CA, USA) at UVic. The laser system was operated with an output of 266 nm that had a maximum energy output of 4 mJ. Optimization was conducted using standard reference material (NIST 613). Material was extracted from otoliths using an NWR 213 laser ablation (LA) module (Elemental Scientific Instruments, Omaha, NE, USA) coupled to an iCAP RQ inductively coupled plasma mass spectrometer (ICP-MS) (Thermo Fisher Scientific, Waltham, MA, USA) at 60% power at TrichAnalytics. An internal reference standard for calcium (Ca) of 40% was used during otolith ablation. The LA-ICP-MS machine was calibrated with a reference standard (NIST 612).

The otolith ablation diameter for both methods was set at 30 μm and tracked across the laser at 5 μm s^{-1}. Isotopes measured in the otoliths were ^{43}Ca, ^{55}Mn, ^{66}Zn, ^{88}Sr, and ^{137}Ba. Elemental concentrations were calculated by correcting the sample signal intensity to the background signal, then to the average intensity standard of the selected element, followed by correcting to the recovery of the internal standard (Ca) in the sample relative to the external standard. All data, therefore, are expressed as a ratio of the element to Ca, but just reported as the individual element. Otolith data were smoothed with a running average for every 5 μm. For natal elemental signatures, data were examined from the line scan approximately 200 μm on either side of the otolith core. Each line was visually assessed for symmetry of elemental signatures and the large spike in Mn at the core. Once the core center was determined, we assessed Sr, Ba, and Mn at a distance greater than 50 μm from the core to limit the potential influence of reservoir-derived maternal signatures, and less than 250 μm to capture elemental signatures for alevin prior to gravel emergence, thus representing the incubation environment [3,27,28]. These three elements were used due to differences characterized among rivers in the Williston watershed [19,20]. Strong relationships also existed between water elemental signatures and otolith microchemistry for Sr and Ba, however, the relationship for Mn was not as strong [19,20].

We calculated average Sr, Ba, and Mn signatures for a 25 μm section of otolith in the region that represented the elemental signature for the intragravel environment post yolk absorption, but prior to feeding and reservoir environmental influence; approximately between 120 μm and 170 μm as derived by [28]. The absolute location where we calculated the natal redd environment signature varied depending on the size of the otolith and cross-sectional length of the line scan. Average values for Sr, Ba, and Mn were calculated for the two sides of the ablation path from the otolith core, but 7 of 64 samples were not symmetrical and only one side was used. Eight samples were excluded as the core was missed. Zinc to calcium was also plotted to estimate age as elemental ratios showed annual oscillations that vary seasonally [21].

Discriminant function analysis using jack-knife resampling was used to estimate fidelity to spawning locations and potential straying among watersheds where Columbia origin Kokanee spawn (SYSTAT version 7). A multivariate combination of Sr, Ba, and Mn to Ca elemental ratios was used to discriminate locations using otolith elemental signatures for the intragravel environment. Jack-knife re-sampling was used to validate the robustness of the discriminant functions. To provide a visualization of the elemental differences for the sections of otoliths prior to emergence from redds, the first two discriminant scores were plotted for each spawning location.

3. Results

3.1. Water Chemistry

Considerable differences existed in water chemistry among the rivers, but also many of the water samples differed from water collected from the Williston Reservoir (Figure 2). Differences within samples collected from an individual river system reflect changes in water chemistry between the headwaters and mouth of the rivers, and differences in elemental signatures for inflowing tributary streams for each river. General patterns were that Sr values were lower for rivers and streams draining the eastern and southern parts of the watershed, and higher for rivers and streams draining the western and northern parts of the watershed. For Ba and Mn, values were generally higher for rivers and streams draining the western parts of the watershed, but not all. Values for Ba were also higher in tributaries to the headwater of the Finlay River, the northernmost region of the watershed. Values for Mn were below detection in many of the tributaries to the Ospika and Finlay Rivers. Another general pattern observed among the larger rivers and their tributaries was a clustering of values for samples from the mainstem, with more variation in signatures from tributary streams and the headwaters.

Discriminant function analysis indicated that samples collected from many of the rivers failed to be classified to the correct location. The total number of cases classified correctly using jack-knife validation was only 35% (Table 1). The five groups with the highest percentage of correctly classified samples were the Omineca River and tributaries (60%), Ingenika River and tributaries (53%), northwestern locations on the Finlay River and tributaries (64%), Ospika River and tributaries (80%), and samples collected from the reservoir (40%). Water chemistry signatures from tributaries within each of these mainstem river systems that showed good separation from the reservoir water chemistry were Stevenson Creek in the Ospika River watershed, Osilinka River in the Omineca River watershed, Pelly Creek in the Ingenika River watershed, and Russel Creek in the lower Finlay River watershed (Figures 1 and 2).

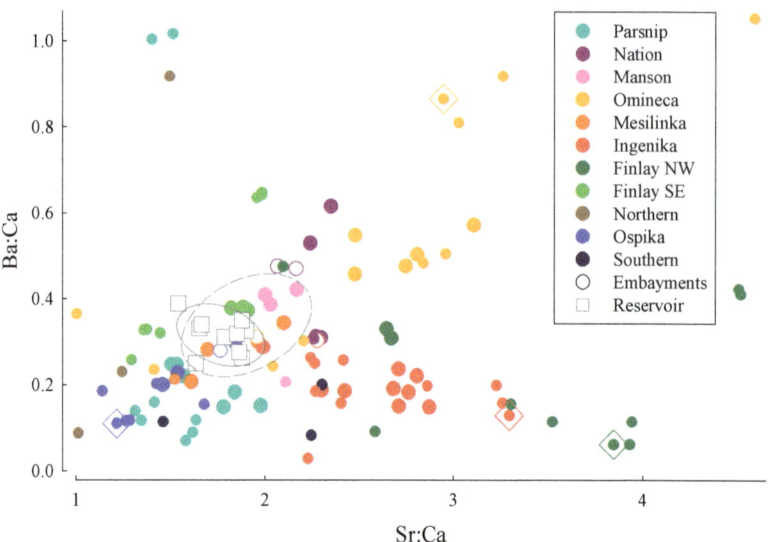

Figure 2. Elemental ratios of Sr and Ba to Ca for water samples collected from rivers and tributaries within the Williston Reservoir watershed. Locations for samples from each watershed are shown in Figure 1. Mainstem rivers are indicated as large, closed circles, tributary streams are indicated as small, closed circles, embayments are indicated as open, large circles with color defining the mainstem river, and samples collected from different locations in the reservoir are indicated as open squares. Ellipses drawn around reservoir (solid line) and embayment (dashed line) data define 95% confidence intervals for elemental signatures for each location. Small, closed circles inside open diamonds indicate water chemistries from tributary streams where mature Kokanee spawners were sampled.

Table 1. The percentage of correct classification determined by discriminant function analysis (DFA) for water samples collected from watersheds within the Williston Reservoir catchment.

	PAR	NAT	MAN	OMI	MES	ING	FIN1	FIN2	NOR	OSP	SOU	EMB	RES	%
PAR	0	0	0	0	0	0	0	1	1	5	3	0	0	0
NAT	0	0	4	2	0	0	0	0	0	0	0	0	0	0
MAN	0	1	0	0	0	0	0	0	0	0	1	2	0	0
OMI	0	1	1	9	0	0	1	0	1	1	1	0	0	60
MES	0	0	2	0	0	0	0	0	0	2	0	0	1	0
ING	0	0	2	0	0	10	3	0	0	0	4	0	0	53
FIN1	0	0	0	0	0	3	7	0	0	0	0	1	0	64
FIN2	0	0	0	0	0	0	0	0	3	0	0	3	1	0
NOR	0	0	0	0	1	0	0	1	0	2	0	0	0	0
OSP	0	0	0	0	2	0	0	0	0	8	0	0	0	80
SOU	0	0	0	0	0	2	0	0	0	1	1	0	0	0
EMB	0	2	2	0	1	0	0	0	0	0	0	0	0	0
RES	2	0	0	0	3	0	0	1	0	0	0	0	4	40
TOT	2	4	11	11	7	15	11	3	5	19	9	6	6	35

Samples were collected from mainstem and tributaries for the Parsnip River (PAR), Nation River (NAT), Manson River (MAN), Omineca River (OMI), Mesilinka River (MES), Ingenika River (ING), northwestern locations on the Finlay River (FIN1), southeastern locations on the Finlay River (FIN2), tributaries flowing into the Finlay Reach (NOR), Ospika River (OSP), tributaries flowing into the Parsnip Reach (SOU), embayments (EMB), and from the reservoir (RES). Data are presented as the number of samples (n) and cross-validation accuracy is expressed as a percentage. Sites where water samples were collected are listed in rows and DFA assignment shown in columns. Elements incorporated into the model were Sr and Ba.

3.2. Otolith Analysis

Elemental line scans of otoliths from mature Kokanee spawners showed considerable variation in Sr, Ba, and Mn elemental ratios to Ca across the otolith. A male spawner from Russell Creek is shown in Figure 3. A distinctive spike in Mn is characteristic of the otolith core, with Ba also high at the core and Sr low at the core. As the embryo developed and hatched, the influence of maternally derived chemical signatures from the reservoir declined, particularly after hatch and ionic uptake by the alevin was increasingly influenced by the water within the redd environment. After emergence and migration downstream to the reservoir, elements incorporated into the otolith reflected water and sources of food from the reservoir. The line scan values beyond the region defined by "F" in Figure 3 likely represent movement to locations with different elemental signatures following emergence from the gravel. There was considerable variation in Sr, Ba, and Mn throughout the life history of Kokanee after emergence, suggesting movement to habitats that appear to differ in elemental composition.

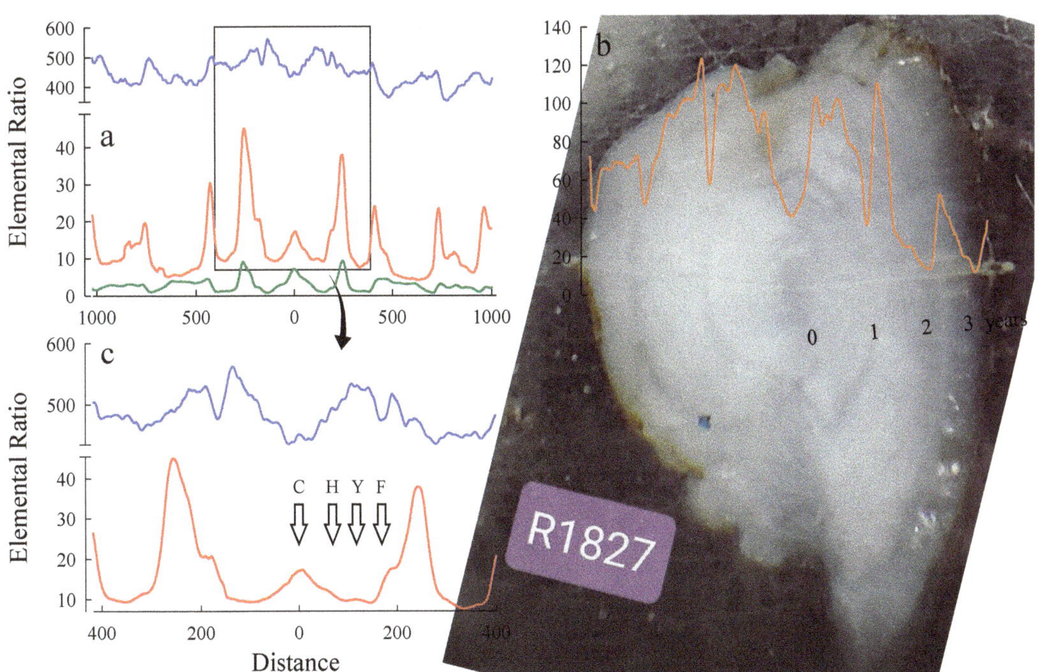

Figure 3. Elemental signatures for an otolith from an adult Kokanee spawner collected in Russell Creek (22.5 cm [L_F] and 99.8 g [M]) on 19 September 2018: (**a**) Elemental signatures from the line scan show changes in Sr (blue), Ba (red), and Mn (green) over the entire life history of the fish; (**b**) Image of otolith showing laser ablation burn to determine elemental composition of the structure. Changes in Zn signatures are superimposed on the image to show how the oscillations align with otolith annuli (photograph courtesy of Jennie Christensen, Trich Analytics, Saanichton, BC, Canada); (**c**) Natal section expanded to show the changes in line scans for Sr (blue) and Ba (red) corresponding to distances defined by [28] from the core on the otolith. The core (C), hatch (H), yolk absorption (Y), and feeding (F) stages for early development stages in Kokanee. Otolith core center is indicated by 0 on the *X*-axis.

Elemental signatures from the section of the otolith that defined the natal redd environment showed considerable variation among samples, but some general patterns were evident. Elemental ratios for Sr, Ba, and Mn were lowest on average for fish from Stevenson

Creek; 374.5 ± 64.4 mmol mol^{-1}, 14.7 ± 4.3 mmol mol^{-1}, and 2.33 ± 1.46 mmol mol^{-1} for Sr, Ba, and Mn, respectively. Fish sampled from the Osilinka River had the highest Ba and Mn levels (24.9 ± 9.6 mmol mol^{-1} and 5.97 ± 3.66 mmol mol^{-1}) and intermediate Sr levels (466.1 ± 39.7 mmol mol^{-1}). Pelly Creek spawners had high values of Sr (511.4 ± 61.5 mmol mol^{-1}), but low Ba (13.2 ± 3.1 mmol mol^{-1}) and Mn (2.66 ± 1.18 mmol mol^{-1}). Russell Creek fish had the highest levels of Sr in the natal section of otolith line scan which was on average (556.0 ± 51.2 mmol mol^{-1}), and the lowest Ba and Mn values on average; 13.1 ± 2.6 mmol mol^{-1} and 2.26 ± 0.72 mmol mol^{-1}, respectively. Significant relationships were found between the natal otolith signatures and surface water chemistry from the spawning streams examined in our study for Sr ($F_{1,62}$ = 94.22, $p = 4.656 \times 10^{-14}$, $R^2 = 0.603$) and Ba ($F_{1,62}$ = 48.41, $p = 2.548 \times 10^{-9}$, $R^2 = 0.438$). Regression analysis was not conducted for Mn as water concentrations for Russel and Stevenson Creeks were below detection. Results from the linear regression for Sr were:

$$Sr_{otolith} = 69.06 \cdot Sr_{water} + 282.5 \tag{1}$$

and for Ba were:

$$Ba_{otolith} = 14.70 \cdot Ba_{water} + 12.15 \tag{2}$$

Discriminant function analysis using jack-knife resampling for natal otolith signatures used the linear combination of the predictor variables and revealed reasonable separation between spawning locations. The percentage of correct classification for each location indicated that the total number of cases classified correctly using jack-knife validation was 73% (Table 2)—suggesting approximately 27% rate of straying. The rate of straying might be higher as some fish were assigned to groups even though elemental signatures did not cluster well with other spawner signatures from that stream. Although elemental signatures showed strong relationships with spatial spawning sites, there was considerable overlap in the confidence ellipses for the discriminant scores of the natal otolith elemental signatures for the four locations where Kokanee spawners were sampled (Figure 4). Discriminant scores tended to cluster for each river system but were not at the center of the ellipse due to values that departed from the cluster suggesting elemental signatures from other river systems.

Table 2. The percentage of correct classification determined by discriminant function analysis (DFA) for otolith elemental signatures characteristic of natal redd environments for Kokanee spawners sampled from Pelly Creek (PEL), Russel Creek (RUS), Osilinka River (OSL), and Stevenson Creek (STV) within the Williston Reservoir watershed.

	RUS	PEL	OSL	STV	%
RUS	14	4	0	0	78
PEL	6	11	0	1	61
OSL	0	2	10	0	83
STV	0	2	2	12	75
TOTAL	22	18	11	15	73

Data are presented as the number of samples (n) and cross-validation accuracy is expressed as a percentage. Sites where Kokanee spawners were collected are listed in rows and DFA assignment shown in columns. Elements incorporated into the model were Sr, Ba, and Mn.

For Pelly Creek and Russel Creek, we also had otolith samples collected in 2006 and 2018. There were 6 Kokanee spawners sampled in 2006 that were not assigned to stream of capture out of 20 samples, suggesting a straying rate of 30%. There were 5 Kokanee spawners sampled in 2018 that were not assigned to stream of capture out of 16 samples, suggesting a straying rate of 31%. Straying did not appear, therefore, to show any temporal difference.

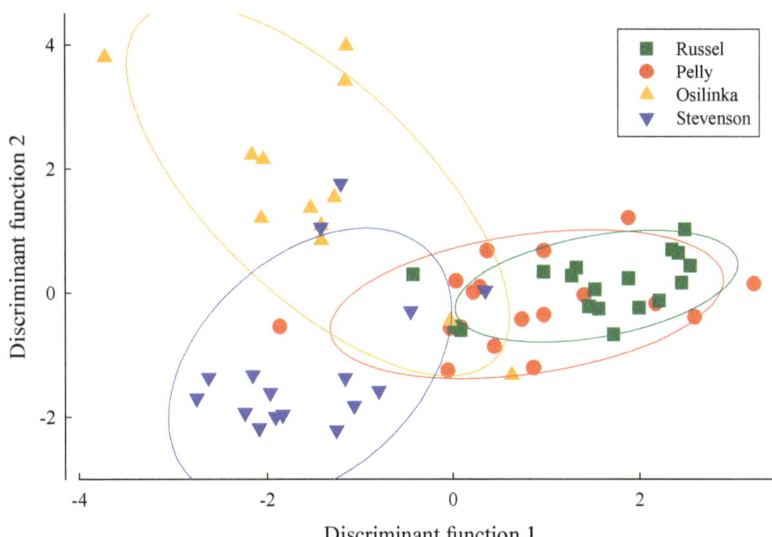

Figure 4. Discriminant function analysis using jack-knife resampling for elemental signatures from sections of otoliths that are characteristic of natal redd environments from spawning Kokanee collected from Pelly Creek, Russel Creek, Osilinka River, and Stevenson Creek. Elemental ratios included in the model were Sr, Ba, and Mn to Ca.

4. Discussion

Determining fidelity to specific locations within the Williston Reservoir by Kokanee for spawning is important to effectively manage this species. We used otolith microchemistry to estimate the proportion of mature Kokanee spawners that home to natal locations and estimated spawning site fidelity to be 73% among the four spawning locations. The success of our study, however, was dependent on several factors. First, spatial variability in water chemistry had to be high. Second, temporal variability in water chemistry had to be low. Third, the section of the otolith corresponding to the embryonic and larval life history period had to be easily distinguished.

4.1. Spatial Variation in Water Elemental Signatures

Our choice of tributaries where Kokanee spawn was based on spatial separation and likelihood of differences in elemental signatures in redd environments, but also availability of archived otoliths from Kokanee spawners collected for multiple cohorts. Elemental signatures differed among tributary streams and from the Williston Reservoir [19,25]. Elemental ratios for Sr, Ba and Mn to Ca were high on the western side of the reservoir and low on the eastern side of the reservoir. Elemental signatures generally varied with latitude, but differences also existed at smaller geographic scales.

Water chemistries from streams on the east side of the reservoir tended to be low in Sr, Ba, and Mn. The Ospika River and tributaries also had the highest cross-validation accuracy (80%). Stevenson Creek had elemental signatures that were lower and differed from the reservoir water samples more than the mainstem of the Ospika River and other Ospika River tributary streams that were sampled. Additionally, we had archived otoliths from Kokanee sampled from Stevenson Creek in 2016 and 2017. Water chemistries from the Omineca River watershed had Sr, Ba, and Mn to Ca ratios higher than values in the reservoir for most locations, but not all (Figure 2). Elemental signatures from water samples collected from the Osilinka River approximately 8 km above the confluence with the Omineca River were much higher in Sr, Ba, and Mn than the reservoir, thereby maximizing potential differences between the maternal and natal otolith signatures. We also had archived otoliths from spawners collected at this location in 2016. There is the potential for Kokanee to migrate above this location on the Osilinka River to spawn along the mainstem or upper Osilinka River tributary streams. Consequently, variability in elemental signatures for Kokanee sampled in the Osilinka River may represent signatures for multiple locations in the Osilinka River watershed. Pelly Creek Sr and Mn signatures were greater than the reservoir, and Ba signatures were lower than the reservoir. The elemental concentrations provided an opportunity to detect differences between the maternal signatures from the reservoir and natal redd signatures. We also had archived otoliths from Kokanee spawners collected from Pelly Creek in 2006 and 2018 available for both spatial and temporal comparisons. Although water chemistry for samples from the lower Finlay River were similar to that of the reservoir, Sr for tributaries differed from the reservoir, particularly Russel Creek where Sr was greater than the reservoir and Ba was slightly lower than the reservoir; Mn was below detection. Other lower Finlay River tributaries draining from the west also had higher levels of Sr than the reservoir. Streams draining from the east, however, had much lower Sr and were similar to the reservoir. We also had archived otoliths from Russel Creek Kokanee spawners collected in 2006 and 2018. Clear separation in water chemistries among the study streams and differences from the reservoir, therefore, provided an opportunity to define natal redd chemistries and assess fidelity to spawning locations in tributary streams to the Williston Reservoir. Stability of water elemental signatures is also necessary to use otolith microchemistry to assess spawner fidelity to natal streams.

4.2. Temporal Stability and Spatial Variation of Water Elemental Signatures

Water stability was important for our study, as large temporal variation in stream chemistry would ultimately confound our ability to differentiate among study locations. While fluctuations in mean elemental concentrations occur seasonally, elemental ratios remain remarkably constant [25]. Elemental ratios were slightly lower during spring freshet (high flow conditions) for the tributaries examined by [25], however, this was of limited concern as most growth for bony structures of fish occurs during base flow conditions [29]—usually summer for temperate regions, but also during winter when larval salmonids are developing. Consistency of elemental signatures within rivers has also been well described by [24]. This finding was also supported by [18], who determined that stream chemistries were stable seasonally and over a two-year duration. Water chemistry was also found to be consistent over two years (2006 and 2007) in the Coldwater River watershed, BC [20]. Water chemistry measured from samples collected at some of the same Coldwater River locations in 2017 indicated that the elemental signatures were consistent after a decade [3].

We also collected water samples in 2020 and 2021 from many of the same locations sampled in 2002 [25]. Water samples collected from Manson River, Omineca River, Osilinka River, Ingenika River, Swannel River, Pelly Creek, and Finlay River at the same locations showed an absolute average difference between the years of 0.156 ± 0.109 mmol mol^{-1}, 0.0291 ± 0.0262 mmol mol^{-1}, and 0.0436 ± 0.0356 mmol mol^{-1}, for Sr, Ba, and Mn, respectively. Chemical signatures, therefore, appear to have been stable over decades for comparisons made for the tributaries sampled in the Williston Reservoir watershed in 2002, 2020, and 2021.

Our work also depended on spatial differences observed in surface waters corresponding to the chemical composition of the intragravel environment, and that the elemental composition of the water would influence the composition of the otolith during the embryonic and alevin stages. In northern areas where freezing occurs, salmon seek areas with upwelling groundwater for spawning [1], which influence the water chemistry within the redd environment. Upwelling groundwater and intragravel permeability within the hyporheic zone results in no significant difference between surface water chemistry and intragravel water [3]. Consequently, the surface water chemistry measurements collected from earlier work and our study adequately represent signatures within the redd environment.

4.3. Distinguishing Natal Otolith Elemental Signatures

Otoliths provide a chronological record of the environmental conditions that fish are exposed to over their lifetime [30,31]. Otoliths are metabolically inert, grow continuously over the lifetime of the fish, and will not be reabsorbed [31]. Elements incorporated into otoliths are also directly proportional to elemental uptake from the surrounding water [19,20,30]. Elemental signatures from the cores of adult otoliths, therefore, are informative of environmental conditions in redds during the early life history of larval salmon. Otolith composition of larval fish, however, is also influenced by maternal nutrients [32]. Determining the end of the maternal signature and the start of the juvenile signature is difficult and complex [27] but can be used to identify natal spawning locations [28] and potential straying among salmonids [3]. Annual otolith increments are wider near the core and become narrower with age. The fixed width of the laser scan, therefore, averages a shorter temporal signature for the natal region of the otolith. Consequently, our ability to detect variation in elemental signatures is greatest near the core of the otolith. The short period of time that post-emergent Kokanee spend in their natal tributaries before migrating downstream to the reservoir, therefore, did not limit the effectiveness of our approach.

4.4. Reliability of Estimates for Fidelity to Spawner Sites

We determined spawning site fidelity to be, on average, 73% among the four spawning locations, but putative homing varied from 61 to 83%. Linear discriminant analysis, however, assigned fish to the best group—even if the elemental signatures did not cluster closely together. Consequently, some of the Kokanee spawners may have been misassigned as they were not from any of the four spawning locations for which we had archived otoliths.

For Stevenson Creek, fidelity was estimated to be 75% with 12 of 16 fish classified to their river of capture. Of the fish assigned to Stevenson Creek, 8 Kokanee spawners clustered with low discriminant function 1 and 2 scores. The other 4 spawners had higher scores, likely due to Sr values similar to the reservoir, but not high enough to cluster with one of the other study streams. Consequently, a more accurate estimate of homing to Stevenson Creek may be 50%. It is likely, however, that these 4 spawners that were assigned to Stevenson Creek but did not cluster with most of the other fish may have been from other Ospika River tributaries that also had low Sr, such as Aley Creek, which has had high escapement [11,12]. Gauvreau Creek is another Ospika River tributary where Kokanee spawn (L. Anderson, personal communication); however, the similarity of water chemistry for Gauvreau Creek to Stevenson Creek made it hard to differentiate these two streams. A more robust interpretation of homing to Stevenson Creek, therefore, may be 50% and an estimate of 25% straying from nearby Ospika River tributaries. The remaining 25% of the spawners likely strayed from spawning locations outside of the Ospika River and potentially from watersheds on the western side or northern region of the reservoir.

Mature Osilinka River Kokanee were intercepted at a site on the mainstem, likely below most of the spawning locations, but 83% of the natal otolith elemental signatures were classified to be from the Osilinka River (10 of 12 fish). Of these 10 fish, however, only 7 clustered tightly and the other 3 were included due to the large size of the confidence ellipse. The high Ba signatures for each of these fish suggest that they likely originated in the Osilinka River watershed, but potentially from an upper tributary or higher in the mainstem. The remaining 2 Kokanee sampled from the Osilinka River had much lower natal Ba signatures and were classified as Pelly Creek fish, although given the overlap of values, there is not much confidence in the assignment. Consequently, homing to the Osilinka River may be as low as 58%.

For Pelly Creek, fidelity was estimated to be 61% (11 of the 18 spawners were correctly assigned). These 11 fish formed a cluster with similar discriminant function scores and likely represent Kokanee homing to Pelly Creek. Homing percentage of 61%, therefore, appears likely for the Pelly Creek spawners. The remaining fish were assigned to Russel Creek (6) and Stevenson Creek (1). The fish assigned to Russel Creek had higher Sr levels and low Ba and Mn levels, which are characteristic of Finlay River tributaries, but we lacked the resolution to define specific natal stream origin. The fish assigned to Stevenson Creek were unlikely to have originated from the Ospika River watershed, due to the magnitude of the Sr and Ba values, and stream origin was unclear for this fish; they may have strayed from other Ingenika River tributaries or even the Ingenika River mainstem.

Russel Creek Kokanee spawners appeared to show high fidelity to a spawning location, 78%. All 4 fish sampled from Russel Creek that were not assigned to the location of capture were classified to Pelly Creek. Two of the fish classified as Pelly Creek clustered with the fish that putatively homed to Pelly Creek and may therefore represent fish straying from Pelly Creek or the Ingenika River watershed more generally. One of the fish classified as Pelly Creek, however, appeared to cluster with the Russel Creek fish, but had the lowest value of Sr of these fish. Based on the similarity of Sr and Ba values for this fish, it likely also homed to Russel Creek. If these fish were assigned to Russel Creek, classification would be 83%. Tributary streams to the Finlay Reach and Omineca Reach have had the highest escapement of Kokanee spawners for aerial enumerations conducted over the last 24 years [10–12], indicating the importance of rivers in this part of the watershed for Kokanee spawning. Five of the Kokanee classified as Russel Creek spawners, however, had Sr values that were higher than those that clustered together. The higher natal Sr signatures for these 5 spawners suggests they did not home to Russel Creek, but likely strayed from other tributaries to the Finlay River—likely tributaries higher in the Finlay River watershed based on patterns observed. Homing percentage of 61% (11 of 18), therefore, appears likely for the Russel Creek spawners—although historically, Russel Creek has had some of the highest escapements based on aerial enumerations [10–12].

4.5. Post-Emergence Movement Patterns

Elemental signatures across the otolith provide information on water composition throughout the life of the fish. Changes in elemental ratios of Sr and Ba to Ca have been useful to define freshwater movement patterns in juvenile Chinook Salmon (*O. tshawytscha*), Coho Salmon [20], Slimy Sculpin, and Arctic Grayling [21]. Although rivers and streams in the Williston Reservoir watershed showed considerable differences in water chemistry, water samples collected from various locations in the reservoir showed consistent elemental signatures. After emergence from their natal redds, Kokanee fry are presumed to migrate immediately downstream and into the reservoir. It was expected, therefore, that line scans would show little variation after equilibrating with waters in the reservoir—but this was not the case. Extensive variation in Sr and Ba along the length of the line scan suggest Kokanee are exploiting aquatic habitat with elemental signatures that differ from the reservoir. Although water is the predominant source of elements to the otolith, other factors, such as dietary shifts, can account for 2 to 41% of elements incorporated into the otolith [33]; however, seasonal or ontogenic patterns would then be expected and these were not observed.

4.6. Factors Affecting Homing/Straying

Columbia origin Kokanee were extensively stocked into tributaries of the Williston Reservoir watershed in the Peace and Parsnip Reaches starting in 1990. The number of returning mature Kokanee to the streams where they were stocked has historically been low. Environmental features of streams or social interactions with conspecifics, however, may have greater influence on spawning site selection than homing [34]. Aerial escapement estimates revealed that Columbia origin Kokanee dispersed rapidly after introduction and colonized to more than 68 new watersheds to streams that were greater than 200 km from the locations where they were stocked [10]. Although the original stocking program anticipated that stocking Kokanee eggs and fry would result in fish homing to natal streams [8], rapid dispersal and exploitation of new habitat following the introduction of a species is common [35]. Colonization of new streams through straying does not follow the typical paradigm for Pacific salmon homing to their natal streams [36], but examples of salmon straying over long distances are quite common. Most fish that stray have been recovered close to their natal streams, often within 25 km, which suggests straying within a watershed [37], but much more extensive straying has been recorded. Recovery of tagged Chinook Salmon from streams over distances up to 480 km has been reported [38]. Chinook Salmon introduced to the South Island of New Zealand colonized rivers 200 km north of the initial transplant site in less than 14 years, or approximately 3 generations [33]. Similar rates of dispersal have been found for Chinook Salmon in South America [39].

Population size, habitat availability, and disturbance contribute to dispersal of fish. Larger populations function as sources for fish that colonize new habitat [40]. The extensive stocking into the watershed of over 3.3 million Columbia origin Kokanee in the 1990s from Hill Creek and then Meadow Creek, created a large population and a source for dispersal [9]. Additionally, by 2008, it was estimated that the Kokanee population in the pelagic zone of the Williston Reservoir was up to 9 million fish [41], with escapement in some years up to 250,000 spawners within a single river [10]. The large population size, therefore, would favor range expansion and colonization of new rivers. There are also numerous tributaries to the reservoir that appear well suited for salmonids to spawn. Consequently, habitat was available, although Kokanee have selected tributaries in the Williston watershed that flow into the northwestern portion of the reservoir—not the regions originally stocked. Higher straying rates have also been reported for altered river systems. High rates of straying in Lower Columbia River Chinook Salmon have been found in response to disturbance [42]. Creation of the Williston Reservoir was clearly a disturbance to the watershed, but at the time of the Kokanee stocking program, productivity was high [43]. Changes associated with flooding the watershed even decades after formation, therefore, would also favor dispersal.

Although rates of straying in the first few generations after transplanting salmon to a novel environment are high, lower rates of straying appear to occur after the populations have been established [35]. Reduced straying and fidelity to spawning locations would create reproductively isolated populations and lead to genetic differentiation. The size of the watershed is clearly large enough for population level differences to exist (see [44] for work done on Arctic Grayling in the Williston Reservoir watershed). Columbia Kokanee, however, have only been in the watershed since 1990 at the earliest, which may not be enough time to establish differences. In an earlier study, we found no indication of genetic differences among the tributary spawning Kokanee of Columbia origin [26]. Although there have been few generations to establish genetic structure, the large scale of the watershed makes this somewhat surprising. The large number of Kokanee stocked [9] and the dynamic nature of the reservoir may continue to favor dispersal over homing to natal streams.

5. Conclusions

Our data indicated that, on average, 54.7% of Columbia origin Kokanee returned to their natal stream, suggesting a high rate of straying. The proportion of fish that returned to the watershed of their origin, however, was higher: 82.8%, suggesting that most fish return to their general region and spawn within a single river system and tributaries. The result of homing to natal streams is reproductive isolation and local adaptation. Local adaptation is beneficial, as locally adapted populations perform better when reared in their own environment compared to being transplanted to different environments [45]. Fine-scale local adaptation at the tributary level, however, will be limited by gene flow [46]. The most likely result of increased levels of migration would be local adaptation at a watershed level rather than a single tributary stream [47]. Our otolith microchemistry results indicated that local adaption might develop in the Williston Reservoir at a regional scale. Such local adaptation may also be maintained due to reduced fitness of migrants. For example, Sockeye Salmon that dispersed from similar habitat types (stream-spawners) had similar reproductive success to those that spawned in their natal stream, whereas Sockeye Salmon that dispersed from lake spawning habitats produced fewer offspring [48].

There are benefits to regional homing rather than stream-specific spawning fidelity. Natural stochastic events or over-exploitation may deplete local populations, but recovery may be facilitated with colonization from nearby donor populations [49]. Additionally, low levels of straying may benefit local adaptation to increase effective population size and maintain genetic variability—something that may be particularly important for small populations [50]. Range expansion of Columbia origin Kokanee continues to be a concern [51]. Whether expansion will continue or whether the fish have defined their "home" ranges will continue to be important to determine.

Author Contributions: Conceived and designed the experiment, J.M.S., P.W.B. and L.A.T.; collected the samples, J.M.S. and P.W.B.; analyzed the data, J.M.S. and L.A.T.; interpreted the results, J.M.S., P.W.B. and L.A.T.; wrote the manuscript, J.M.S.; writing—review and editing, J.M.S., P.W.B. and L.A.T.; project administration and funding acquisition, J.M.S., P.W.B. and L.A.T. All authors have read and agreed to the published version of the manuscript.

Funding: This project was funded by the Fish and Wildlife Compensation Program (FWCP), grant number PEA-F21-F-3176. The FWCP is a partnership between BC Hydro, the Province of BC, Fisheries and Oceans Canada, First Nations, and public stakeholders to conserve and enhance fish and wildlife in watersheds impacted by BC Hydro dams.

Institutional Review Board Statement: The study was conducted following the guidelines for experimental procedures in animal research from the Animal Care and Use Committee of the University of Northern British Columbia, protocols 2016-05, 2017-06, and 2018-10.

Data Availability Statement: Data from this study are available from the corresponding author upon request (J.M.S.; mark.shrimpton@unbc.ca).

Acknowledgments: We thank Chelsea Coady (FWCP), Matt Casselman (BC Hydro), Carmen Richter (Saulteau First Nations), Nikolaus Gantner (BCMoF), Randy Zemlak (BC Hydro), and Kristen Peck (DFO) for technical advice on the project, Arne Langston and Randy Zemlak (BC Hydro) for collecting fish in 2006, Lindi Anderson and Mike Tilson (Chu Cho Environmental) for collecting fish in 2016 and 2017, Andrew McDermot-Fouts and Jesse Laframboise (DWB Consulting) for collecting fish in 2018 under contract to BC MoFLNRORD funded by FWCP—Peace Region, Mike Stamford (Stamford Environmental) for providing unpublished water chemistry data and collecting water samples, Adrian Clarke (FFSBC) for providing unpublished water chemistry data, Claire Shrimpton and Phillip Shrimpton for assistance in collecting water samples, Erwin Rehl and Charles Bradshaw (Northern Analytical Lab Services) for ICP-OES analysis of water samples, Jody Spence and Nicole LaForge (UVic) for LA-ICP-MS analysis of otolith samples, Jennie Christensen and Dwayne Smith (TrichAnalytics) for LA-ICP-MS analysis of otolith samples, and two anonymous reviewers for their constructive comments.

Conflicts of Interest: The authors declare no conflict of interest.

References

1. Quinn, T.P. *The Behavior and Ecology of Pacific Salmon and Trout*; UBC Press: Vancouver, BC, Canada, 2005.
2. Quinn, T.P.; Stewart, I.J.; Boatright, C.P. Experimental evidence of homing to site of incubation by mature sockeye salmon, *Oncorhynchus nerka*. *Anim. Behav.* **2006**, *72*, 941–949. [CrossRef]
3. Turcotte, L.A.; Shrimpton, J.M. Assessment of spawning site fidelity in Interior Fraser River Coho Salmon *Oncorhynchus kisutch* using otolith microchemistry, in British Columbia, Canada. *J. Fish. Biol.* **2020**, *97*, 1833–1841. [CrossRef] [PubMed]
4. Baxter, C.V.; Hauer, F.R. Geomorphology, hyporheic exchange, and selection of spawning habitat by bull trout (*Salvelinus confluentus*). *Can. J. Fish. Aquat. Sci.* **2000**, *57*, 1470–1481. [CrossRef]
5. McRae, C.J.; Warren, K.D.; Shrimpton, J.M. Spawning site selection in Interior Fraser River coho salmon (*Oncorhynchus kisutch*): An imperiled population of anadromous salmon from an interior, snow-dominated watershed. *Endang. Species Res.* **2012**, *16*, 249–260. [CrossRef]
6. Tuor, K.M.F.; Shrimpton, J.M. Differences in water temperatures experienced by embryo and larval Coho Salmon (*Oncorhynchus kisutch*) across their geographic range in British Columbia. *Environ. Biol. Fish.* **2019**, *102*, 955–967. [CrossRef]
7. Shrimpton, J.M.; Heath, D.D. Census vs. effective population size in chinook salmon: Large- and small-scale environmental perturbation effects. *Mol. Ecol.* **2003**, *12*, 2571–2583. [CrossRef] [PubMed]
8. Blackman, B.G.; Jesson, D.A.; Ableson, D.; Down, T. *Williston Lake Fisheries Compensation Program Management Plan*; Peace/Williston Fish & Wildlife Compensation Program Report No. 58; BC Hydro: Prince George, BC, Canada, 1990; 38p.
9. Langston, A.R.; Murphy, E.B. *The History of Fish Introductions (to 2005) in the Peace/Williston Fish and Wildlife Compensation Program Area*; Peace/Williston Fish and Wildlife Compensation Program Report No. 325; BC Hydro: Prince George, BC, Canada, 2008; 59p.
10. Langston, A.R. *Williston Watershed Kokanee Spawner Distribution and Enumeration Surveys (2002–2006)*; Fish and Wildlife Compensation Program–Peace Region Report No. 357; BC Hydro: Prince George, BC, Canada, 2012; 11p.

11. McDermot-Fouts, A. *Williston Watershed Kokanee Spawner Distribution and Aerial Enumeration Surveys (2018)*; FWCP Project No. PEA-F19-F-2895-DC-103364; DWB Consulting Services Ltd.: Prince George, BC, Canada, 2019; 23p.
12. Robinson, M. *Williston Watershed Kokanee Spawner Distribution and Aerial Enumeration Surveys (2019)*; FWCP Project No. PEA-F20-F-3359-DC-106374; DWB Consulting Services Ltd.: Prince George, BC, Canada, 2020; 34p.
13. Langston, A.R.; Zemlak, R.J. *Williston Reservoir Stocked Kokanee Spawning Assessment, 1994*; Peace/Williston Fish and Wildlife Compensation Program, Report No. 176; BC Hydro: Prince George, BC, Canada, 1998; 13p.
14. Burgner, R.L. Life history of sockeye salmon (*Oncorhynchus nerka*). In *Pacific Salmon Life Histories*; Groot, C., Margolis, L., Eds.; UBC Press: Vancouver, BC, Canada, 1991; pp. 3–101.
15. Kennedy, B.P.; Blum, J.D.; Folt, C.L. Natural isotope markers in salmon. *Nature* **1997**, *387*, 766–767. [CrossRef]
16. Rutter, N.W. *Multiple Glaciations in the Area of Williston Lake, British Columbia*; Geological Survey of Canada Bulletin no. 273; Energy, Mines and Resources Canada: Ottawa, ON, Canada, 1976; 31p.
17. Armstrong, A.T. *1979 Report of Exploration Activities on North Carbon Creek Coal Property*; Utah Mines Ltd., Exploration Division: Vancouver, BC, Canada, 1979; 23p.
18. Wells, B.K.; Rieman, B.E.; Clayton, J.L.; Horan, D.L.; Jones, C.M. Relationships between water, otolith, and scale chemistries of westslope cutthroat trout from the Coeur d'Alene River, Idaho: The potential application of hard-part chemistry to describe movements in freshwater. *Tran. Am. Fish. Soc.* **2003**, *132*, 409–424. [CrossRef]
19. Clarke, A.D.; Telmer, K.H.; Shrimpton, J.M. Using natural elemental signatures to determine habitat use and population structure for a fluvial species, the Arctic grayling, in a watershed impacted by a large reservoir. *J. Appl. Ecol.* **2007**, *44*, 1156–1165. [CrossRef]
20. Shrimpton, J.M.; Warren, K.D.; Todd, N.L.; McRae, C.J.; Glova, G.J.; Telmer, K.H.; Clarke, A.D. Freshwater movement patterns in juvenile Pacific salmon *Oncorhynchus* spp. before they migrate to the ocean: Oh the places you'll go! *J. Fish Biol.* **2014**, *85*, 987–1004. [CrossRef]
21. Clarke, A.D.; Telmer, K.H.; Shrimpton, J.M. Movement patterns of fish revealed by otolith microchemistry: A comparison of putative migratory and resident species. *Environ. Biol. Fish.* **2015**, *98*, 1583–1597. [CrossRef]
22. Shiller, A.M. Syringe filtration methods for examining dissolved and colloidal trace element distribution in remote field locations. *Environ. Sci. Technol.* **2003**, *37*, 3953–3957. [CrossRef] [PubMed]
23. Freeze, R.A.; Cherry, J.A. *Groundwater*; Prentice-Hall: Englewood Cliffs, NJ, USA, 1979.
24. Taylor, B.R.; Hamilton, H.R. Comparison of methods for determination of total solutes in flowing waters. *J. Hydrol.* **1994**, *154*, 291–300. [CrossRef]
25. Clarke, A.D.; Telmer, K.; Shrimpton, J.M. *Discrimination of Habitat Use by Slimy Sculpin (Cottus cognatus) in Tributaries of the Williston Reservoir Using Natural Elemental Signatures*; Peace/Williston Fish and Wildlife Compensation Program Report No. 288; BC Hydro: Prince George, BC, Canada, 2004; 33p.
26. Wilson, P.N.; Shrimpton, J.M. *Genetic Population Structure and Demographics of Kokanee Introduced to the Williston Reservoir*; Fish & Wildlife Compensation Program—Peace Region Project Report PEA-F29-F-3143-DCA; BC Hydro: Prince George, BC, Canada, 2021; 41p.
27. Hegg, J.C.; Kennedy, B.P.; Chittaro, P. What did you say about my mother? The complexities of maternally derived chemical signatures in otoliths. *Can. J. Fish. Aquat. Sci.* **2019**, *76*, 81–94. [CrossRef]
28. Janak, J.M.; Linley, T.J.; Harnish, R.A.; Shen, S.D. Partitioning maternal and exogenous diet contributions to otolith $^{87}Sr/^{86}Sr$ in Kokanee Salmon (*Oncorhynchus nerka*). *Can. J. Fish. Aquat. Sci.* **2021**, *78*, 1146–1157. [CrossRef]
29. Kennedy, B.P.; Folt, C.L.; Blum, J.D.; Nislow, K.H. Using natural strontium isotopic signatures as fish markers: Methodology and application. *Can. J. Fish. Aquat. Sci.* **2000**, *57*, 2280–2292. [CrossRef]
30. Campana, S.; Thorrold, S. Otoliths, increments, and elements: Keys to a comprehensive understanding of fish populations? *Can. J. Fish. Aquat. Sci.* **2001**, *58*, 30–38. [CrossRef]
31. Campana, S.E. Chemistry and composition of fish otoliths: Pathways, mechanisms and applications. *Mar. Ecol. Prog. Ser.* **1999**, *188*, 263–297. [CrossRef]
32. Volk, E.C.; Blakley, A.; Schroder, S.L.; Kuehner, S.M. Otolith chemistry reflects migratory characteristics of Pacific salmonids: Using otolith core chemistry to distinguish maternal associations with sea and freshwaters. *Fish. Res.* **2000**, *46*, 251–266. [CrossRef]
33. Doubleday, Z.; Izzo, C.; Gillanders, B.; Woodcock, S. Relative contribution of water and diet to otolith chemistry in freshwater fish. *Aquat. Biol.* **2013**, *18*, 271–280. [CrossRef]
34. Dittman, A.H.; May, D.; Larson, D.A.; Moser, M.L.; Johnston, M.; Fast, D. Homing and spawning site selection by supplemented hatchery- and natural-origin Yakima River Chinook Salmon. *Trans. Am. Fish. Soc.* **2010**, *139*, 1014–1028. [CrossRef]
35. Quinn, T.P.; Kinnison, M.T.; Unwin, M.J. Evolution of chinook salmon (*Oncorhynchus tshawytscha*) populations in New Zealand: Pattern, rate, and process. *Genetica* **2001**, *112–113*, 493–513. [CrossRef]
36. Groot, C.; Margolis, L. *Pacific Salmon Life Histories*; UBC Press: Vancouver, BC, USA, 1991.
37. Hard, J.J.; Heard, W.R. Analysis of straying variation in Alaskan hatchery chinook salmon (*Oncorhynchus tshawytscha*) following transplantation. *Can. J. Fish. Aquat. Sci.* **1999**, *56*, 578–589. [CrossRef]
38. Candy, J.R.; Beacham, T.D. Patterns of homing and straying in southern British Columbia coded-wire tagged chinook salmon (*Oncorhynchus tshawytscha*) populations. *Fish. Res.* **2000**, *47*, 41–56. [CrossRef]
39. Correa, C.; Gross, M.R. Chinook salmon invade southern South America. *Biol. Invasions* **2008**, *10*, 615–639. [CrossRef]

40. Palstra, F.P.; O'Connell, M.F.; Ruzzante, D.E. Population structure and gene flow reversals in Atlantic salmon (*Salmo salar*) over contemporary and long-term temporal scales: Effects of population size and life history. *Mol. Ecol.* **2007**, *16*, 4504–4522. [CrossRef]
41. Sebastian, D.; Andrusak, G.; Scholten, G.; Langston, A. *Peace Project Water Use Plan Williston Fish Index, GMSMON-13. An Index of Fish Distribution and Abundance in Peace Arm of Williston*; Ministry of Environment: Victoria, BC, Canada, 2009; 58p.
42. Quinn, T.P.; Nemeth, R.S.; McIsaac, D.O. Homing and straying patterns of fall chinook salmon in the lower Columbia River. *Trans. Am. Fish. Soc.* **1991**, *120*, 150–156. [CrossRef]
43. Sebastian, D.C.; Scholten, G.H.; Woodruff, P.E. *Williston Reservoir Fish Assessment: Results of Hydroacoustic, Trawl and Gill Net Surveys in August 2000*; Stock Management Report No. 16; Ministry of Water Land and Air Protection: Victoria, BC, Canada, 2003; 34p.
44. Stamford, M.D.; Taylor, E.B. Population subdivision and genetic signatures of demographic changes in Arctic grayling (*Thymallus arcticus*) from an impounded watershed. *Can. J. Fish. Aquat. Sci.* **2005**, *62*, 2548–2559. [CrossRef]
45. Fraser, D.J.; Weir, L.K.; Bernatchez, L.; Hansen, M.M.; Taylor, E.B. Extent and scale of local adaptation in salmonid fishes: Review and meta-analysis. *Heredity* **2011**, *106*, 404–420. [CrossRef]
46. Hendry, A.P.; Taylor, E.B.; McPhail, J.D. Adaptive divergence and the balance between selection and gene flow: Lake and stream stickleback in the Misty system. *Evolution* **2002**, *56*, 1199–1216. [CrossRef]
47. Vähä, J.-P.; Erkinaro, J.; Niemelä, E.; Primmer, C.R. Temporally stable genetic structure and low migration in an Atlantic salmon population complex: Implications for conservation and management. *Evol. Applic.* **2008**, *1*, 137–154. [CrossRef] [PubMed]
48. Peterson, D.A.; Hilborn, R.; Hauser, L. Local adaptation limits lifetime reproductive success of dispersers in a wild salmon metapopulation. *Nat. Commun.* **2014**, *5*, 3696. [CrossRef] [PubMed]
49. Heath, D.D.; Busch, C.; Kelly, J.; Atagi, D.Y. Temporal change in genetic structure and effective population size in steelhead trout (*Oncorhynchus mykiss*). *Mol. Ecol.* **2002**, *11*, 197–214. [CrossRef] [PubMed]
50. Walter, R.P.; Aykanat, T.; Kelly, D.W.; Shrimpton, J.M.; Heath, D.D. Gene flow increases temporal stability of Chinook salmon (*Oncorhynchus tshawytscha*) populations in the Upper Fraser River, British Columbia, Canada. *Can. J. Fish. Aquat. Sci.* **2009**, *66*, 167–176. [CrossRef]
51. Pearce, T.; Morgan, J.; Case, S. *First Nations Information Gathering on Kokanee, Bull Trout and Arctic Grayling: Kwadacha Nation*; Fish & Wildlife Compensation Progrm—Peace Region Project No. PEA-F19-F-2866; BC Hydro: Prince George, BC, Canada, 2019; 18p.

MDPI

Article

Revealing Population Connectivity of the Estuarine Tapertail Anchovy *Coilia nasus* in the Changjiang River Estuary and Its Adjacent Waters Using Otolith Microchemistry

Tao Jiang [1], Hongbo Liu [1], Yuhai Hu [2], Xiubao Chen [1] and Jian Yang [1,2,*]

[1] Laboratory of Fishery Microchemistry, Freshwater Fisheries Research Center, Chinese Academy of Fishery Sciences, Wuxi 214081, China; jiangt@ffrc.cn (T.J.); liuhb@ffrc.cn (H.L.); chenxb@ffrc.cn (X.C.)
[2] Wuxi Fisheries College, Nanjing Agricultural University, Wuxi 214081, China; 2020113023@stu.njau.edu.cn
* Correspondence: jiany@ffrc.cn; Tel./Fax: +86-510-8555-7823

Abstract: The estuarine tapertail anchovy, *Coilia nasus*, is a migratory fish with high economic value in China. We collected fish from the Changjiang River (the Yangtze River) estuary, the Qiantang River estuary, and the southern Yellow Sea, and studied their relationships using otolith elemental and stable isotopic microchemistry signatures to assess the population connectivity of *C. nasus*. Results show that, in addition to Ca, other elements were present in the otolith core. The $\delta^{18}O$, Na/Ca, Fe/Ca, and Cu/Ca values of the Qiantang population were significantly higher than those of the others, whereas its $\delta^{13}C$ and Ba/Ca values were found to be significantly lower. Otolith multi-element composition and stable isotope ratios differed significantly between the Qiantang and Changjiang estuary groups ($p < 0.05$); however, no difference was observed between the latter and the Yellow Sea group. Cluster analysis, linear discriminant analysis, and a self-organizing map strongly suggest possible connectivity between the fish populations of the Changjiang estuary and Yellow Sea, while the population of the Qiantang River estuary appears to be independent. Notably, results suggest a much closer connectivity between the fish populations of the Changjiang River and the Yellow Sea.

Keywords: *Coilia nasus*; otolith; river estuary; Yellow Sea; elemental signature; stable isotopic signature

Citation: Jiang, T.; Liu, H.; Hu, Y.; Chen, X.; Yang, J. Revealing Population Connectivity of the Estuarine Tapertail Anchovy *Coilia nasus* in the Changjiang River Estuary and Its Adjacent Waters Using Otolith Microchemistry. *Fishes* **2022**, *7*, 147. https://doi.org/10.3390/fishes7040147

Academic Editor: Josipa Ferri

Received: 5 May 2022
Accepted: 22 June 2022
Published: 23 June 2022

1. Introduction

The connectivity between and identification of different populations of migratory fishes have been the focus of relevant fish surveys and conservation efforts [1–4]. Migration promotes gene exchange among different fish populations and hinders the development of their genetic structuring [5]. In contrast, it builds a bridge for resource flow among different populations. As a result, populations distributed differently in time and space tend to be connected [6,7]. In fact, revealing migration characteristics at a large spatial scale is quite difficult with regard to fishes with a complex habitat history [8,9]. Meanwhile, population connectivity and structure are also difficult to study in this field. Even genetic approaches may not have adequate resolutions, unless straying is negligible over evolutionary time scales [10].

Fortunately, with the development of interdisciplinary fields, otolith, which used to be an important material for the analysis of the age of fish, has increasingly become an important material for studying fish migration ecology. Otoliths, including lapillus, sagitta, and asteriscus, are types of hard tissues located in the fish inner ear, and play an important role in the senses of balance and hearing because of their unique shapes and microcrystalline structures [11]. Although an otolith was originally used as a tool to estimate the age of fish [12,13], analyses of its shape and elemental/isotopic microchemistry have recently demonstrated enormous potential for its use in studies of species, population, or ecosystem-based management [14–19]. Meanwhile, otolith elemental/isotopic chemistry, also termed as otolith microchemistry, is a powerful technique for revealing migration history, population connectivity of the research object, temporal and spatial distribution characteristics (using elemental signatures) [20–24], dietary changes, and environmental temperature in the species' life-history using stable isotopic signatures [25–28], even with limited (e.g., 2–10) numbers of otolith samples [21,23,29].

The chemical characteristics of otoliths can reflect biographical information of fish habitats throughout their life-history because otoliths are formed before the fish hatches, and elements deposited on them are not metabolized by the fish during their life cycle [30]. In particular, the nucleus region, defined as a small area in the central region of an otolith [31,32], which usually represents its initial growth period (i.e., larvae and juveniles) [33,34], is important as it corresponds to the fish's incubation and early life-history stage; therefore, the chemical characteristics of the otolith nucleus region can be used to determine the characteristics of hatchery/spawning sites and population origin, which can retrospectively determine the origin of the population [35–37]. In these studies, the elemental "fingerprint" and stable isotope characteristics are often used to determine the physical and chemical nature of the aqueous environment during the incubation and early life-history of research objects for identification and connectivity studies of populations [38,39]. In addition, there are three main channels for element deposition into the otolith: binding with protein in the otolith, inclusion into the calcium carbonate lattice, and substitution of calcium in calcium carbonate crystals [40–42]. The modes of deposition, as observed for the elements, are: (a) Cu and Zn through channel 1, (b) Mg through channel 2, and (c) Sr and Ba through channel 3, which reflect environmental concentrations [42]. The Sr/Ca and Ba/Ca ratios were used as common elemental ratios to study the habitat history and migration patterns of fish because of their direct relationships with the contents of the elements in waterbodies [24,43,44].

The estuarine tapertail anchovy *Coilia nasus* (Temminck et Schlegel, 1846) (junior synonym *Coilia ectenes* (Jordan et Seale, 1905) is a small migratory fish found in large rivers, such as the Changjiang River (the Yangtze River), the Yellow River, and the Qiantang River, and has a high economic value in China. However, its resources are facing a precarious situation because of overfishing and habitat destruction. By contrast, the values of two other ecotype populations, which are *C. nasus taihuensis* (with similar long supermaxilla) and *C. brachygnathus* (with a shorter supermaxilla), are significantly lower, in spite of their relatively substantial resources. As an anadromous fish, the adults migrate from the Yellow Sea and East China Sea upstream along the Changjiang River to spawn in affiliate lakes (e.g., the Poyang Lake), mainly from February to September every year [45]. Notably, the migration patterns of *C. nasus* populations in the Qiantang River are significantly different from those in the Changjiang River [46], indicating that they might be independent populations. In addition, *C. nasus* can be found around the Zhoushan Islands, outside of the mouth of the Qiantang River [47]. The juveniles possibly leave the Changjiang River and move southward into the adjacent East China Sea, and northward into the adjacent Yellow Sea [45]. However, the migration of *C. nasus* from the Qiantang estuary northward to the Yellow Sea has not been reported. Therefore, we hypothesize that *C. nasus* in the Qiantang estuary belong to a resource population with no connectivity to the populations found in the Yellow Sea and the Changjiang River.

The aim of the present study was to assess the population connectivity of *C. nasus* in the Changjiang River and Qiantang River estuaries and the southern Yellow Sea, using an analysis of elemental and stable isotopic microchemistry signatures of the otolith nucleus region of *C. nasus*. The results of the present study could provide a technical basis for the conservation and objective evaluation of *C. nasus* populations in the abovementioned waterbodies.

2. Materials and Methods

2.1. Sampling Sites

The Changjiang river is the main distribution area of *C. nasus*, and the affiliated lakes used to be the spawning grounds. Presently, many lakes along the Changjiang River are blocked by water gates, and only the Dongting Lake, the Poyang Lake, and the Shiju

Lake are believed to be still connected to the river (Figure 1). In our previous work, we confirmed that there was connectivity among *C. nasus* populations in the Poyang Lake, the Changjiang estuary, and the adjacent Yellow Sea [48]. Presently, three waterbodies, including two estuary waters and one sea water, were selected to discover the population connectivity and migration routes of *C. nasus* around the Changjiang estuary (Figure 1). The Jiuduansha Shoal, which used to be a fishing ground of *C. nasus*, is the distributary sandbar located in the southern branch of the Changjiang estuary, with an average depth of 0–6 m. The Qiantang estuary, which is adjacent to the Changjiang estuary, was also selected, to compare the difference in populations with those in the Changjiang River. In addition to these two estuary waters, the coastal waters of the Yellow Sea near Nantong City, which is named the Lvsi fishing ground and is one of the popular fishing grounds in China, was also chosen to detect the possible migration routes of *C. nasus*.

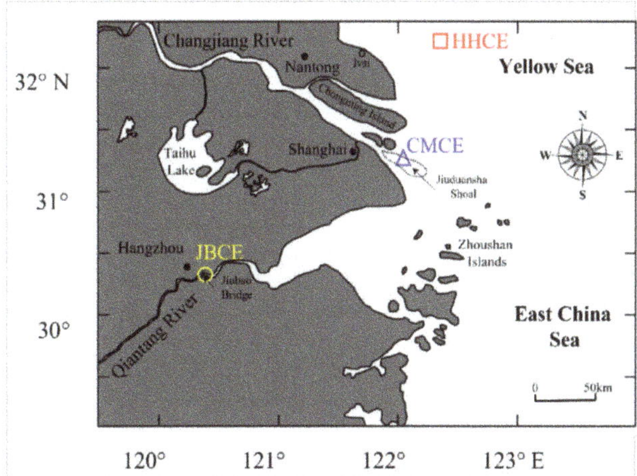

Figure 1. Sampling sites for *Coilia nasus*. Different markers indicate sampling sites in the Nantong coastal area of the Yellow Sea (HHCE, red square), the Changjiang estuary near the Jiuduansha Shoal (CMCE, blue triangle), and the Jiubao bridge waters in the Qiantang estuary (JBCE, yellow circle), from north to south.

2.2. Experimental Fish

Thirty *C. nasus* specimens were collected from the Qiantang estuary near the Jiubao Bridge waters (JBCE, 120.3° E, 30.3° N) in April 2016, the Changjiang estuary near the Jiuduansha Shoal (CMCE, 122.1° E, 31.2° N) in April 2016, and the coastal waters of the Yellow Sea near Nantong City (HHCE, 122.2° E, 32.2° N) in May 2016 (Figure 1). Sagittal otoliths were extracted from these specimens after measuring their total length and body weight (Table 1). After cleaning, the otoliths were embedded in epoxy resin (Epofix; Struers, Copenhagen, Denmark) with the proximal surface upside. The otoliths were then ground by a grinding machine (Discoplan-TS, Struers, Copenhagen, Denmark) to expose the cores and polished using an automated polishing machine (Roto Pol-35; Struers, Copenhagen, Denmark) with a polishing solution (OP-S NonDry, Struers, Copenhagen, Denmark) to remove scratches on the surface. Following this, a flat surface through the core of each otolith was obtained. After checking the annulus and determining the age of each otolith, the samples were cleaned in an ultrasonic bath, rinsed with Milli-Q water (Millipore, Molsheim, France), and oven-dried at 38 °C overnight.

Table 1. Sampling details of *Coilia nasus* in the present study.

Sampling Site	Sample Code	Sampling Date	Sample Size (*N*)	Total Length (mm)	Body Weight (g)	Age [a] (Year)	Gonad Stage [b]
Coastal water in the Yellow Sea near Nantong City	HHCE	13 April 2016	10	261 ± 14	49.00 ± 15.56	2	II
Changjiang River estuary near Jiuduansha Shoal	CMCE	1–5 April 2016	10	284 ± 10	64.61 ± 7.75	2	II (2 ind.) III (8 ind.)
Qiantang River estuary near Jiubao bridge waters	JBCE	10 May 2016	10	271 ± 12	46.85 ± 7.79	2	III

Note: [a] Fish age was determined by the annuli in the otolith [49]. [b] Gonad stages of maturation were determined by visual examination of the gonads [50,51], i.e., stage II (developmental stage), stage III (early mature stage), etc.

2.3. Elemental Analysis

Elemental signature analysis of the otolith nucleus region, which was estimated at 0–1 month [52], was performed at the Laboratory of Fishery Microchemistry at the Freshwater Fisheries Research Center, Wuxi, China, using an NW213 laser ablation system (New Wave Research, Fremont, CA, USA) coupled with an Agilent 7500ce ICPMS (Agilent Technologies, Wilmington, DE, USA). Nineteen elements, including Li, Na, Mg, Al, K, Ca, Sc, Ti, V, Cr, Mn, Fe, Co, Ni, Cu, Zn, Sr, Cd, and Ba, were analyzed with a laser wavelength of 213 nm. The applied voltage, pulse rate, energy intensity, and residence time were 10 kV, 10 Hz, 11.76 J/cm^2, and 5 s, respectively. Because the nucleus region was defined as the round area that spreads at a radius of 190–230 μm from the core [53], a smaller beam width of 80 μm, which was estimated at approximately 10 d based on our unpublished work and previous literature [52], was selected for ablating the nucleus regions of all samples.

NIST612 and MACS-3 were used as the standards and analyzed after every 10 samples; 100-s blank data were used as the background signal to calculate the limits of detection (LOD) before and after the analysis. All results were expressed as the molecular weight ratio (mmol/mol) of the element to Ca. During the analysis, the relative standard deviation (RSD) of the corresponding elemental ratio in the signal of the standard sample was calculated to determine the signal stability (RSD% < 10).

2.4. Stable Isotope Analysis

Stable isotope analysis was also performed at the Laboratory of Fishery Microchemistry, Freshwater Fisheries Research Center, Wuxi, China. After elemental analysis, the otoliths were repolished to remove the laser ablation burn scars and cleaned in an ultrasonic bath, rinsed with Milli-Q water (Millipore, Molsheim, France), and oven-dried at 38 °C overnight. A MicroMill (New Wave Research, Fremont, CA, USA) microsampler was used to drill the nucleus region (see Figure 2), and powdered samples were collected. For adequate signals, the depth was set as 300 μm, while the diameter of the drilling hole was 198.1 ± 2.5 μm in the central parts of the aforementioned otolith core areas [53], which were estimated at approximately 25 d based on our unpublished data and a previous report [52]. Otolith isotope analysis was performed using a Delta V Advantage stable isotope ratio mass spectrometer equipped with a Gas Bench II Gas generator (Thermo Fisher Scientific, Waltham, MA, USA). The samples were placed at the bottom of glass test tubes (QSA00303, Thermo Fisher Scientific) before analysis. The natural abundance of carbon/oxygen stable isotope was expressed as follows:

$$\delta X\ (‰) = [(R_{otolith} - R_{standard})/R_{standard}] \times 1000$$

where X represents $\delta^{13}C$ and $\delta^{18}O$ and R represents the $^{13}C/^{12}C$ or $^{18}O/^{16}O$ ratio. The accuracies of $\delta^{13}C$ and $\delta^{18}O$ detection were monitored against an isotope standard (NBS 18, International Atomic Energy Agency, Vienna, Austria) and expressed using Vienna Pee Dee Belemnite (VPDB) as an international standard unit. The precision of the analyses based on the measurements of this standard was within 0.03%.

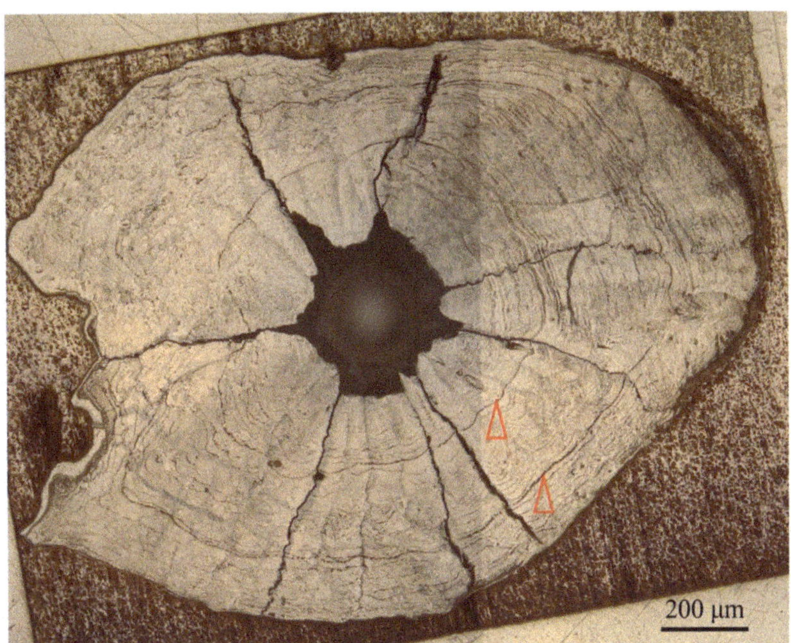

Figure 2. Otolith after drilling of the nucleus region. The red triangles show the annuli.

2.5. Statistical Analysis

The homogeneity of variance and the normality of the distribution of all data were checked using SPSS 23.0 (IBM SPSS Statistics Inc., Chicago, IL, USA). Multiple analysis of variance (MANOVA) and one-way analysis of variance (ANOVA) were carried out to assess the differences among all three groups and between independent groups using SPSS 23.0 (IBM SPSS Statistics Inc.). All elemental and stable isotopic ratios were centered using the Z-Score and displayed using the pheatmap (v 1.0.12) package in R (Ver. 4.1.2). A cluster analysis was also performed based on the complete linkage method, with Euclidean distance as the distance measuring method using the same package. The ratios were further examined using principal component analysis (PCA) and linear discriminant analysis (LDA) in SPSS 23.0 and self-organizing map (SOM) analysis, with the help of the Toolbox2.0 toolkit in Matlab 9.0 (Mathworks, Inc., El Segundo, CA, USA), to study the relationships between the groups. The map size was set according to $5 \times (\text{number of samples})^{\frac{1}{2}}$ [54], and then based on the minimizing quantization error (QE) and topographic error (TE) [55]. The number of groups was set mainly based on the clustering analysis. The cells were then divided into different groups according to the similarity of the weight vectors of the neurons by Ward's linkage method [56].

3. Results

3.1. Elemental Ratios in the Otolith Nucleus Regions of C. nasus

In addition to Ca, we detected Na, Mn, Fe, Ni, Cu, Sr, and Ba in the samples with %RSD ranging from 2.05 (Sr/Ca) to 8.5 (Ni/Ca), and LODs (mmol/mol) for Na/Ca (0.031), Mn/Ca (6.2×10^{-4}), Fe/Ca (0.0051), Ni/Ca (6.4×10^{-5}), Cu/Ca (1.1×10^{-4}), Sr/Ca (6.6×10^{-4}), and Ba/Ca (3.2×10^{-6}), based on the standards (Table 2). Apart from Mn/Ca and Sr/Ca, for which there were no significant differences among the three groups, there was no significant difference between the specimens of HHCE and CMCE in the elemental ratios ($p > 0.05$). Meanwhile, four elemental ratios determined in the present study- Na/Ca,

Fe/Ca, Ni/Ca, and Ba/Ca- were significantly different between the specimens of CMCE and JBCE ($p < 0.05$) (Table 2).

Table 2. Elemental ratios in the otolith nucleus regions of *Coilia nasus*.

Item	Elemental Ratio (mmol/mol)						
	Na/Ca	Mn/Ca	Fe/Ca	Ni/Ca	Cu/Ca	Sr/Ca	Ba/Ca
RSD (%)	7.33	5.42	2.31	8.50	6.25	2.05	3.14
LOD	0.031	6.20×10^{-4}	0.0051	6.40×10^{-5}	1.10×10^{-4}	6.60×10^{-6}	3.20×10^{-6}
HHCE	8.73 ± 0.3^{a}	0.012 ± 0.008^{a}	0.43 ± 0.01^{a}	0.0077 ± 0.0011^{ab}	0.00030 ± 0.00007^{a}	0.69 ± 0.14^{a}	0.013 ± 0.006^{ab}
CMCE	8.51 ± 0.36^{a}	0.011 ± 0.01^{a}	0.42 ± 0.01^{a}	0.0072 ± 0.0007^{b}	0.00035 ± 0.00015^{ab}	0.70 ± 0.10^{a}	0.017 ± 0.005^{b}
JBCE	9.15 ± 0.3^{b}	0.005 ± 0.005^{a}	0.46 ± 0.01^{b}	0.0082 ± 0.0009^{a}	0.00043 ± 0.00008^{b}	0.74 ± 0.21^{a}	0.009 ± 0.004^{a}

Note: Different superscript letters in each column indicate significant differences between groups ($p < 0.05$). *C. nasus* collected in the Nantong coastal region of the Yellow Sea (HHCE), *C. nasus* collected in the Changjiang estuary near the Jiuduansha Shoal (CMCE), and *C. nasus* collected in the Jiubao Bridge waters in the Qiantang estuary (JBCE). Data are mean ± standard deviation.

3.2. Stable Isotopic Ratios in the Otolith Nucleus Regions of *C. nasus*

The $\delta^{13}C$ and $\delta^{18}O$ in the otolith nucleus region of *C. nasus* showed no significant differences between HHCE and CMCE ($p > 0.05$). The $\delta^{13}C$ of JBCE was lower than that of the other groups, while $\delta^{18}O$ was higher to a very significant extent ($p < 0.05$) (Table 3).

Table 3. Stable isotopic ratios in the otolith nucleus regions of *Coilia nasus*.

Group	Stable Isotopic Ratios (‰, VPDB)	
	$\delta^{13}C$	$\delta^{18}O$
HHCE	-10.42 ± 1.81^{a}	-7.72 ± 1.79^{a}
CMCE	-10.31 ± 1.49^{a}	-7.32 ± 0.93^{a}
JBCE	-13.12 ± 1.00^{b}	-4.08 ± 0.72^{b}

Note: Different superscript letters in each column indicate significant differences ($p < 0.05$). *C. nasus* collected in the Nantong coastal region of the Yellow Sea (HHCE), *C. nasus* collected in the Changjiang estuary near the Jiuduansha Shoal (CMCE), and *C. nasus* collected in the Jiubao Bridge waters in the Qiantang estuary (JBCE). Data are mean ± standard error.

3.3. Multivariate Statistics Based on Elemental and Stable Isotopic Ratios

Results of the MANOVA indicated that there was no significant difference between the HHCE and CMCE specimens ($p > 0.05$). However, significant differences were observed between JBCE and the other specimens ($p < 0.05$).

The $\delta^{18}O$, Na/Ca, Fe/Ca, and Cu/Ca values of *C. nasus* in JBCE were found to be higher (red boxes) than those in the other populations, while $\delta^{13}C$ and Ba/Ca values were found to be lower (blue box) (Figure 3). According to these elemental and stable isotope ratios, the results of the cluster analysis showed that the *C. nasus* of JBCE formed a distinct single group, while the individuals of HHCE and CMCE showed a mixed pattern. Consistent with this result, PCA also clearly showed that the individuals of HHCE and CMCE were mixed well, while those of JBCE were independent from the others (Figure 4).

Cross-validation through LDA showed that the discriminant accuracy of *C. nasus* of HHCE and CMCE was 60%. The remaining 40% were misjudged as individuals of each other's groups. The discriminant accuracy for the JBCE group was 100%, which was different from that of the other two groups (Table 4 and Figure 5).

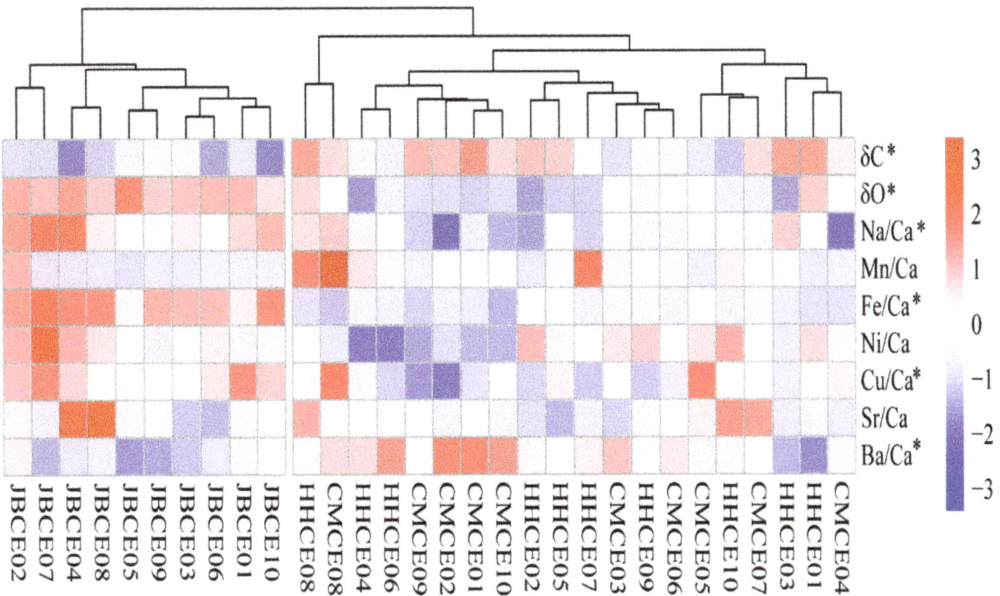

Figure 3. Microchemical characteristics of *Coilia nasus* from the Nantong coastal area of the Yellow Sea (HHCE), the Changjiang estuary near the Jiuduansha Shoal (CMCE), and the Jiubao bridge waters in the Qiantang estuary (JBCE), based on the Pheatmap. The different shades of blue and red reflect different values. The (*) symbol represents a significant difference ($p < 0.05$, one-way ANOVA).

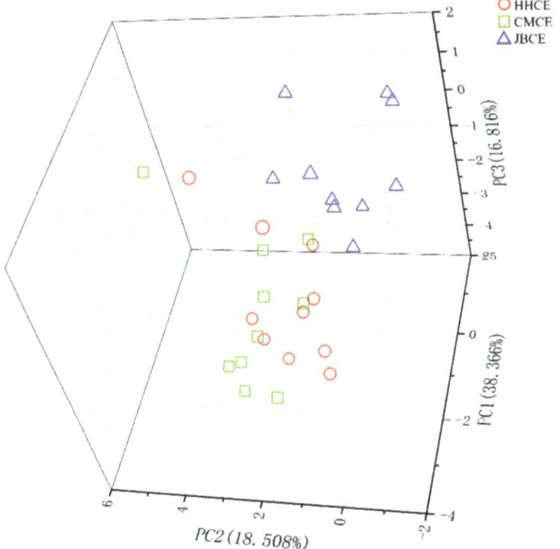

Figure 4. Principal component analysis of *Coilia nasus* from the Nantong coastal area of the Yellow Sea (HHCE), the Changjiang estuary near the Jiuduansha Shoal (CMCE), and the Jiubao bridge waters in the Qiantang estuary (JBCE), based on microchemical characteristics of the otolith core. Those of HHCE and CMCE were mixed well, while those of JBCE were independent from the others.

Table 4. Linear discriminant analysis of *Coilia nasus* populations in the Yellow Sea (HHCE), the Changjiang estuary (CMCE), and the Qiantang estuary (JBCE) based on $\delta^{13}C$, $\delta^{18}O$, and elemental ratio characteristics in the otolith nucleus region.

Group		Predicted Group (Original/Cross-Validated) HHCE	CMCE	JBCE
Count	HHCE	6 (6)	4 (4)	0 (0)
	CMCE	3 (4)	7 (6)	0 (0)
	JBCE	0 (0)	0 (0)	10 (10)
%	HHCE	60 (60)	40 (40)	0 (0)
	CMCE	30 (40)	70 (60)	0 (0)
	JBCE	0 (0)	0 (0)	100 (100)

Note: HHCE: *C. nasus* collected in the Nantong coastal region of the Yellow Sea. CMCE: *C. nasus* collected in the Changjiang estuary near the Jiuduansha Shoal. JBCE: *C. nasus* collected in the Jiubao Bridge waters in the Qiantang estuary.

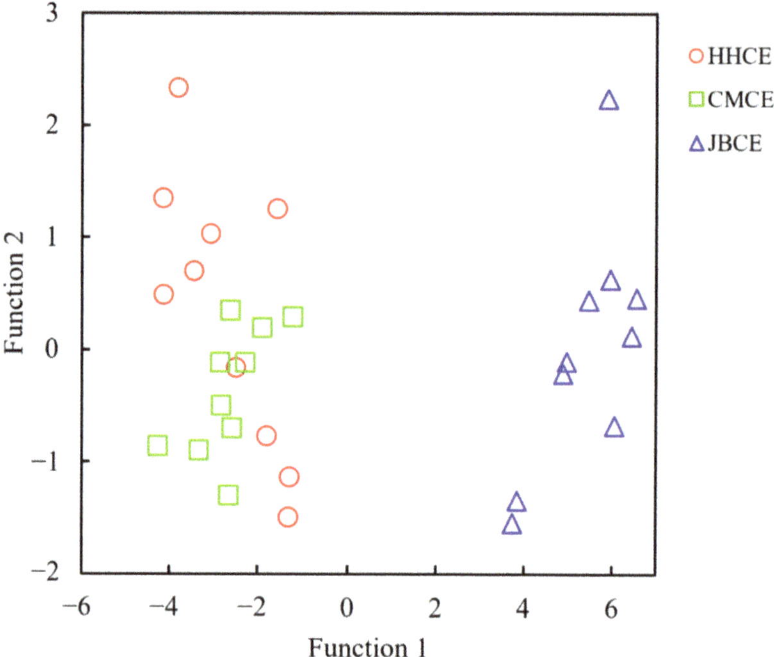

Figure 5. Discriminant analysis of *Coilia nasus* from the Nantong coastal area of the Yellow Sea (HHCE), the Changjiang estuary near the Jiuduansha Shoal (CMCE), and the Jiubao bridge waters in the Qiantang estuary (JBCE), based on microchemical characteristics of the otolith core. Those of HHCE and CMCE were mixed, while those of JBCE were independent from the others.

The SOM analysis also clearly indicated a similar relationship. Results showed that the JBCE was clearly distinguished from the other two populations (CMCE and HHCE). The latter two populations gathered into a large group with 25 (5 × 5) output cells, corresponding to 30 samples from three waterbodies, which could obtain the best network training effect with QE and TE values of 0.345 and 0.024, respectively (Figure 6).

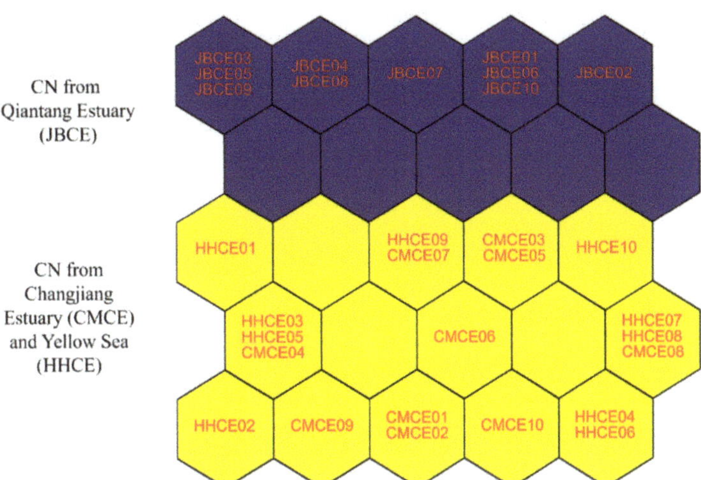

Figure 6. Self-organizing map (SOM) analysis of *Coilia nasus* from the Nantong coastal area of the Yellow Sea (HHCE), the Changjiang estuary near the Jiuduansha Shoal (CMCE), and the Jiubao bridge waters in the Qiantang estuary (JBCE). Those of HHCE and CMCE were mixed in yellow cells, while those of JBCE were independent of the others in blue cells.

4. Discussion

The deposition mechanisms of different elements were distinct and were derived from ambient waterbodies. Consequently, the interannual changes in otolith elements were smaller than their spatial differences [57–59]. Therefore, in addition to Sr/Ca and Ba/Ca ratios, the ratios of other elements to Ca in special areas of otoliths (such as core areas corresponding to hatcheries) may also reflect the differences in the research objects, and these may be used to understand population structures and connectivity characteristics [37,58]. The elemental ratios Na/Ca, Mn/Ca, Fe/Ca, Ni/Ca, Cu/Ca, Sr/Ca, and Ba/Ca were selected according to a previous study [48], and the present results suggested that those elemental characteristics could be used as natural tags in discriminating *C. nasus* of the Changjiang Estuary and adjacent waterbodies.

The stable isotopic ratios were significantly different between the *C. nasus* from the Qiantang estuary and those from the Changjiang estuary and the Yellow Sea. Notably, juveniles of *C. nasus* usually live in freshwater during their first year of growth [60]. The length the otolith of *C. nasus* during its early life history in the freshwater was 980 ± 111 μm [47], which means that the materials extracted from a hole with a diameter of 200 μm (corresponding to an age of 25 d) in the nucleus region mainly reflect the early freshwater life-history in their natal river. Therefore, the respective region should also be named as the "natal region." In addition to regional differences, the differences in $\delta^{18}O$ between otoliths and water was negatively correlated with water temperature, and therefore it can be used as an indicator of ambient water temperature [25,61]. The significant differences between the populations of *C. nasus* in the Qiantang estuary and those in the Yellow Sea and the Changjiang estuary may mainly reflect the differences in the average water temperatures during the early life of *C. nasus*. The higher $\delta^{18}O$ of *C. nasus* in the Qiantang estuary, the lower the temperature of this waterbody compared to that of the Changjiang River and the Yellow Sea. This may result in earlier breeding of *C. nasus* in the Qiantang River. The Qiantang River is located south of the Changjiang River. These facts are consistent with the report that the populations of *C. nasus* in the northern waterbodies start migration in May–June, while the populations in the southern region, such as the Changjiang estuary, start in March [62]. Future studies can advantageously use our results for the determination of exact spawning times and locations. Furthermore, future studies should consider more diverse environments, in order to reveal the relationship between chemical information obtained from otoliths and environmental conditions.

Both water temperature and $\delta^{13}C$ in ambient water can partly influence the content of $\delta^{13}C$ in the otolith [63,64]; however, the diet condition of the fish will continue to have the most important role [65,66]. Results indicated that the populations of *C. nasus* in the Yellow Sea and the Changjiang estuary were significantly greater than that in the Qiantang estuary. Previous studies have shown that the Poyang Lake is one of the important spawning grounds for the migratory population in the Changjiang River, while there is no large lake to serve as a spawning ground along the Qiantang River. Therefore, the food conditions that *C. nasus* experiences during its early life are expected to be different between the Yellow Sea and Changjiang estuary, that have a higher $\delta^{13}C$, and the Qiantang estuary, that has a lower $\delta^{13}C$.

In the present study, the gonad development of most *C. nasus* fish of both the Changjiang and Qiantang populations were at gonad stage III (Table 2), indicating that both populations were reproductive [50] and that their spawning sites were located in the corresponding rivers, as they had already migrated into the estuary waters. As previously mentioned, the analyzed nucleus region represented approximately 10-d and 25-d life histories in the larva stage [52]. Although there was no report on how fast the larvae can move or drift from their originating locations (i.e., spawning site and hatchery) until now, results of both the trace elemental and isotopic ratios reflect the conditions prevailing in their originating rivers during such a short period, and in addition, during their first year of growth in freshwater [60]. Results of multivariate statistics, based on the elemental and stable isotopic ratios, clearly reveal the possibility of connectivity among the populations of the three waterbodies. Although the sample size was relatively low in the present study, the cluster analysis, PCA, LDA, and SOM clearly show that *C. nasus* in the Qiantang estuary is a population that is independent of those in the Changjiang and the Yellow Sea, which is consistent with previous studies that have revealed significant differences in the life history of the Qiantang population with those of the Changjiang and the Yellow Sea [46,47,60]. Meanwhile, the population of the Yellow Sea mixed well with that of the Changjiang, indicating the possibility of connectivity between these two populations. In addition, the Qiantang and Changjiang Rivers are adjacent to each other. Findings of previous [46,47] and the present study indicated that some individuals can be "mixed" or "lost" in the two adjacent rivers during the migratory season, which strongly suggests that the adults can recognize and return to their natal river for spawning, which highlights a strong "natal homing" feature in the species, like in the case of salmon [67].

Results of the present study clearly indicate the connectivity between the fish populations in the Changjiang estuary and the Yellow Sea. This is in accordance with the results of our previous study, in which samples were gathered in 2011–2012 [48]. Therefore, the results of both these studies suggest that some juveniles from the Changjiang River swam northeast into the Yellow Sea after almost one year of freshwater life history [60]. By contrast, there was no connectivity between the JBCE and HHCE populations, which were randomly sampled during the migratory season, as suggested by their different life histories and migratory patterns [46,60]. Because of the low sample size, the ability of the Qiantang population to move north into the Yellow Sea could not be confirmed; however, the results imply a stronger connectivity between the Yellow Sea and Changjiang populations than with the Qiantang population.

Knowledge on the fine scale natal origin of the anadromous fish *C. nasus* is difficult to obtain because of gaps in knowledge on their distribution in spawning sites and in surveys of resources during their early growth stage. Fortunately, we discovered a spawning site in the Poyang Lake, the largest freshwater lake in China, which is affiliated to the Changjiang River [68]. We expect that other spawning sites are likely to exist along the river [48]. The elemental ratios or $\delta^{13}C$ and $\delta^{18}O$ in the otoliths provide adequate information on the connectivity among different populations, compared with establishing connectivity based on the identification of original locations (i.e., spawning/hatchery site) without considering a special population, such as fully mature adults within the spawning site [48] or weakly swimming larvae and juveniles within the natal waters [10]. However, the strontium isotope ratio ($^{87}Sr/^{86}Sr$) in the otolith seems to accurately delineate the natal origin at a fine scale, as it has a high correlation with that of ambient waters with spatial features at a fine scale [69]. In the future, a robust framework should be built based on the strontium isotope ratios of both the otoliths and habitat waters of *C. nasus*, which will play an important role in assessing natal sources at a fine spatial scale and delineate its detailed migratory route during different developmental stages.

In addition, the results of this study show that the *C. nasus* population in the Yellow Sea enters the Qiantang estuary, and that from the Qiantang estuary it enters into the Yellow Sea or the Changjiang River estuary. This indicates that *C. nasus* migration may be very accurate, like that of salmon migration. Salmon primarily relies on smell to locate hatching/spawning sites accurately [70]. A similar olfactory gene was also found in *C. nasus* [71]. Therefore, *C. nasus* may also use "olfactory" memories to return to their hatcheries for breeding. This behavior of migratory fish, in which a population can return to respective spawning ground with a high success rate of hatching, is known as natal homing, and it guarantees a considerable population reproduction success rate [5]. It is likely that *C. nasus* displays similar behavior.

5. Conclusions

The aforementioned results of otolith microchemistry strongly suggest that there are two natal original populations in the Qiantang and Changjiang Rivers. They also suggest that the population in the Yellow Sea has little connectivity with that of the Qiantang River, but that it had a supplementary relationship with that of the Changjiang River. Nevertheless, whether the population of the Qiantang can move north into the Yellow Sea remains to be confirmed because of the limitation of the low sample size of the present study and the knowledge gaps on the marine life histories and fine scale natal origins of these three populations. Consequently, future research should focus on the connectivity of populations of *C. nasus*, which inhabit the South–East China Sea and the adjacent river estuaries, based on otolith microchemistry and assessment of environmental conditions (i.e., water temperature, salinity, etc.), to determine the migratory dynamics of the Changjiang, Qiantang, and other southern populations. Furthermore, the elemental characteristics of the otolith from the core to the edge, and a robust framework based on strontium isotope ratios of both otolith and waters, must be studied to arrive at a more accurate migration route, in order to assess the finer-scale natal origin of *C. nasus*. In addition, as there is the possibility of natal homing behavior, resource estimation and conservation work should be carried out for different populations of *C. nasus*.

Author Contributions: Conceptualization, T.J., J.Y., H.L., and X.C.; Methodology, T.J., H.L., Y.H., X.C., and J.Y.; Investigation, T.J., H.L., Y.H., and J.Y.; Resources, T.J. and J.Y.; Funding acquisition, T.J. and J.Y.; Writing—Original Draft, T.J., H.L., Y.H., and X.C.; Writing—Review & Editing, J.Y.; Supervision, J.Y. All authors have read and agreed to the published version of the manuscript.

Funding: This research was funded by the Central Public-interest Scientific Institution Basal Research Fund, Freshwater Fisheries Research Center, CAFS (2021JBFM14), and the Central Public-interest Scientific Institution Basal Research Fund, CAFS (2021GH08).

Institutional Review Board Statement: The study was conducted according to the guidelines of the Declaration of Helsinki. All *Coilia nasus* samples were dead individuals from legal commercial fisheries. Therefore, there was no need to apply for an approval by the Animal Ethics Committee.

Informed Consent Statement: Not applicable.

Data Availability Statement: The data that support the findings of this study are available from the corresponding author upon reasonable request.

Conflicts of Interest: The authors declare no conflict of interest.

References

1. Avigliano, E.; Carvalho, B.; Velasco, G.; Tripodi, P.; Vianna, M.; Volpedo, A.V. Nursery areas and connectivity of the adult anadromous catfish (*Genidens barbus*) revealed by otolith-core microchemistry in the south-western Atlantic Ocean. *Mar. Freshw. Res.* **2017**, *68*, 931–940. [CrossRef]
2. Rohtla, M.; Matetski, L.; Svirgsden, R.; Kesler, M.; Taal, I.; Saura, A.; Vaittinen, M.; Vetemaa, M. Do sea trout *Salmo trutta* parr surveys monitor the densities of anadromous or resident maternal origin parr, or both? *Fish. Manag. Ecol.* **2017**, *24*, 156–162. [CrossRef]
3. Koeberle, A.L.; Arismendi, I.; Crittenden, W.; Di Prinzio, C.; Gomez-Uchida, D.; Noakes, D.L.G.; Richardson, S. Otolith shape as a classification tool for Chinook salmon (*Oncorhynchus tshawytscha*) discrimination in native and introduced systems. *Can. J. Fish. Aquat. Sci.* **2020**, *77*, 1172–1188. [CrossRef]
4. Beacham, T.D.; Wallace, C.G.; Jonsen, K.; McIntosh, B.; Candy, J.R.; Horst, K.; Lynch, C.; Willis, D.; Luedke, W.; Kearey, L.; et al. Parentage-based tagging combined with genetic stock identification is a cost-effective and viable replacement for coded-wire tagging in large-scale assessments of marine Chinook salmon fisheries in British Columbia, Canada. *Evol. Appl.* **2021**, *14*, 1365–1389. [CrossRef] [PubMed]
5. McDowall, R.M. Diadromy, diversity and divergence: Implications for speciation processes in fishes. *Fish Fish.* **2001**, *2*, 278–285. [CrossRef]
6. Torniainen, J.; Vuorinen, P.J.; Jones, R.I.; Keinänen, M.; Palm, S.; Vuori, K.A.M.; Kiljunen, M. Migratory connectivity of two Baltic Sea salmon populations: Retrospective analysis using stable isotopes of scales. *ICES J. Mar. Sci.* **2014**, *71*, 336–344. [CrossRef]
7. Alshwairikh, Y.A.; Kroeze, S.L.; Olsson, J.; Stephens-Cardenas, S.A.; Swain, W.L.; Waits, L.P.; Horn, R.L.; Narum, S.R.; Seaborn, T. Influence of environmental conditions at spawning sites and migration routes on adaptive variation and population connectivity in Chinook salmon. *Ecol. Evol.* **2021**, *11*, 16890–16908. [CrossRef]
8. Arai, T. Migration ecology in the freshwater eels of the genus *Anguilla* Schrank, 1798. *Trop. Ecol.* **2022**, *6*, e05176. [CrossRef]
9. Miller, M.J.; Yoshinaga, T.; Aoyama, J.; Otake, T.; Mochioka, N.; Kurogi, H.; Tsukamoto, K. Offshore spawning of *Conger myriaster* in the western North Pacific: Evidence for convergent migration strategies of anguilliform eels in the Atlantic and Pacific. *Naturwissenschaften* **2011**, *98*, 537–543. [CrossRef]
10. Thorrold, S.R.; Latkoczy, C.; Swart, P.K.; Jones, C.M. Natal homing in a marine fish metapopulation. *Science* **2001**, *291*, 297–299. [CrossRef]
11. Schulz Mirbach, T.; Ladich, F.; Plath, M.; Heß, M. Enigmatic ear stones: What we know about the functional role and evolution of fish otoliths. *Biol. Rev.* **2019**, *94*, 457–482. [CrossRef] [PubMed]
12. Burton, M.L.; Potts, J.C.; Ostrowski, A.D.; Shertzer, K.W. Age, growth, and natural mortality of Graysby, *Cephalophilis cruentata*, from the southeastern United States. *Fishes* **2019**, *4*, 36. [CrossRef]
13. Burton, M.L.; Potts, J.C.; Ostrowski, A.D. Preliminary estimates of age, growth and natural mortality of margate, *Haemulon album*, and black margate, *Anisotremus surinamensis*, from the Southeastern United States. *Fishes* **2019**, *4*, 44. [CrossRef]
14. Geffen, A.J.; Morales-Nin, B.; Gillanders, B.M. Fish otoliths as indicators in ecosystem based management: Results of the 5th International Otolith Symposium (IOS2014). *Mar. Freshw. Res.* **2016**, *67*, i–iv. [CrossRef]
15. D'Iglio, C.; Albano, M.; Famulari, S.; Savoca, S.; Panarello, G.; Di Paola, D.; Perdichizzi, A.; Rinelli, P.; Lanteri, G.; Spano, N.; et al. Intra- and interspecific variability among congeneric *Pagellus* otoliths. *Sci. Rep.* **2021**, *11*, 16315. [CrossRef]
16. D'Iglio, C.; Natale, S.; Albano, M.; Savoca, S.; Famulari, S.; Gervasi, C.; Lanteri, G.; Panarello, G.; Spanò, N.; Capillo, G. Otolith analyses highlight morpho-functional differences of three species of mullet (Mugilidae) from transitional water. *Sustainability* **2022**, *14*, 398. [CrossRef]
17. Mitsui, S.; Strüssmann, C.A.; Yokota, M.; Yamamoto, Y. Comparative otolith morphology and species identification of clupeids from Japan. *Ichthyol. Res.* **2020**, *67*, 502–513. [CrossRef]
18. D'Iglio, C.; Albano, M.; Tiralongo, F.; Famulari, S.; Rinelli, P.; Savoca, S.; Spanò, N.; Capillo, G. Biological and ecological aspects of the blackmouth catshark (*Galeus melastomus* Rafinesque, 1810) in the Southern Tyrrhenian Sea. *J. Mar. Sci. Eng.* **2021**, *9*, 967. [CrossRef]
19. Polito, M.J.; Trivelpiece, W.Z.; Karnovsky, N.J.; Ng, E.; Patterson, W.P.; Emslie, S.D. Integrating stomach content and stable isotope analyses to quantify the diets of pygoscelid penguins. *PLoS ONE* **2011**, *6*, e26642. [CrossRef]
20. Chino, N.; Arai, T. Migratory history of the fourspine sculpin *Rheopresbe kazika*, a national monument species in Japan. *Thalassa* **2020**, *36*, 395–398. [CrossRef]
21. Tran, N.T.; Labonne, M.; Chung, M.T.; Wang, C.H.; Huang, K.F.; Durand, J.D.; Grudpan, C.; Chan, B.; Hoang, H.D.; Panfili, J. Natal origin and migration pathways of Mekong catfish (*Pangasius krempfi*) using strontium isotopes and trace element concentrations in environmental water and otoliths. *PLoS ONE* **2021**, *16*, e0252769. [CrossRef] [PubMed]
22. Rohtla, M.; Matetski, L.; Taal, I.; Svirgsden, R.; Kesler, M.; Paiste, P.; Vetemaa, M. Quantifying an overlooked aspect of partial migration using otolith microchemistry. *J. Fish Biol.* **2020**, *97*, 1582–1585. [CrossRef] [PubMed]
23. Avigliano, E.; Pisonero, J.; Mendez, A.; Tombari, A.; Volpedo, A.V. Habitat use of the amphidromous catfish *Genidens barbus*: First insights at its southern distribution limit. *N. Z. J. Mar. Fresh.* **2021**, *263*, 107637. [CrossRef]
24. Xiong, Y.; Yang, J.; Jiang, T.; Liu, H.B.; Zhong, X.M.; Xiong, Y.; Yang, J.; Jiang, T.; Liu, H.; Zhong, X.; et al. Temporal stability in the otolith Sr:Ca ratio of the yellow croaker, *Larimichthys polyactis* (Actinopterygii, Perciformes, Sciaenidae), from the southern Yellow Sea. *Acta Ichthyol. Piscat.* **2021**, *51*, 59–65. [CrossRef]

25. Nakamura, M.; Yoneda, M.; Ishimura, T.; Shirai, K.; Tamamura, M.; Nishida, K. Temperature dependency equation for chub mackerel (*Scomber japonicus*) identified by a laboratory rearing experiment and microscale analysis. *Mar. Freshw. Res.* **2020**, *71*, 1384–1389. [CrossRef]
26. Sanchez, P.; Rooker, J.R.; Zapp Sluis, M.; Pinsky, J.; Dance, M.; Falterman, B.; Allman, R. Application of otolith chemistry at multiple life history stages to assess population structure of Warsaw grouper in the Gulf of Mexico. *Mar. Ecol. Prog. Ser.* **2020**, *651*, 111–123. [CrossRef]
27. Pereira, N.S.; Sial, A.N.; Pinheiro, P.B.; Freitas, F.L.; Silva, A.M.C. Carbon and oxygen stable isotopes of freshwater fish otoliths from the São Francisco River, northeastern Brazil. *An. Acad. Bras. Cienc.* **2021**, *93*, e20191050. [CrossRef]
28. Shiao, J.C.; Hsu, J.; Cheng, C.C.; Tsai, W.Y.; Lu, H.; Tanaka, Y.; Wang, P. Contribution rates of different spawning and feeding grounds to adult Pacific bluefin tuna (*Thunnus orientalis*) in the northwestern Pacific Ocean. *Deep Sea Res. I* **2021**, *169*, 103453. [CrossRef]
29. Tripp, A.; Murphy, H.M.; Davoren, G.K. Otolith chemistry reveals natal region of larval capelin in coastal Newfoundland, Canada. *Front. Mar. Sci.* **2020**, *7*, 258. [CrossRef]
30. Campana, S.E.; Thorrold, S.R. Otoliths, increments, and elements: Keys to a comprehensive understanding of fish populations? *Can. J. Fish. Aquat. Sci.* **2001**, *58*, 30–38. [CrossRef]
31. Zhang, Z.; Moksness, E. A chemical way of thinning otoliths of adult Atlantic herring (*Clupea harengus*) to expose the microstructure in the nucleus region. *ICES J. Mar. Sci.* **1993**, *50*, 213–217. [CrossRef]
32. Campana, S.E.; Fowler, A.J. Otolith elemental fingerprinting for stock identification of Atlantic Cod (*Gadus morhua*) using laser ablation ICPMS. *Can. J. Fish. Aquat. Sci.* **1994**, *51*, 1942–1950. [CrossRef]
33. Itoh, T.; Tsuji, S. Age and growth of juvenile southern bluefin tuna *Thunnus maccoyii* based on otolith microstructure. *Fish. Sci.* **1996**, *62*, 892–896. [CrossRef]
34. Busbridge, T.A.J.; Marshall, C.T.; Arkhipkin, A.I.; Shcherbich, Z.; Marriott, A.L.; Brickle, P. Can otolith microstructure and elemental fingerprints elucidate the early life history stages of the gadoid southern blue whiting (*Micromesistius australis australis*)? *Fish. Res.* **2020**, *228*, 105572. [CrossRef]
35. Brickle, P.; Schuchert, P.C.; Arkhipkin, A.I.; Reid, M.R.; Randhawa, H.S. Otolith trace elemental analyses of south American austral hake, *Merluccius australis* (Hutton, 1872) indicates complex salinity structuring on their spawning/larval grounds. *PLoS ONE* **2016**, *11*, e0145479. [CrossRef]
36. Mu, X.X.; Zhang, C.; Zhang, C.L.; Yang, J.; Ren, Y. Age-structured otolith chemistry profiles revealing the migration of *Conger myriaster* in China Seas. *Fish. Res.* **2021**, *239*, 105938. [CrossRef]
37. Lee, Y.C.; Chang, P.H.; Shih, C.H.; Shiao, J.C.; Tzeng, T.D.; Chang, W.C. The impact of religious release fish on conservation. *Glob. Ecol. Conser.* **2021**, *27*, e01556. [CrossRef]
38. Lazartigues, A.; Girard, C.; Brodeur, P.; Lecomte, F.; Mingelbier, M.; Sirois, P. Otolith microchemistry to identify sources of larval yellow perch in a fluvial lake: An approach towards freshwater fish management. *Can. J. Fish. Aquat. Sci.* **2018**, *75*, 474–487. [CrossRef]
39. Fitzpatrick, R.M.; Winkelman, D.L.; Johnson, B.M. Using isotopic data to evaluate *Esox lucius* (Linnaeus, 1758) natal origins in a hydrologically complex river basin. *Fishes* **2021**, *6*, 67. [CrossRef]
40. Izzo, C.; Doubleday, Z.A.; Gillanders, B.M. Where do elements bind within the otoliths of fish? *Mar. Freshw. Res.* **2016**, *67*, 1072–1076. [CrossRef]
41. Thomas, O.R.B.; Ganio, K.; Roberts, B.R.; Swearer, S.E. Trace element-protein interactions in endolymph from the inner ear of fish: Implications for environmental reconstructions using fish otolith chemistry. *Metallomics* **2017**, *9*, 239–249. [CrossRef] [PubMed]
42. Hüssy, K.; Limburg, K.E.; De Pontual, H.; Thomas, O.R.B.; Cook, P.K.; Heimbrand, Y.; Blass, M.; Sturrock, A.M. Trace element patterns in otoliths: The role of biomineralization. *Rev. Fish. Sci. Aquac.* **2020**, *29*, 1–33. [CrossRef]
43. Jessop, B.M.; Wang, C.H.; Tzeng, W.N.; You, C.F.; Shiao, J.C.; Lin, S.H. Otolith Sr:Ca and Ba:Ca may give inconsistent indications of estuarine habitat use for American eels (*Anguilla rostrata*). *Environ. Biol. Fish.* **2012**, *93*, 193–207. [CrossRef]
44. Menezes, R.; Moura, P.E.S.; Santos, A.C.A.; Moraes, L.E.; Condini, M.V.; Rosa, R.S.; Albuquerque, C.Q. Habitat use plasticity by the dog snapper (*Lutjanus jocu*) across the Abrolhos Bank shelf, eastern Brazil, inferred from otolith chemistry. *Estuar. Coast. Shelf Sci.* **2021**, *263*, 107637. [CrossRef]
45. Yuan, C.M.; Qin, A.L.; Liu, R.H.; Lin, J.B. On the classification of the anchovics, *Coilia*, from the lower Yangtze River and the southeast coast of China. *J. Nanjing Univ. (Nat. Sci.)* **1980**, *3*, 67–77, (In Chinese with English Abstract).
46. Khumbanyiwa, D.D.; Li, M.M.; Jiang, T.; Liu, H.B.; Yang, J. Unraveling habitat use of *Coilia nasus* from Qiantang River of China by otolith microchemistry. *Reg. Stud. Mar. Sci.* **2018**, *18*, 122–128. [CrossRef]
47. Jiang, T.; Liu, H.B.; Shen, X.Q.; Shimasaki, Y.; Ohshima, Y.; Yang, J. Life history variations among different populations of *Coilia nasus* along the Chinese Coast, inferred from otolith microchemistry. *J. Fac. Agric. Kyushu Univ.* **2014**, *59*, 383–389. [CrossRef]
48. Jiang, T.; Liu, H.B.; Lu, M.J.; Chen, T.T.; Yang, J. A possible connectivity among estuarine tapertail anchovy (*Coilia nasus*) populations in the Yangtze River, Yellow Sea, and Poyang Lake. *Estuar. Coast.* **2016**, *39*, 1762–1768. [CrossRef]
49. Moura, A.; Dias, E.; López, R.; Antunes, C. Regional population structure of the European eel at the southern limit of its distribution revealed by otolith shape signature. *Fishes* **2022**, *7*, 135. [CrossRef]
50. Li, Y.; Xie, S.; Li, Z.; Gong, W.; He, W. Gonad development of an anadromous fish *Coilia ectenes* (Engraulidae) in lower reach of Yangtze River, China. *Fish. Sci.* **2007**, *73*, 1224–1230. [CrossRef]

51. Xu, G.C.; Nie, Z.J.; Zhang, C.X.; Wei, G.L.; Xu, P.; Gu, R.B. Histological studies on testis development of *Coilia nasus* under artificial farming conditions. *J. Huazhong Agric. Univ.* **2012**, *31*, 247–252, (In Chinese with English Abstract). [CrossRef]
52. Ge, K.K.; Zhong, J.S. Daily-age structure and growth characteristics of *Coilia nasus* larvae and juveniles in the surf zone of Yangtze River Estuary. *Acta Hydro. Sin.* **2010**, *34*, 716–721. [CrossRef]
53. Dou, S.Z.; Amano, Y.; Yu, X.; Cao, L.; Kotaro, S.; Otake, T.; Tsukamoto, K. Elemental signature in otolith nuclei for stock discrimination of anadromous tapertail anchovy (*Coilia nasus*) using laser ablation ICPMS. *Environ. Biol. Fish.* **2012**, *95*, 431–443. [CrossRef]
54. Vesanto, J. Neural network tool for data mining: SOM Toolbox. In Proceedings of the 5th International Symposium on Tool Environments and Development Methods for Intelligent Systems (TOOLMET2000), Oulu, Finland, 13–14 April 2000; Oulun Yliopistopaino: Oulu, Finland, 2000; pp. 184–196.
55. Tsai, W.P.; Huang, S.P.; Cheng, S.T.; Shao, K.T.; Chang, F.J. A data-mining framework for exploring the multi-relation between fish species and water quality through self-organizing map. *Sci. Total Environ.* **2017**, *579*, 474–483. [CrossRef]
56. Wang, C.; Li, X.H.; Lai, Z.N.; Li, Y.F.; Dauta, A.; Lek, S. Patterning and predicting phytoplankton assemblages in a large subtropical river. *Fund. Appl. Limnol.* **2014**, *185*, 263–279. [CrossRef]
57. Schaffler, J.J.; Young, S.P.; Herrington, S.; Ingram, T.; Tannehill, J. Otolith chemistry to determine within-river origins of Alabama Shad in the Apalachicola-Chattahoochee-Flint River basin. *Trans. Am. Fish. Soc.* **2015**, *144*, 1–10. [CrossRef]
58. Aschenbrenner, A.; Ferreira, B.P.; Rooker, J.R. Spatial and temporal variability in the otolith chemistry of the Brazilian snapper *Lutjanus alexandrei* from estuarine and coastal environments. *J. Fish Biol.* **2016**, *89*, 753–769. [CrossRef] [PubMed]
59. Bouchoucha, M.; Pécheyran, C.; Gonzalez, J.L.; Lenfant, P.; Darnaude, A.M. Otolith fingerprints as natural tags to identify juvenile fish life in ports. *Estuar. Coast. Shelf Sci.* **2018**, *212*, 210–218. [CrossRef]
60. Jiang, T.; Yang, J.; Liu, H.; Shen, X.Q. Life history of *Coilia nasus* from the Yellow Sea inferred from otolith Sr:Ca ratios. *Environ. Biol. Fish.* **2012**, *95*, 503–508. [CrossRef]
61. Hane, Y.; Kimura, S.; Yokoyama, Y.; Miyairi, Y.; Ushikubo, T.; Ishimura, T.; Ogawa, N.; Aono, T.; Nishida, K. Reconstruction of temperature experienced by Pacific bluefin tuna *Thunnus orientalis* larvae using SIMS and microvolume CF-IRMS otolith oxygen isotope analyses. *Mar. Ecol. Prog. Ser.* **2020**, *649*, 175–188. [CrossRef]
62. Yuan, C.M.; Qin, A.L. Ecological habits and distribution of *Coilia* along the Chinese Coast and its changes of output. *Mar. Sci.* **1984**, *5*, 35–37, (In Chinese with English Abstract).
63. Martino, J.C.; Doubleday, Z.A.; Chung, M.T.; Gillanders, B.M. Experimental support towards a metabolic proxy in fish using otolith carbon isotopes. *J. Exp. Biol.* **2020**, *223*, jeb217091. [CrossRef] [PubMed]
64. Chung, M.T.; Trueman, C.N.; Godiksen, J.A.; Grønkjær, P. Otolith $\delta^{13}C$ values as a metabolic proxy: Approaches and mechanical underpinnings. *Mar. Freshw. Res.* **2019**, *70*, 1747. [CrossRef]
65. Elsdon, T.S.; Ayvazian, S.; McMahon, K.W.; Thorrold, S.R. Experimental evaluation of stable isotope fractionation in fish muscle and otoliths. *Mar. Ecol. Prog. Ser.* **2010**, *408*, 195–205. [CrossRef]
66. Von Biela, V.R.; Newsome, S.D.; Zimmerman, C.E. Examining the utility of bulk otolith $\delta^{13}C$ to describe diet in wild-caught black rockfish *Sebastes melanops*. *Aquat. Biol.* **2015**, *23*, 201–208. [CrossRef]
67. Dittman, A.H.; Quinn, T.P. Homing in Pacific salmon: Mechanisms and ecological basis. *J. Exp. Biol.* **1996**, *199*, 83–91. [CrossRef]
68. Jiang, T.; Yang, J.; Lu, M.J.; Liu, H.B.; Chen, T.T.; Gao, Y.W. Discovery of a spawning area for anadromous *Coilia nasus* temminck et schlegel, 1846 in Poyang Lake, China. *J. Appl. Ichthyol.* **2017**, *33*, 189–192. [CrossRef]
69. Brennan, S.R.; Zimmerman, C.E.; Fernandez, D.P.; Cerling, T.E.; Mcphee, M.V.; Wooller, M.J. Strontium isotopes delineate fine-scale natal origins and migration histories of Pacific salmon. *Sci. Adv.* **2015**, *1*, e1400124. [CrossRef]
70. Yano, K.; Nakamura, A. Observations on the effect of visual and olfactory ablation on the swimming behavior of migrating adult chum salmon, *Oncorhynchus keta*. *Jpn. J. Ichthyol.* **1992**, *39*, 67–83. [CrossRef]
71. Zhu, G.; Wang, L.; Tang, W.; Wang, X.; Wang, C. Identification of olfactory receptor genes in the Japanese grenadier anchovy *Coilia nasus*. *Genes Genom.* **2017**, *39*, 521–532. [CrossRef]

MDPI
St. Alban-Anlage 66
4052 Basel
Switzerland
Tel. +41 61 683 77 34
Fax +41 61 302 89 18
www.mdpi.com

Fishes Editorial Office
E-mail: fishes@mdpi.com
www.mdpi.com/journal/fishes

www.ingramcontent.com/pod-product-compliance
Lightning Source LLC
LaVergne TN
LVHW070141170726
843515LV00007B/1841